Information Theory and Machine Learning

Information Theory and Machine Learning

Editors

Lizhong Zheng
Chao Tian

MDPI • Basel • Beijing • Wuhan • Barcelona • Belgrade • Manchester • Tokyo • Cluj • Tianjin

Editors
Lizhong Zheng
Massachusetts Institute of Technology
USA

Chao Tian
Texas A&M University
USA

Editorial Office
MDPI
St. Alban-Anlage 66
4052 Basel, Switzerland

This is a reprint of articles from the Special Issue published online in the open access journal *Entropy* (ISSN 1099-4300) (available at: https://www.mdpi.com/journal/entropy/special_issues/inf_learn).

For citation purposes, cite each article independently as indicated on the article page online and as indicated below:

LastName, A.A.; LastName, B.B.; LastName, C.C. Article Title. *Journal Name* **Year**, *Volume Number*, Page Range.

ISBN 978-3-0365-5307-8 (Hbk)
ISBN 978-3-0365-5308-5 (PDF)

Contents

Preface to "Information Theory and Machine Learning" vii

Leighton Pate Barnes, Alex Dytso and Harold Vincent Poor
Improved Information-Theoretic Generalization Bounds for Distributed, Federated, and Iterative Learning
Reprinted from: *Entropy* **2022**, *24*, 1178, doi:10.3390/e24091178 1

Elyas Sabeti, Sehong Oh, Peter X.K. Song and Alfred O. Hero
A Pattern Dictionary Method for Anomaly Detection
Reprinted from: *Entropy* **2022**, *24*, 1095, doi:10.3390/e24081095 15

Joshua Lee, Yuheng Bu, Prasanna Sattigeri, Rameswar Panda, Gregory W. Wornell, Leonid Karlinsky, Rogerio Schmidt Feris
A Maximal Correlation Framework for Fair Machine Learning
Reprinted from: *Entropy* **2022**, *24*, 461, doi:10.3390/e24040461 41

Xili Dai, Shengbang Tong, Mingyang Li, Ziyang Wu, Michael Psenka, Kwan Ho Ryan Chan, Pengyuan Zhai, Yaodong Yu, Xiaojun Yuan, Heung-Yeung Shum and Yi Ma
CTRL: Closed-Loop Transcription to an LDR via Minimaxing Rate Reduction
Reprinted from: *Entropy* **2022**, *24*, 456, doi:10.3390/e24040456 57

Xiangxiang Xu, Shao-Lun Huang, Lizhong Zheng and Gregory W. Wornell
An Information Theoretic Interpretation to
Neural Networks
Reprinted from: *Entropy* **2022**, *24*, 135, doi:10.3390/e24010135 97

Yuheng Bu, Weihao Gao, Shaofeng Zou and Venugopal V. Veeravalli
Population Risk Improvement with Model Compression: An Information-Theoretic Approach
Reprinted from: *Entropy* **2021**, *23*, 1255, doi:10.3390/e23101255 125

Shuangming Yang, Jiangtong Tan and Badong Chen
Robust Spike-Based Continual Meta-Learning Improved by Restricted Minimum Error Entropy Criterion
Reprinted from: *Entropy* **2022**, *24*, 455, doi:10.3390/e24040455 145

Sarah E. Marzen and James P. Crutchfield
Probabilistic Deterministic Finite Automata andRecurrent Networks, Revisited
Reprinted from: *Entropy* **2022**, *24*, 90, doi:10.3390/e24010090 163

Shunki Kyoya and Kenji Yamanishi
Summarizing Finite Mixture Model with Overlapping Quantification
Reprinted from: *Entropy* **2021**, *23*, 1503, doi:10.3390/e23111503 177

Matthew Dixon and Tyler Ward
Information-Corrected Estimation: A Generalization Error Reducing Parameter Estimation Method
Reprinted from: *Entropy* **2021**, *23*, 1419, doi:10.3390/e23111419 201

Farzad Shahrivari and Nikola Zlatanov
On Supervised Classification of Feature Vectors with Independent and Non-Identically Distributed Elements
Reprinted from: *Entropy* **2021**, *23*, 1045, doi:10.3390/e23081045 221

Preface to "Information Theory and Machine Learning"

The recent successes of machine learning, especially regarding systems based on deep neural networks, have encouraged further research activities and raised a new set of challenges in understanding and designing complex machine learning algorithms. New applications require learning algorithms to be distributed, have transferable learning results, use computation resources efficiently, convergence quickly on online settings, have performance guarantees, satisfy fairness or privacy constraints, incorporate domain knowledge on model structures, etc. A new wave of developments in statistical learning theory and information theory has set out to address these challenges. This Special Issue, "Machine Learning and Information Theory", aims to collect recent results in this direction reflecting a diverse spectrum of visions and efforts to extend conventional theories and develop analysis tools for these complex machine learning systems. We would like to thank all contributing authors and the MDPI team for their support.

Lizhong Zheng and Chao Tian

Editors

entropy

MDPI

Article

Improved Information-Theoretic Generalization Bounds for Distributed, Federated, and Iterative Learning †

Leighton Pate Barnes [1,*], Alex Dytso [2] and Harold Vincent Poor [1]

1 Department of Electrical and Computer Engineering, Princeton University, Princeton, NJ 08544, USA
2 Department of Electrical and Computer Engineering, New Jersey Institute of Technology, Newark, NJ 07102, USA
* Correspondence: leightonbarnes@gmail.com
† This paper is an extended version of our paper published in Proceedings of the 2022 IEEE International Symposium on Information Theory, Espoo, Finland, 26 June–1 July 2022.

Abstract: We consider information-theoretic bounds on the expected generalization error for statistical learning problems in a network setting. In this setting, there are K nodes, each with its own independent dataset, and the models from the K nodes have to be aggregated into a final centralized model. We consider both simple averaging of the models as well as more complicated multi-round algorithms. We give upper bounds on the expected generalization error for a variety of problems, such as those with Bregman divergence or Lipschitz continuous losses, that demonstrate an improved dependence of $1/K$ on the number of nodes. These "per node" bounds are in terms of the mutual information between the training dataset and the trained weights at each node and are therefore useful in describing the generalization properties inherent to having communication or privacy constraints at each node.

Keywords: generalization error; information-theoretic bounds; distribution and federated learning

Citation: Barnes, L.P.; Dytso, A.; Poor, H.V. Improved Information-Theoretic Generalization Bounds for Distributed, Federated, and Iterative Learning. *Entropy* **2022**, *24*, 1178. https://doi.org/10.3390/e24091178

Academic Editors: Chao Tian and Lizhong Zheng

Received: 25 July 2022
Accepted: 20 August 2022
Published: 24 August 2022

1. Introduction

A key feature of machine learning systems is their ability to generalize new and unknown data. Such a system is trained on a particular set of data but must then perform well even on new data points that have not previously been considered. This ability, deemed generalization, can be formulated in the language of statistical learning theory by considering the generalization error of an algorithm (i.e., the difference between the population risk of a model trained on a particular dataset and the empirical risk for the same model and dataset). We say that a model generalizes well if it has a small generalization error, and because models are often trained by minimizing empirical risk or some regularized version of it, a small generalization error also implies a small population risk, which is the average loss over new samples taken randomly from the population. It is therefore of interest to find an upper bound on the generalization error and understand which quantities control it so that we can quantify the generalization properties of a machine learning system and offer guarantees about its performance.

In recent years, it has been shown that information-theoretic measures such as mutual information can be used for generalization error bounds under the assumption of the tail of the distribution of the loss function [1–4]. In particular, when the loss function is sub-Gaussian, the expected generalization error can scale at most with the square root of the mutual information between the training dataset and the model weights [2]. Such bounds offer an intuitive explanation for generalization and overfitting: if an algorithm uses only limited information from its training data, then this will bound the expected generalization error and prevent overfitting. Conversely, if an algorithm uses all of the information from its training set, in the sense that the model is a deterministic function of the training set, then this mutual information can be infinite, and there is the possibility of overfitting.

Another modern focus of machine learning systems has been that of distributed and federated learning [5–7]. In these systems, data are generated and processed in a distributed network of machines. The main differences between the distributed and centralized settings are the information constraints imposed by the network. There has been considerable interest in understanding the impact of both communication constraints [8,9] and privacy constraints [10–13] on the performance of machine learning systems, as well as designing protocols that efficiently train the systems under these constraints.

Since both communication and local differential privacy constraints can be thought of as special cases of mutual information constraints, they should pair naturally with some form of information theoretic generalization bounding in order to induce control over the generalization error of the distributed machine learning system. The information constraints inherent to the network can themselves give rise to tighter bounds on generalization error and thus provide better guarantees against overfitting. Along these lines, in a recent work [14], a subset of the present authors introduced the framework of using information theoretic quantities for bounding both the expected generalization error and a measure of privacy leakage in distributed and federated learning systems. The generalization bounds in this work, however, are essentially the same as those obtained by thinking of the entire system, from the data at each node in the network to the final aggregated model, as a single, centralized algorithm. Any improved generalization guarantees from these bounds would remain implicit in the mutual information terms involved.

In this work, we develop improved bounds on the expected generalization error for distributed and federated learning systems. Instead of leaving the differences between these systems and their centralized counterparts implicit in the mutual information terms, we bring analysis of the structure of the systems directly to the bounds. By working with the contribution from each node separately, we are able to derive upper bounds on the expected generalization error that scale with the number of nodes K as $O\left(\frac{1}{K}\right)$ instead of $O\left(\frac{1}{\sqrt{K}}\right)$. This improvement is shown to be tight for certain examples, such as learning the mean of a Gaussian distribution with quadratic loss. We develop bounds that apply to distributed systems in which the submodels from K different nodes are averaged together, as well as bounds that apply to more complicated multi-round stochastic gradient descent (SGD) algorithms, such as in federated learning. For linear models with Bregman divergence losses, these "per node" bounds are in terms of the mutual information between the training dataset and the trained weights at each node and are therefore useful in describing the generalization properties inherent to having communication or privacy constraints at each node. For arbitrary nonlinear models that have Lipschitz continuous losses, the improved dependence of $O\left(\frac{1}{K}\right)$ can still be recovered but without a description in terms of mutual information. We demonstrate the improvements given by our bounds over the existing information theoretic generalization bounds via simulation of a distributed linear regression example. A preliminary conference version of this paper was presented in [15]. The present paper completes the work by including all of the missing proof details as well as providing new bounds for noisy SGD in Corollary 4.

Technical Preliminaries

Suppose we have independent and identically distributed (i.i.d.) data $Z_i \sim \pi$ for $i = 1, \dots, n$, and let $S = (Z_1, \dots, Z_n)$. Suppose further that $W = \mathcal{A}(S)$ is the output of a potentially stochastic algorithm. Let $\ell(W, Z)$ be a real-valued loss function and define

$$L(w) = \mathbb{E}_\pi[\ell(w, Z)]$$

to be the population risk for weights (or model) w. We similarly define

$$L_s(w) = \frac{1}{n}\sum_{i=1}^{n} \ell(w, z_i)$$

to be the empirical risk on dataset s for model w. The generalization error for dataset s is then

$$\Delta_{\mathcal{A}}(s) = L(\mathcal{A}(s)) - L_s(\mathcal{A}(s))$$

In addition, the expected generalization error is

$$\mathbb{E}_{S\sim\pi^n}[\Delta_{\mathcal{A}}(S)] = \mathbb{E}_{S\sim\pi^n}[L(\mathcal{A}(S)) - L_S(\mathcal{A}(S))] \tag{1}$$

where the expectation is also over any randomness in the algorithm. Below, we present some standard results for the expected generalization error that will be needed:

Theorem 1 (Leave-One-Out Expansion; Lemma 11 in [16]). *Let $S^{(i)} = (Z_1, \ldots, Z_i', \ldots, Z_n)$ be a version of S with Z_i replaced by an i.i.d. copy Z_i'. Denote $S' = (Z_1', \ldots, Z_n')$. Then, we have*

$$\mathbb{E}_{S\sim\pi^n}[\Delta_{\mathcal{A}}(S)] = \frac{1}{n}\sum_{i=1}^{n} \mathbb{E}_{S,S'}[\ell(\mathcal{A}(S), Z_i') - \ell(\mathcal{A}(S^{(i)}), Z_i')]\,.$$

Proof. Observe that

$$\mathbb{E}_{S\sim\pi^n}[L(\mathcal{A}(S))] = \mathbb{E}_{S,S'}[\ell(\mathcal{A}(S), Z_i')] \tag{2}$$

for each i and that

$$\begin{aligned}
\mathbb{E}_{S\sim\pi^n}[L_S(\mathcal{A}(S))] &= \frac{1}{n}\sum_{i=1}^{n} \mathbb{E}_{S\sim\pi^n}[\ell(\mathcal{A}(S), Z_i)] \\
&= \frac{1}{n}\sum_{i=1}^{n} \mathbb{E}_{S,S'\sim\pi^n}\left[\ell(\mathcal{A}(S^{(i)}), Z_i')\right]\,.
\end{aligned} \tag{3}$$

Putting Equations (2) and (3) together with (1) yields the result. □

In many of the results in this paper, we will use one of the two following assumptions:

Assumption 1. *The loss function $\ell(\widetilde{W}, \widetilde{Z})$ satisfies*

$$\log \mathbb{E}\left[\exp\left(\lambda\left(\ell(\widetilde{W}, \widetilde{Z}) - \mathbb{E}[\ell(\widetilde{W}, \widetilde{Z})]\right)\right)\right] \le \psi(-\lambda)$$

for $\lambda \in (b, 0]$, $\psi(0) = \psi'(0) = 0$, where $\widetilde{W}$ and $\widetilde{Z}$ are taken independently from the marginals for W and Z, respectively,

The next assumption is a special case of the previous one with $\psi(\lambda) = \frac{R^2\lambda^2}{2}$:

Assumption 2. *The loss function $\ell(\widetilde{W}, \widetilde{Z})$ is sub-Gaussian with parameter R^2 in the sense that*

$$\log \mathbb{E}\left[\exp\left(\lambda\left(\ell(\widetilde{W}, \widetilde{Z}) - \mathbb{E}[\ell(\widetilde{W}, \widetilde{Z})]\right)\right)\right] \le \frac{R^2\lambda^2}{2}\,.$$

Theorem 2 (Theorem 2 in [3]). *Under Assumption 1, we have*

$$\mathbb{E}_{S\sim\pi^n}[\Delta_{\mathcal{A}}(S)] \leq \frac{1}{n}\sum_{i=1}^{n}\psi^{*-1}(I(W;Z_i))$$

where $\psi^{*-1}(y) = \inf_{\lambda\in[0,b)}\left(\frac{y+\psi(\lambda)}{\lambda}\right)$.

For a continuously differentiable and strictly convex function $F:\mathbb{R}^m \to \mathbb{R}$, we define the associated Bregman divergence [17,18] between two points $p,q \in \mathbb{R}^m$ to be

$$D_F(p,q) = F(p) - F(q) - \langle \nabla F(q), p-q \rangle\,,$$

where $\langle \cdot,\cdot \rangle$ denotes the usual inner product.

2. Distributed Learning and Model Aggregation

Now suppose that there are K nodes each having n samples. Each node $k = 1,\ldots,K$ has a dataset $S_k = (Z_{1,k},\ldots,Z_{n,k})$, with $Z_{i,k}$ taken i.i.d. from π. We use $S = (S_1,\ldots,S_K)$ to denote the entire dataset of size nK. Each node locally trains a model $W_k = \mathcal{A}_k(S_k)$ with algorithm $\mathcal{A}_k$. After each node locally trains its model, the models W_k are then combined to form the final model $\widehat{W}$ using an aggregation algorithm $\widehat{W} = \widehat{\mathcal{A}}(W_1,\ldots,W_K)$ (see Figure 1). In this section, we will assume that $W_k \in \mathbb{R}^d$ and that the aggregation is performed by simple averaging (i.e., $\widehat{W} = \frac{1}{K}\sum_{k=1}^{K} W_k$). Define $\mathcal{A}$ to be the total algorithm from the data S to the final weights $\widehat{W}$ such that $\widehat{W} = \mathcal{A}(S)$. In this section, if we say that Assumption 1 or 2 holds, we mean that it holds for each algorithm $\mathcal{A}_k$. As in Theorem 1, we use $S^{(i,k)}$ to denote the entire dataset S with sample $Z_{i,k}$ replaced by an independent copy $Z'_{i,k}$, and similarly, we use $S_k^{(i)}$ to refer to the sub-dataset at node k, with sample $Z_{i,k}$ replaced by an independent copy $Z'_{i,k}$:

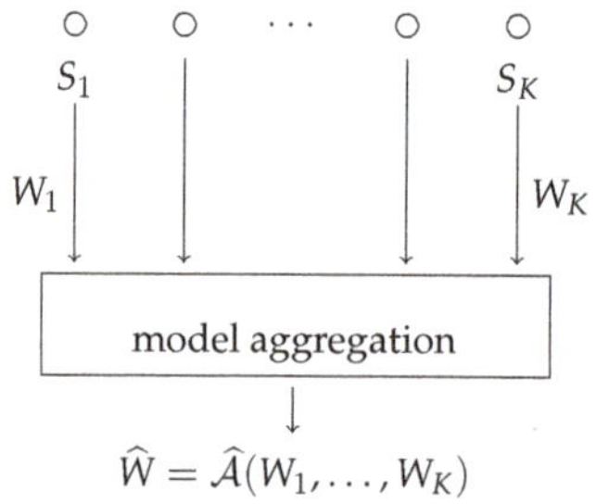

Figure 1. The distributed learning setting with model aggregation.

Theorem 3. *Suppose that $\ell(\cdot,z)$ is a convex function of $w \in \mathbb{R}^d$ for each z and that $\mathcal{A}_k$ represents the empirical risk minimization algorithm on local dataset S_k in the sense that*

$$W_k = \mathcal{A}_k(S_k) = \operatorname*{argmin}_{w}\sum_{i=1}^{n}\ell(w,Z_{i,k})\,.$$

Then, we have

$$\Delta_{\mathcal{A}}(s) \leq \frac{1}{K}\sum_{k=1}^{K}\Delta_{\mathcal{A}_k}(s_k)\,.$$

Proof.

$$\begin{aligned}
\Delta_{\mathcal{A}}(s) &= \mathbb{E}_{Z\sim\pi}[\ell(\mathcal{A}(s),Z)] - \frac{1}{nK}\sum_{i,k}\ell(\mathcal{A}(s),z_{i,k}) \\
&= \mathbb{E}_{Z\sim\pi}\left[\ell\left(\frac{1}{K}\sum_{k=1}^{K}w_k, Z\right)\right] - \frac{1}{nK}\sum_{i,k}\ell(\mathcal{A}(s),z_{i,k}) \\
&\leq \frac{1}{K}\sum_{k=1}^{K}\mathbb{E}_{Z\sim\pi}[\ell(w_k,Z)] - \frac{1}{nK}\sum_{i,k}\ell(\mathcal{A}(s),z_{i,k}) \qquad (4) \\
&\leq \frac{1}{K}\sum_{k=1}^{K}\mathbb{E}_{Z\sim\pi}[\ell(w_k,Z)] - \frac{1}{K}\sum_{k=1}^{K}\min_{w}\frac{1}{n}\sum_{i=1}^{n}\ell(w,z_{i,k}) \qquad (5) \\
&= \frac{1}{K}\sum_{k=1}^{K}\Delta_{\mathcal{A}_k}(s_k).
\end{aligned}$$

In the above display, Equation (4) follows by the convexity of ℓ via Jensen's inequality, and Equation (5) follows by minimizing the empirical risk over each node's local dataset, which exactly corresponds to what each node's local algorithm $\mathcal{A}_k$ does. □

While Theorem 3 seems to be a nice characterization of the generalization bounds for the aggregate model (in that the aggregate generalization error cannot be any larger than the average generalization errors over each node), it does not offer any improvement in the expected generalization error that one might expect when given nK total samples instead of just n samples. A naive application of the generalization bounds from Theorem 2, followed by the data processing inequality $I(\widehat{W};Z_{i,k}) \leq I(W_k;Z_{i,k})$, runs into the same problem.

2.1. Improved Bounds

In this subsection, we demonstrate bounds on the expected generalization error that remedy the above shortcomings. In particular, we would like to demonstrate the following two properties:

(1) The bound should decay with the number of nodes K in order to take advantage of the total dataset from all K nodes.

(2) The bound should be in terms of the information theoretic quantities $I(W_k;S_k)$, which can represent (or be bounded from above by) the capacities of the channels over which the nodes are communicating. This can, for example, represent a communication or local differential privacy constraint for each node.

At a high level, we will improve on the bound from Theorem 3 by taking into account the fact that a small change in S_k will only change $\widehat{W}$ by a fraction $\frac{1}{K}$ of the amount that it will change W_k. In the case where W is a linear or location model, and the loss ℓ is a Bregman divergence, we can obtain an upper bound on the expected generalization error that satisfies properties (1) and (2) as follows:

Theorem 4 (Linear or Location Models with Bregman Loss). *Suppose the loss ℓ takes the form of one of the following:*

(i) $\ell(w,(x,y)) = D_F(w^T x, y)$;

(ii) $\ell(w,z) = D_F(w,z)$.

In addition, assume that Assumption 1 holds. Then, we have

$$\mathbb{E}_{S\sim\pi^{nK}}[\Delta_{\mathcal{A}}(S)] = \frac{1}{K^2}\sum_{k=1}^{K}\mathbb{E}_{S_k\sim\pi^n}[\Delta_{\mathcal{A}_k}(S_k)]$$

and

$$\begin{aligned}\mathbb{E}_{S\sim\pi^{nK}}[\Delta_{\mathcal{A}}(S)] &\leq \frac{1}{nK^2}\sum_{i,k}\psi^{*-1}(I(W_k;Z_{i,k})) \\ &\leq \frac{1}{K^2}\sum_{k=1}^{K}\psi^{*-1}\left(\frac{I(W_k;S_k)}{n}\right).\end{aligned}$$

Proof. Here, we restrict our attention to case (ii), but the two cases have nearly identical proofs. Using Theorem 1, we have

$$\begin{aligned}
&\mathbb{E}_{S\sim\pi^{nK}}[\Delta_{\mathcal{A}}(S)] \\
&= \frac{1}{nK}\sum_{i,k}\mathbb{E}_{S,S'}\left[\ell(\mathcal{A}(S),Z'_{i,k}) - \ell(\mathcal{A}(S^{(i,k)}),Z'_{i,k})\right] \\
&= \frac{1}{nK}\sum_{i,k}\mathbb{E}_{S,S'}\Big[F(\mathcal{A}(S)) - F(Z'_{i,k}) - \langle\nabla F(Z'_{i,k}),\mathcal{A}(S)-Z'_{i,k}\rangle \\
&\qquad - F(\mathcal{A}(S^{(i,k)})) + F(Z'_{i,k}) + \langle\nabla F(Z'_{i,k}),\mathcal{A}(S^{(i,k)})-Z'_{i,k}\rangle\Big]
\end{aligned}$$

$$=\frac{1}{nK}\sum_{i,k}\mathbb{E}_{S,S'}\left[\langle\nabla F(Z'_{i,k}),\mathcal{A}(S^{(i,k)})-\mathcal{A}(S)\rangle\right] \tag{6}$$

$$=\frac{1}{nK}\sum_{i,k}\mathbb{E}_{S,S'}\left[\langle\nabla F(Z'_{i,k}),\frac{1}{K}W_k^{(i)}+\frac{1}{K}\sum_{j\neq k}W_j-\frac{1}{K}\sum_j W_j\rangle\right]$$

$$=\frac{1}{nK^2}\sum_{i,k}\mathbb{E}_{S,S'}\left[\langle\nabla F(Z'_{i,k}),W_k^{(i)}-W_k\rangle\right]. \tag{7}$$

In Equation (7), we use $W_k^{(i)}$ to denote $\mathcal{A}_k(S_k^{(i)})$. Equation (6) follows the linearity of the inner product and cancels the higher order terms $F(\mathcal{A}(S))$ and $F(\mathcal{A}(S^{(i,k)}))$, which have the same expected values. The key step in Equation (7) then follows by noting that $\mathcal{A}(S^{(i,k)})$ only differs from $\mathcal{A}(S)$ in the submodel coming from node k, which is multiplied by a factor of $\frac{1}{K}$ when averaging all of the submodels. By backing out of Equation (6) and re-adding the appropriate canceled terms, we get

$$\mathbb{E}_{S\sim\pi^{nK}}[\Delta_{\mathcal{A}}(S)] = \frac{1}{K^2}\sum_{k=1}^{K}\mathbb{E}_{S_k\sim\pi^n}[\Delta_{\mathcal{A}_k}(S_k)].$$

By applying Theorem 2, this yields

$$\mathbb{E}_{S\sim\pi^{nK}}[\Delta_{\mathcal{A}}(S)] \leq \frac{1}{nK^2}\sum_{i,k}\psi^{*-1}(I(W_k;Z_{i,k})).$$

Then, by noting that ψ^{*-1} is non-decreasing and concave, we have

$$\frac{1}{nK^2}\sum_{i,k}\psi^{*-1}(I(W_k;Z_{i,k})) \leq \frac{1}{K^2}\sum_{k=1}^{K}\psi^{*-1}\left(\sum_{i=1}^{n}\frac{I(W_k;Z_{i,k})}{n}\right).$$

Using the property that conditioning decreases entropy yields

$$\sum_{i=1}^{n} I(W_k;Z_{i,k}) \leq I(W_k;S_k),$$

and we have

$$\frac{1}{K^2}\sum_{k=1}^{K}\psi^{*-1}\left(\sum_{i=1}^{n}\frac{I(W_k;Z_{i,k})}{n}\right) \leq \frac{1}{K^2}\sum_{k=1}^{K}\psi^{*-1}\left(\frac{I(W_k;S_k)}{n}\right)$$

as desired. □

The result in Theorem 4 is general enough to apply to many problems of interest. For example, if $F(p) = \|p\|_2^2$, then the Bregman divergence D_F gives the ubiquitous squared ℓ^2 loss (i.e., $D_F(p,q) = \|p-q\|_2^2$). For a comprehensive list of realizable loss functions, the interested reader is referred to [19]. Using F above, Theorem 4 can be applied to ordinary least squares regression, which we will examine in greater detail in Section 4. Other regression models such as logistic regression have loss functions that cannot be described with a Bregman divergence without the inclusion of additional nonlinearity. However, the result in Theorem 4 is agnostic to the algorithm that each node uses to fit its individual model. In this way, each node could fit a logistic model to its data, and the total aggregate model would then be an average over these logistic models. Theorem 4 would still control the expected generalization error for the aggregate model with the extra $\frac{1}{K}$ factor. However, critically, the upper bound would only be for the generalization error that is with respect to a loss of the form $D_F(w^T x, y)$, such as quadratic loss.

In order to show that the dependence on the number of nodes K from Theorem 4 is tight for certain problems, consider the following example from [3]. Suppose that $Z\sim\pi = \mathcal{N}(\mu,\sigma^2 I_d)$ and $\ell(w,z) = \|w-z\|_2^2$ so that we are trying to learn the mean μ of a Gaussian distribution. An obvious algorithm for each node to use is simple averaging of its dataset:

$$w_k = \mathcal{A}_k(s_k) = \frac{1}{n}\sum_{i=1}^{n} z_{i,k}\,.$$

For this algorithm, it can be shown that

$$I(\widehat{W};Z_{i,k}) = \frac{d}{2}\log\frac{nK}{nK-1}$$

and

$$\psi^{*-1}(y) = 2\sqrt{d\left(1+\frac{1}{nK}\right)^2\sigma^4 y}$$

See Section IV.A. in [3] for further details. If we apply the existing information theoretic bounds from Theorem 2 in an end-to-end way, such as in the approach from [14], we would get

$$\begin{aligned}\mathbb{E}_{S\sim\pi^{nK}}[\Delta_{\mathcal{A}}(S)] &\leq \sigma^2 d\sqrt{2\left(1+\frac{1}{nK}\right)^2\log\frac{nK}{nK-1}}\\ &= O\left(\frac{1}{\sqrt{nK}}\right).\end{aligned}$$

However, for this choice of algorithm at each node, the true expected generalization error can be computed to be

$$\mathbb{E}_{S\sim\pi^{nK}}[\Delta_{\mathcal{A}}(S)] = \frac{2\sigma^2 d}{nK}\,.$$

By applying our new bound from Theorem 4, we get

$$\begin{aligned}\mathbb{E}_{S\sim\pi^{nK}}[\Delta_{\mathcal{A}}(S)] &\leq \frac{\sigma^2 d}{K}\sqrt{2\left(1+\frac{1}{n}\right)^2 \log\frac{n}{n-1}} \\ &\leq O\left(\frac{1}{K\sqrt{n}}\right)\end{aligned}$$

which shows the correct dependence on K and improves upon the $O\left(\frac{1}{\sqrt{K}}\right)$ result from prior information theoretic methods.

2.2. General Models and Losses

In this section, we briefly describe some results that hold for more general classes of models and loss functions, such as deep neural networks and other nonlinear models:

Theorem 5 (Lipschitz Continuous Loss). *Suppose that $\ell(w,z)$ is Lipschitz continuous as a function of w in the sense that*

$$|\ell(w,z) - \ell(w',z)| \leq C\|w - w'\|_2$$

for any z and that $\mathbb{E}[\|W_k - \mathbb{E}[W_k]\|_2] \leq \sigma_0$ for each k. Then, we have

$$\mathbb{E}_{S\sim\pi^{nK}}[\Delta_{\mathcal{A}}(S)] \leq \frac{2C\sigma_0}{K}.$$

Proof. Starting with Theorem 1, we have

$$\begin{aligned}&\mathbb{E}_{S\sim\pi^{nK}}[\Delta_{\mathcal{A}}(S)] \\ &\quad= \frac{1}{nK}\sum_{i,k}\mathbb{E}_{S,S'}\Big[\ell(\mathcal{A}(S), Z'_{i,k}) - \ell(\mathcal{A}(S^{(i,k)}), Z'_{i,k})\Big] \\ &\quad\leq \frac{1}{nK}\sum_{i,k}\mathbb{E}_{S,S'}\left[C\left\|\mathcal{A}(S) - \mathcal{A}(S^{(i,k)})\right\|_2\right] &(8)\\ &\quad= \frac{1}{nK^2}\sum_{i,k}\mathbb{E}_{S,S'}\left[C\left\|W_k - W_k^{(i)}\right\|_2\right] \\ &\quad\leq \frac{C}{nK^2}\sum_{i,k}\mathbb{E}_{S,S'}[\|W_k - \mathbb{E}[W_k]\|_2] + \mathbb{E}_{S,S'}\left[\left\|W_k^{(i)} - \mathbb{E}[W_k]\right\|_2\right] &(9)\\ &\quad\leq \frac{2C\sigma_0}{K}, &(10)\end{aligned}$$

where Equation (8) follows from Lipschitz continuity, Equation (9) uses the triangle inequality, and Equation (10) is assumed. □

The bound in Theorem 5 is not in terms of the information theoretic quantities $I(W_k; S_k)$, but it does show that the $O\left(\frac{1}{K}\right)$ upper bound can be shown for much more general loss functions and arbitrary nonlinear models.

2.3. Privacy and Communication Constraints

Both communication and local differential privacy constraints can be thought of as special cases of mutual information constraints. Motivated by this observation, Theorem 4 immediately implies corollaries for these types of systems:

Corollary 1 (Privacy Constraints). *Suppose each node's algorithm $\mathcal{A}_k$ is an ε-local, differentially private mechanism in the sense that $\frac{p(w_k|s_k)}{p(w_k|s_k')} \le e^{\varepsilon}$ for each w_k, s_k, s_k'. Then, for losses ℓ of the form in Theorem 4, and under Assumption 2, we have*

$$\mathbb{E}_{S\sim\pi^{nK}}[\Delta_{\mathcal{A}}(S)] \le \frac{1}{K}\sqrt{\frac{2R^2\min\{\varepsilon,(e-1)\varepsilon^2\}}{n}}\,.$$

Proof. Note that

$$\begin{aligned} I(W_k;S_k) &= \sum_{w_k,s_k} p(w_k,s_k)\log\frac{p(w_k|s_k)}{\sum_{s_k'}p(w_k|s_k')p(s_k')} \\ &\le \sum_{w_k,s_k} p(w_k,s_k)\log\frac{p(w_k|s_k)}{\inf_{s_k'}p(w_k|s_k')} \\ &\le \sum_{w_k,s_k} p(w_k,s_k)\varepsilon = \varepsilon\,. \end{aligned}$$

Similarly, it is true that

$$\begin{aligned} I(W_k;S_k) &= \mathsf{KL}(P_{W_kS_k}\|P_{S_k}P_{W_k}) \\ &\le \mathsf{KL}(P_{W_kS_k}\|P_{S_k}P_{W_k}) + \mathsf{KL}(P_{S_k}P_{W_k}\|P_{W_kS_k}) \\ &= \sum_{w_k,s_k} p(w_k)p(s_k)\left(\frac{p(w_k|s_k)}{p(w_k)}-1\right)\log\frac{p(w_k|s_k)}{p(w_k)} \\ &\le \sum_{w_k,s_k} p(w_k)p(s_k)(e^{\varepsilon}-1)\varepsilon \le (e-1)\varepsilon^2 \end{aligned}$$

where the last inequality is only true for $\varepsilon \le 1$. Putting these two displays together gives $I(W_k;S_k) \le \min\{\varepsilon,(e-1)\varepsilon^2\}$, and the result follows from Theorem 4. □

Corollary 2 (Communication Constraints). *Suppose each node can only transit B bits of information to the model aggregator, meaning that each W_k can only take 2^B distinct possible values. Then, for losses ℓ of the form in Theorem 4, and under Assumption 2, this yields*

$$\mathbb{E}_{S\sim\pi^{nK}}[\Delta_{\mathcal{A}}(S)] \le \frac{1}{K}\sqrt{\frac{2(\log 2)R^2B}{n}}\,.$$

Proof. The corollary follows immediately from Theorem 4 and

$$I(W_k;S_k) \le H(W_k) \le (\log 2)B\,.$$

□

3. Iterative Algorithms

We now turn to considering more complicated multi-round and iterative algorithms. In this setting, after T rounds, there is a sequence of weights $W^{(T)} = (W^1,\ldots,W^T)$, and the final model $\widehat{W}_T = f_T(W^{(T)})$ is a function of that sequence, where f_T gives a linear combination of the T vectors $W^1,\ldots,W^T$. The function f_T could represent, for example, averaging over the T iterates, choosing the last iterate W^T or some weighted average over the iterates. For each round t, each node k produces an updated model W_k^t based on its local dataset S_k and the previous timestep's global model W^{t-1}. The global model is then updated via an average over all K updated submodels:

$$W^t = \frac{1}{K}\sum_{k=1}^{K} W_k^t\,.$$

The particular example that we will consider is that of a distributed SGD, where each node constructs its updated model W_k^t by taking one or more gradient steps starting from W^{t-1} with respect to random minibatches of its local data. Our model is general enough to account for multiple local gradient steps, as are used in so-called federated learning [5–7], as well as noisy versions of SGDs, such as in [20,21]. If only one local gradient step is taken for each iteration, then the update rule for this example could be written as

$$W_k^t = W^{t-1} - \eta_t \nabla_w \ell(W^{t-1}, Z_{t,k}) + \xi_t \tag{11}$$

where $Z_{t,k}$ is a data point (or minibatch) sampled from S_k on timestep t, η_t is the learning rate, and ξ_t is some potential added noise. We assume that the data points $Z_{t,k}$ are sampled without replacement so that the samples are distinct across different values of t. We will also assume, for notational simplicity, that $\widehat{W}_T = W^T$, although the more general result follows in a straightforward manner.

For this type of iterative algorithm, we will consider the following timestep-averaged empirical risk quantity:

$$\frac{1}{KT} \sum_{t=1}^{T} \sum_{k=1}^{K} \ell(W^t, Z_{t,k}) \,,$$

and the corresponding generalization error, expressed as

$$\Delta_{\mathsf{sgd}}(S) = \frac{1}{T} \sum_{t=1}^{T} \left(\mathbb{E}_{Z \sim \pi}[\ell(W^t, Z)] - \frac{1}{K} \sum_{k=1}^{K} \ell(W^t, Z_{t,k}) \right) . \tag{12}$$

Note that Equation (12) is slightly different from the end-to-end generalization error that we would get from considering the final model W^T and whole dataset S. It is instead an average over the generalization error we would get from each model, stopping at iteration t. We perform this so that when we apply the leave-one-out expansion from Theorem 1, we do not have to account for the dependence of W_k^t on past samples $Z_{t',k'}$ for $t' < t$ and $k' \neq k$. Since we expect the generalization error to decrease as we use more samples, this quantity should result in a more conservative upper bound and be a reasonable surrogate object to study. The next bound follows as a corollary to Theorem 4:

Corollary 3. *For losses ℓ of the form in Theorem 4, and under Assumption 2 (for each W_k^t), we have*

$$\mathbb{E}\left[\Delta_{\mathsf{sgd}}(S)\right] \le \frac{1}{T} \sum_{t=1}^{T} \frac{1}{K^2} \sum_{k=1}^{K} \sqrt{2R^2 I(W_k^t; Z_{t,k})} \,.$$

In the particular example described in Equation (11), where Gaussian noise $\xi_t \sim \mathcal{N}(0, I_d \sigma_t^2)$ is added to each iterate, Corollary 3 yields the following. As in [20], we assume that the updates are magnitude-bounded (i.e., $\sup_{w,x} \|\nabla_w \ell(w,z)\|_2 \le L$), the stepsizes satisfy $\eta_t = \frac{c}{t}$ for a constant $c > 0$, and that $\sigma_t = \sqrt{\eta_t}$:

Corollary 4. *Under the assumptions above, we have*

$$\mathbb{E}\left[\Delta_{\mathsf{sgd}}(S)\right] \le \frac{2RL}{K} \sqrt{\frac{c}{T}} \,.$$

Proof. The mutual information terms in Corollary 3 satisfy

$$I(W_k^t; Z_{t,k}) \leq I(W_k^t, W^{t-1}; Z_{t,k}) \tag{13}$$
$$= I(W_k^t; Z_{t,k}|W^{t-1}) + I(W^{t-1}; Z_{t,k}) \tag{14}$$
$$= I(W_k^t; Z_{t,k}|W^{t-1}) \tag{15}$$
$$\leq \frac{d}{2}\log\left(1 + \frac{\eta_t^2 L^2}{d\sigma_t^2}\right) \tag{16}$$
$$\leq \frac{\eta_t^2 L^2}{2\sigma_t^2} = \frac{cL^2}{2t}. \tag{17}$$

Equation (13) follows from the data-processing inequality, Equation (14) is the chain rule for mutual information, and Equation (15) follows from the independence of $Z_{t,k}$ and W^{t-1}. Equation (16) is due to the capacity of the additive white Gaussian noise channel, and Equation (17) just uses the approximation $\log(1+x) \leq x$. Thus, we have

$$\mathbb{E}[\Delta_{\mathsf{sgd}}(S)] \leq \frac{1}{TK}\sum_{t=1}^{T} RL\sqrt{\frac{c}{t}} \leq \frac{2RL}{K}\sqrt{\frac{c}{T}}.$$

□

4. Simulations

We simulated a distributed linear regression example in order to demonstrate the improvement in our bounds over the existing information-theoretic bounds. To accomplish this, we generated $n = 10$ synthetic datapoints at each of K different nodes for various values of K. Each datapoint consisted of a pair (x, y), where $y = xw_0 + n$ with $x, n \sim \mathcal{N}(0,1)$, and $w_0 \sim \mathcal{N}(0,1)$ was the randomly generated true weight that was common to all datapoints. Each node constructed an estimate $\widehat{w}_k$ of w_0 using the well-known normal equations which minimize the quadratic loss (i.e., $\widehat{w}_k = \operatorname{argmin}_w \sum_{i=1}^{n}(wx_{i,k} - y_{i,k})^2$). The aggregate model was then the average $\widehat{w} = \frac{1}{K}\sum_{k=1}^{K}\widehat{w}_k$. In order to estimate the old and new information-theoretic generalization bounds (i.e., the bounds from Theorems 2 and 4, respectively), this procedure was repeated $M = 10^6$ times, and the datapoint and model values were binned in order to estimate the mutual information quantities. The value of M was increased until the mutual information estimates were no longer particularly sensitive to the number and widths of the bins. In order to estimate the true generalization error, the expectations for both the population risk and the dataset were estimated by Monte Carlo experimentation, with 10^4 trials each. The results can be seen in Figure 2, where it is evident that the new information theoretic bound is much closer to the true expected generalization error and decays with an improved rate as a function of K.

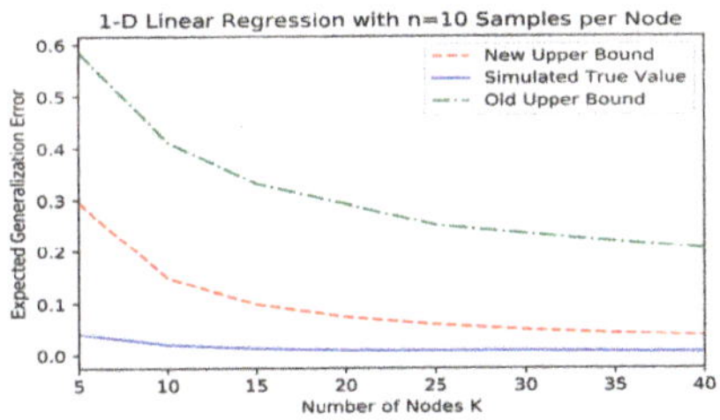

Figure 2. *Cont.*

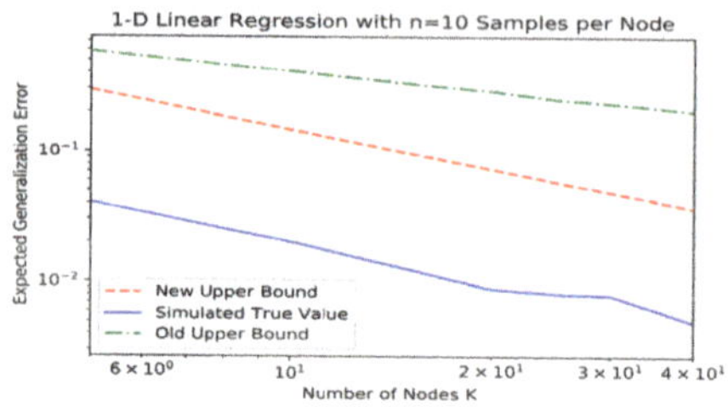

Figure 2. Information-theoretic upper bounds and expected generalization error for a simulated linear regression example in linear (**top**) and log (**bottom**) scales.

Author Contributions: Conceptualization, L.P.B., A.D. and H.V.P.; Formal analysis, L.P.B., A.D. and H.V.P.; Investigation, L.P.B., A.D. and H.V.P.; Methodology, L.P.B., A.D. and H.V.P.; Supervision, H.V.P.; Writing—original draft, L.P.B.; Writing—review & editing, L.P.B., A.D. and H.V.P. All authors have read and agreed to the published version of the manuscript.

Funding: This research was funded by National Science Foundation grant number CCF-1908308.

Institutional Review Board Statement: Not applicable.

Informed Consent Statement: Not applicable.

Data Availability Statement: Not applicable.

Conflicts of Interest: The authors declare no conflict of interest.

References

1. Russo, D.; Zou, J. How Much Does Your Data Exploration Overfit? Controlling Bias via Information Usage. *IEEE Trans. Inf. Theory* **2020**, *66*, 302–323. [CrossRef]
2. Xu, A.; Raginsky, M. Information-Theoretic Analysis of Generalization Capability of Learning Algorithms. *Adv. Neural Inf. Process. Syst.* **2017**, *30*, 2521–2530.
3. Bu, Y.; Zou, S.; Veeravalli, V.V. Tightening Mutual Information-Based Bounds on Generalization Error. *IEEE J. Sel. Areas Inf. Theory* **2020**, *1*, 121–130. [CrossRef]
4. Aminian, G.; Bu, Y.; Wornell, G.W.; Rodrigues, M.R. Tighter Expected Generalization Error Bounds via Convexity of Information Measures. In Proceedings of the 2022 IEEE International Symposium on Information Theory (ISIT), Espoo, Finland, 26 June–1 July 2022.
5. McMahan, H.B.; Moore, E.; Ramage, D.; Hampson, S.; y Arcas, B.A. Communication-Efficient Learning of Deep Networks from Decentralized Data. In Proceedings of the 20th International Conference on Artificial Intelligence and Statistics, Fort Lauderdale, FL, USA, 20–22 April 2017.
6. Konecný, J.; McMahan, H.B.; Ramage, D.; Richtárik, P. Federated Optimization: Distributed Machine Learning for On-Device Intelligence. *arXiv* **2016**, arXiv:1610.02527.
7. Konečný, J.; McMahan, H.B.; Yu, F.X.; Richtarik, P.; Suresh, A.T.; Bacon, D. Federated Learning: Strategies for Improving Communication Efficiency. *arXiv* **2016**, arXiv:1610.05492.
8. Lin, Y.; Han, S.; Mao, H.; Wang, Y.; Dally, W.J. Deep Gradient Compression: Reducing the Communication Bandwidth for Distributed Training. In Proceedings of the 6th International Congress on Learning Representations (ICLR), Vancouver, BC, Canada, 30 April–3 May 2018.
9. Barnes, L.P.; Inan, H.A.; Isik, B.; Ozgur, A. rTop-k: A Statistical Estimation Approach to Distributed SGD. *IEEE J. Sel. Areas Inf. Theory* **2020**, *1*, 897–907. [CrossRef]
10. Warner, S.L. Randomized Response: A Survey Technique for Eliminating Evasive Answer Bias. *J. Am. Stat. Assoc.* **1965**, *60*, 63–69. [CrossRef] [PubMed]
11. Dwork, C.; McSherry, F.; Nissim, K.; Smith, A. Calibrating Noise to Sensitivity in Private Data Analysis. In *Theory of Cryptography Conference*; Halevi, S., Rabin, T., Eds.; Springer: Berlin/Heidelberg, Geramny, 2006.
12. Kasiviswanathan, S.P.; Lee, H.K.; Nissim, K.; Raskhodnikova, S.; Smith, A. What Can We Learn Privately? *SIAM J. Comput.* **2011**, *40*, 793–826. [CrossRef]
13. Cuff, P.; Yu, L. Differential Privacy as a Mutual Information Constraint. In Proceedings of the 2016 ACM SIGSAC Conference on Computer and Communications Security, Vienna, Austria, 24–28 October 2016; pp. 43–54.
14. Yagli, S.; Dytso, A.; Poor, H.V. Information-Theoretic Bounds on the Generalization Error and Privacy Leakage in Federated Learning. In Proceedings of the 2020 IEEE 21st International Workshop on Signal Processing Advances in Wireless Communications (SPAWC), Atlanta, GA, USA, 26–29 May 2020; pp. 1–5. [CrossRef]

15. Barnes, L.P.; Dytso, A.; Poor, H.V. Improved Information Theoretic Generalization Bounds for Distributed and Federated Learning. *arXiv* **2022**, arXiv:2202.02423.
16. Shalev-Shwartz, S.; Shamir, O.; Srebro, N.; Sridharan, K. Learnability, Stability and Uniform Convergence. *J. Mach. Learn. Res.* **2010**, *11*, 2635–2670.
17. Bregman, L.M. The Relaxation Method of Finding the Common Point of Convex Sets and Its Application to the Solution of Problems in Convex Programming. *USSR Comput. Math. Math. Phys.* **1967**, *7*, 200–217. [CrossRef]
18. Dytso, A.; Fauß, M.; Poor, H.V. Bayesian Risk With Bregman Loss: A Cramér–Rao Type Bound and Linear Estimation. *IEEE Trans. Inf. Theory* **2022**, *68*, 1985–2000. [CrossRef]
19. Banerjee, A.; Merugu, S.; Dhillon, I.S.; Ghosh, J.; Lafferty, J. Clustering with Bregman Divergences. *J. Mach. Learn. Res.* **2005**, *6*, 1705–1749.
20. Pensia, A.; Jog, V.; Loh, P.L. Generalization Error Bounds for Noisy, Iterative Algorithms. In Proceedings of the 2018 IEEE International Symposium on Information Theory (ISIT), Vail, CO, USA, 17–22 June 2018; pp. 546–550.
21. Wang, H.; Gao, R.; Calmon, F.P. Generalization Bounds for Noisy Iterative Algorithms Using Properties of Additive Noise Channels. *arXiv* **2021**, arXiv:2102.02976.

entropy

Article

A Pattern Dictionary Method for Anomaly Detection

Elyas Sabeti [1], Sehong Oh [2], Peter X. K. Song [3] and Alfred O. Hero [2,*]

1 Michigan Institute for Data Science, University of Michigan, Ann Arbor, MI 48109, USA
2 Department of Electrical Engineering and Computer Science, University of Michigan, Ann Arbor, MI 48109, USA
3 Department of Biostatistics, University of Michigan, Ann Arbor, MI 48109, USA; pxsong@umich.edu
* Correspondence: hero@umich.edu

Abstract: In this paper, we propose a compression-based anomaly detection method for time series and sequence data using a pattern dictionary. The proposed method is capable of learning complex patterns in a training data sequence, using these learned patterns to detect potentially anomalous patterns in a test data sequence. The proposed pattern dictionary method uses a measure of complexity of the test sequence as an anomaly score that can be used to perform stand-alone anomaly detection. We also show that when combined with a universal source coder, the proposed pattern dictionary yields a powerful atypicality detector that is equally applicable to anomaly detection. The pattern dictionary-based atypicality detector uses an anomaly score defined as the difference between the complexity of the test sequence data encoded by the trained pattern dictionary (typical) encoder and the universal (atypical) encoder, respectively. We consider two complexity measures: the number of parsed phrases in the sequence, and the length of the encoded sequence (codelength). Specializing to a particular type of universal encoder, the Tree-Structured Lempel–Ziv (LZ78), we obtain a novel non-asymptotic upper bound, in terms of the Lambert W function, on the number of distinct phrases resulting from the LZ78 parser. This non-asymptotic bound determines the range of anomaly score. As a concrete application, we illustrate the pattern dictionary framework for constructing a baseline of health against which anomalous deviations can be detected.

Keywords: pattern dictionary; atypicality; Lempel–Ziv algorithm; lossless compression; anomaly detection

Citation: Sabeti, E.; Oh, S.; Song, P.X.K.; Hero, A.O. A Pattern Dictionary Method for Anomaly Detection. *Entropy* **2022**, *24*, 1095. https://doi.org/10.3390/e24081095

Academic Editors: Chao Tian and Lizhong Zheng

Received: 14 July 2022
Accepted: 7 August 2022
Published: 9 August 2022

1. Introduction

Anomaly detection and outlier detection are used for detecting data samples that are inconsistent with normal data samples. Early methods did not take the sequential structure of the data into consideration [1]. However, many real world applications involve data collected as a sequence or time series. In such data, anomalous samples are better characterized as subsequences of time series. Anomaly detection is a challenging task due to the uncertain nature of anomalies. Anomaly detection in time series and sequence data is particularly difficult since both length and occurrence frequency of potentially anomalous subsequences are unknown. Additionally, algorithmic computational complexity can be a challenge, especially for streaming data with large alphabet sizes.

In this paper, we propose a universal nonparametric model-free anomaly detection method for time series and sequence data based on a pattern dictionary (PD). Given training and test data sequences, a pattern dictionary is created from the sets of all the patterns in the training data. This dictionary is then used to sequentially parse and compress (in a lossless manner) the test data sequence. Subsequently, we interpret the number of parsed phrases or the codelength of the test data as anomaly scores. The smaller the number of parsed phrases or the shorter the compressed codelength of the test data, the more similarity between training and test data patterns. This sequential parsing and lossless compression procedure leads to detection of anomalous test sequences and their potential anomalous patterns (subsequences).

The proposed pattern dictionary method has the following properties: (i) it is nonparametric since it does not rely on a family of parametric distributions; (ii) it is universal in the sense that the detection criterion does not require any prior modeling of the anomalies or nominal data; (iii) it is non-Bayesian as the detection criterion is model-free; and (iv) as it depends on data compression, data discretization is required prior to building the dictionary. While the proposed pattern dictionary can be used as a stand-alone anomaly detection method (Pattern Dictionary for Detection (PDD)), we show how it can be utilized in the atypicality framework [2,3] for more general data discovery problems. This results in a method we call PDA (Pattern Dictionary based Atypicality), in which the proposed pattern dictionary is contrasted against a universal source coder which is the Tree-Structured Lempel–Ziv (LZ78) [4,5]. We use the LZ78 as the universal encoder since its compression procedure is similar to our proposed pattern dictionary, and it is (asymptotically) optimal [4,5].

The main contributions of this paper are as follows. First, we propose the pattern dictionary method for anomaly detection and characterize its properties. We show in Theorem 1 that using a multi-level dictionary that separates the patterns by their depth results in a shorter average indexing codelength in comparison to a uni-level dictionary that uses a uniform indexing approach. Second, we develop novel non-asymptotic lower and upper bounds of the LZ78 parser in Theorem 2 and further analyze its non-asymptotic properties. We demonstrate that the non-asymptotic upper bound on the number of distinct phrases resulting from the LZ78 parsing of an $|\mathcal{X}|$-ary sequence of length l can be explicitly expressed by the Lambert W function [6]. To the best of our knowledge, such characterization has not previously appeared in the literature. Then, we show in Lemma 1 that the achieved non-asymptotic upper bound on the number of distinct phrases resulting from the LZ78 parsing converges to the optimal upper bound $\frac{l}{\log l}$ of the LZ78 parser as $l \to \infty$. Third, we show how the pattern dictionary and LZ78 can be used together in an atypicality detection framework. We demonstrate that the achieved non-asymptotic lower and upper bounds on both LZ78 and pattern dictionary determine the range of the anomaly score. Consequently, we show how these bounds can be used to analyze the effect of dictionary depth on the anomaly score. Furthermore, the bounds are used to set the anomaly detection threshold. Finally, we compare our proposed methods with the competing methods, including nearest neighbors-based similarity [7], threshold sequence time-delay embedding [8–11], and compression-based dissimilarity measure [12–15,15,16], that are designed for anomaly detection in sequence data and time series. We conclude our paper with an experiment that details how the proposed framework can be used to construct a baseline of health against which anomalous deviations are detected.

The paper is organized as follows. In Section 2, we briefly review the relevant literature in anomaly detection (readers who are familiar with anomaly detection can skip this section). Section 3 introduces the detection framework and the notation used in this paper. Section 4 presents our proposed pattern dictionary method and its properties. In Section 5, we show how the proposed pattern dictionary can be used in an atypicality framework alongside LZ78, and we analyze the non-asymptotic properties of the LZ78 parser. Section 6 presents experiments that illustrate the proposed pattern dictionary anomaly detection procedure. Finally, Section 7 concludes our paper.

2. Related Works

Anomaly detection has a vast literature. Anomaly detection procedures can be categorized into parametric and nonparametric methods. Parametric methods rely on a family of parametric distributions to model the normal data. The slippage problem [17], change detection [18–21], concept drift detection [19–22], minimax quickest change detection (MQCD) [23–25], and transient detection [26–29] are examples of parametric anomaly detection problems. The main difference between our proposed pattern dictionary method and the aforementioned techniques is that our method is a model-free nonparametric method. The main drawback of the parametric anomaly detection procedure is that it is difficult to accurately specify the parametric distribution for the data under investigation.

Nonparametric anomaly detection approaches do not assume any explicit parameterized model for the data distributions. An example is an adaptive nonparametric anomaly detection approach called geometric entropy minimization (GEM) [30,31] that is based on the minimal covering properties of K-point entropic graphs constructed on N training samples from a nominal probability distribution. The main difference between GEM-based methods and our proposed pattern dictionary is that former techniques are designed to detect outliers and cannot easily incorporate the temporal information regarding anomaly in a data stream. Another nonparametric detection method is sequential nonparametric testing that considers data as online stream and addresses the growing data storage problem by sequentially testing every new data samples [32,33]. A key difference between sequential nonparametric testing and our proposed pattern dictionary method is that our method is based on coding theory instead of statistical decision theory.

Information theory and universal source coding have been used previously in anomaly detection [34–45]. The detection criteria in these approaches are based on comparing metrics such as complexity or similarity distances that depend on entropy rate. An issue with these approaches is that there are many completely dissimilar sources with the same entropy rate, reducing outlier sensitivity. Another related problem is universal outlier detection [46,47]. In these works, different levels of knowledge about nominal and outlier distributions and number of outliers are incorporated. Unlike these methods, our proposed pattern dictionary approach does not require any prior knowledge about outliers and anomalies. In [48], a measure of empirical informational divergence between two individual sequences generated from two finite-order, finite-alphabet, stationary Markov processes is introduced and used for a simple universal classification. While the parsing procedure used in [48] is similar to the pattern dictionary used in this paper, there are important differences. The empirical measure proposed in [48] is a stand alone score function that is designed for two-class classification, while our measure is a direct byproduct of the LZ78 encoding algorithm designed for single-class classification, i.e., anomaly detection. In addition, the theoretical convergence of the empirical measure to the relative entropy between the class conditioned distributions, shown in [48], is only guaranteed when the sequences satisfy the finite-order Markov property, a condition that may be difficult to satisfy in practice. In [2,3], an information theoretic data discovery framework called *atypicality* has been introduced in which the detection criterion is based on a descriptive codelength comparison of an optimum encoder or a training-based fixed source coder, namely a data-dependent source coder introduced in [2]) with a universal source coder. In this paper, we show how our proposed pattern dictionary method can be used as a training-based fixed source coder in an atypicality framework.

Anomaly and outlier detection for time series has also been extensively studied [49]. Various time series modeling techniques such as regression [50], auto regression [51], auto regression moving average [52], auto regressive integrated moving average [53], support vector regression [54], and Kalman filters [55] have been used to detect anomalous observations by comparing the estimated residuals to a threshold. Many of these methods depend on a statistical assumption on the residuals, e.g., an assumption of Gaussian distribution, while the pattern dictionary method is model-free.

The proposed pattern dictionary method is closely related to the anomaly detection methods that are designed for sequence data. Many of these methods are focused on specific applications. For instance, detection of mutations in DNA sequences [7,56], detection of cyberattacks in computer network [57], and detection of irregular behaviors in online banking [58] are all application-specific examples of anomaly detection for discrete sequences. In the recent years, multiple sequence data anomaly detection methods have been developed specifically for graphs [59], dynamic networks [60], and social networks [61]. Chandola et al. [34] summarized many anomaly detection methods for discrete sequences and identified three general approaches to this problem. These anomaly detection formulations are unique in the way that anomalies are defined, but similar in their reliance on comparison between a test (sub)sequence and normal sequences in the training data. For example, kernel-based techniques such as nearest neighbor-based similarity (NNS) [7] are designed to detect anomalous sequences that are dissimilar to the training data. As another example, threshold sequence time-delay embedding (t-STIDE) [8–11] is established to detect anomalous sequences that contain subsequences with anomalous occurrence frequencies. The compression-based dissimilarity measure (CDM) is proposed for discord detection [12–15,15,16] to detect anomalous subsequences within a long sequence. Chandola et al. [34] also showed how various techniques developed for one problem formulation can be changed and applied to other problem formulations. While our pattern dictionary method shares similarity with NNS, CDM, and t-STIDE, our proposed method is generally applicable to any of the categories of anomaly detection identified in [34]. Furthermore, our detection criterion does not depend on the specific type of anomaly. Note that while CDM is also a compression-based method, its anomaly score is based on a dissimilarity measure that might fail to detect atypical subsequences [2]. For instance, using CDM method, a binary i.i.d. uniform training sequence is equally dissimilar to another binary i.i.d. uniform test sequence or to a test sequence drawn from some other distribution. In Section 6, the detection performance of our proposed pattern dictionary method is compared with NNS, CDM, t-STIDE, and the Ziv–Merhav method of [48].

It is worth mentioning that since the proposed pattern dictionary method is based on lossless source coding, it requires discretization of time series prior to deployment. In fact, many anomaly detection approaches require discretization of continuous data prior to applying inference techniques [62–65]. Note that discretization is also a requirement in other problem settings such as continuous optimization in genetic algorithms [66], image pattern recognition [67], and nonparametric histogram matching over codebooks in computer vision [68].

3. Framework and Notation

In the anomaly detection literature for sequence data and time series, the following three general formulations are considered [34]: (i) an entire test sequence is anomalous if it is notably different from normal training sequences; (ii) a subsequence within a long test sequence is anomalous if it is notably different from other subsequences in the same test sequence or the subsequences in a given training sequence; and (iii) a given test subsequence or pattern is anomalous if its occurrence frequency in a test sequence is notably different from its occurrence frequency in a normal training sequence. In this paper, we consider a unified formulation in which we determine if a (sub)sequence is anomalous with respect to a training sequence (or training sequence database) if any of the aforementioned three conditions are met. In other words, given a training sequence or a training sequence database, a test sequence is anomalous if it is significantly different from training sequences, or it contains a subsequence that is significantly different from subsequences in the training sequence, or it contains a subsequence whose occurrence frequency is significantly different from its occurrence frequency in the training data.

Notation

We use x to denote a sequence and x_n^m to denote a subsequence of x: $x_n^m = \{x_i, i = n, n+1, \ldots, m\}$, and x^l represents a sequence of length l, i.e., $\{x_n, n = 1, \ldots, l\}$. $\mathcal{X}$ denotes a finite set, and $\mathcal{D}$ represents a dictionary of subsequences. Throughout this paper:

- All logarithms are base 2 unless otherwise is indicated.
- In the encoding process, we always adhere to lossless compression and strict decodability at the decoder.
- While adhering to strict decodability, we only care about the codelength, not the codes themselves.

4. Pattern Dictionary: Design and Properties

Consider a long sequence, called the training data, $\{x_n, n = 1, \ldots, L\}$ of length L drawn from a finite alphabet $\mathcal{X}$. The goal is to *learn* the patterns (subsequences) of this sequence by creating a dictionary that contains all distinct patterns of maximum length (depth) $D_{max} \ll L$ that are embedded in the sequence. We call this dictionary a *pattern dictionary* $\mathcal{D}$ with the maximum depth D_{max} and the set of observed patterns $\mathcal{S}_{\mathcal{D}}(x_1^L)$.

Example 1. *Suppose $D_{max} = 2$, the alphabet is $\mathcal{X} = \{A, B, C, D\}$ and the training sequence is $x = ABACADABBACCADDABABACADAB$. The set of patterns with depth $d \leq D_{max}$ in this sequence is $\mathcal{S}_{\mathcal{D}}(x) = \{A, B, C, D, AB, BA, AC, CA, AD, DA, BB, CC, DD\}$.*

Since the pattern dictionary is going to be used as a training-based fixed source coder (a data-dependent source coder as defined in [2]), an efficient structure for the pattern representation that minimizes the indexing codelength is of interest. The simplest approach is to consider all the patterns of length $1 \leq d \leq D_{max}$ in one set $\mathcal{S}_{\mathcal{D}}$ and use a uniform indexing approach. This approach is called a *uni-level dictionary*. Another approach is to separate all the patterns by their depth (pattern length) and arrange them in D_{max} sets $\mathcal{S}_{\mathcal{D}}^{(1)}, \mathcal{S}_{\mathcal{D}}^{(2)}, \ldots, \mathcal{S}_{\mathcal{D}}^{(\mathrm{D_{max}})}$, and define $\mathcal{S}_{\mathcal{D}} = \bigcup_{d=1}^{D_{max}} \mathcal{S}_{\mathcal{D}}^{(\mathrm{d})}$, which we call a *multi-level dictionary*. In the following sections, we show that the latter results in a shorter average indexing codelength. It is worth mentioning that since a multi-level dictionary results in a depth-dependent indexing codelength, the average over the depth is considered. A relevant question is if the average of indexing codelength over all the patterns independent of depth should be used as an alternative. Since such pattern dictionaries are used to sequentially parse test data, patterns at smaller depth are more likely to be matched, even if they are anomalous. Thus, the average of indexing codelength over depth can better differentiate depth-dependent anomalies.

4.1. A Special Case

Suppose all the possible patterns of depth $d \leq D_{max}$ exist in the training sequence $\{x_n, n = 1, \ldots, L\}$. That is, the cardinality of $\mathcal{S}_{\mathcal{D}}^{(\mathrm{d})}$ is $\left|\mathcal{S}_{\mathcal{D}}^{(\mathrm{d})}\right| = |\mathcal{X}|^d$ for $1 \leq d \leq D_{max}$. Then, the total number of patterns is

$$\begin{aligned}
\left|\mathcal{S}_{\mathcal{D}}\left(x_1^L\right)\right| &= \sum_{d=1}^{D_{max}} \left|\mathcal{S}_{\mathcal{D}}^{(\mathrm{d})}\left(x_1^L\right)\right| \\
&= \sum_{d=1}^{D_{max}} |\mathcal{X}|^d \\
&= \frac{|\mathcal{X}|\left(|\mathcal{X}|^{D_{max}} - 1\right)}{|\mathcal{X}| - 1}.
\end{aligned}$$

Hence, a uni-level dictionary results in a uniform indexing codelength of

$$\begin{aligned}
L^{uni} &= \log\left(\frac{|\mathcal{X}|\left(|\mathcal{X}|^{D_{max}} - 1\right)}{|\mathcal{X}| - 1}\right) \\
&\approx D_{max} \log(|\mathcal{X}|).
\end{aligned}$$

On the other hand, a multi-level dictionary requires a two-stage description of index. The first stage is the index of the depth d (using $\log D_{max}$ bits), and the second stage is the index of the pattern among all the patterns with the same depth (using $d \log(|\mathcal{X}|)$ bits). This two-stage description of the index leads to a non-uniform indexing of codelength: the minimum indexing codelength occurring for the patterns of depth $d = 1$ equals to $L_{min}^{multi} = \log D_{max} + \log(|\mathcal{X}|)$ bits, while the maximum indexing codelength occurring for the patterns of depth $d = D_{max}$ equals to L_{max}^{multi} $=\log D_{max} + D_{max}\log(|\mathcal{X}|)$ bits. Thus, the average indexing codelength of a multi-level dictionary is given by

$$\begin{aligned} L^{multi} &= \frac{1}{D_{max}}\sum_{d=1}^{D_{max}}(\log D_{max} + d\log(|\mathcal{X}|)) \\ &= \log D_{max} + \frac{\log(|\mathcal{X}|)}{D_{max}}\sum_{d=1}^{D_{max}} d \\ &\approx \log D_{max} + \frac{1}{2}D_{max}\log(|\mathcal{X}|). \end{aligned}$$

Figures 1 and 2 graphically compare the indexing codelength between a uni-level dictionary and a multi-level dictionary for a fixed alphabet size and a fixed D_{max}, respectively. As seen, the average indexing codelength of a multi-level dictionary results in a shorter indexing codelength.

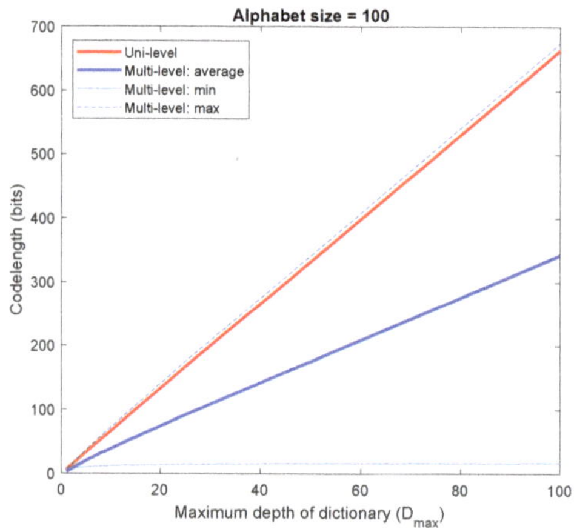

Figure 1. Comparison of indexing codelength between a uni-level dictionary and a multi-level dictionary (fixed alphabet size $|\mathcal{X}| = 100$).

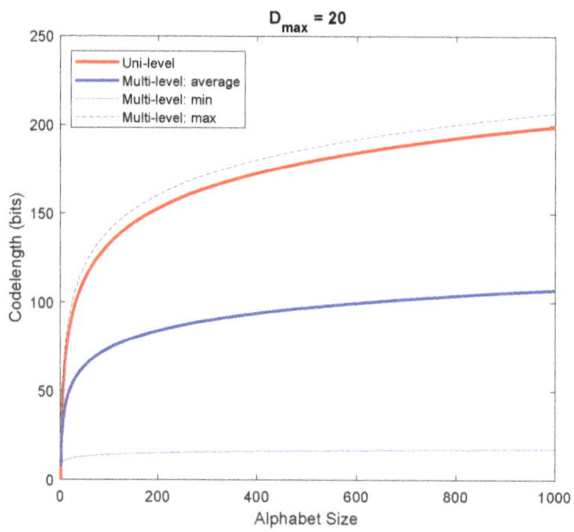

Figure 2. Comparison of indexing codelength between a uni-level dictionary and a multi-level dictionary (fixed $D_{max} = 20$).

4.2. The General Case

Given the training sequence $\{x_n, n = 1, \ldots, L\}$, suppose there are $a_d = \left|\mathcal{S}_{\mathcal{D}}^{(d)}\right| \leq |\mathcal{X}|^d$ patterns of depth $d \leq D_{max}$ (a_1 patterns of depth one, a_2 patterns of depth two, etc.). The following Theorem 1 shows that the average indexing codelength using a multi-level dictionary is always less than the indexing codelength of a uni-level dictionary.

Theorem 1. *Assume there are embedded $a_d = \left|\mathcal{S}_{\mathcal{D}}^{(d)}\right| \leq |\mathcal{X}|^d$ patterns of depth $1 \leq d \leq D_{max}$ in a training sequence of length $L \gg D_{max}$. Let L^{uni} and L^{multi} be the indexing codelength of a uni-level dictionary and the average indexing codelength of a multi-level dictionary, respectively. Then,*

(1) $L^{multi} \leq L^{uni}$; and

(2) $\log\left(1 + \frac{\left(\sqrt{a_{D_{max}}} - \sqrt{a_1}\right)^2}{D_{max} a_{D_{max}}}\right) \leq L^{uni} - L^{multi} \leq \log\left(1 + w + (1-w)\frac{a_{D_{max}}}{a_1} - a_1^{w-1} a_{D_{max}}^{1-w}\right)$,

where

$$w = \frac{\ln\left[\left(\frac{a_{D_{max}}}{a_{D_{max}} - a_1}\right) \ln \frac{a_{D_{max}}}{a_1}\right]}{\ln \frac{a_{D_{max}}}{a_1}}.$$

Proof. Since $L \gg D_{max}$, clearly $0 < a_1 \leq a_2 \leq \cdots \leq a_{D_{max}}$. Using a uni-level dictionary, the indexing codelength is

$$\begin{aligned} L^{uni} &= \log\left(\sum_{d=1}^{D_{max}} a_d\right) \\ &= \log D_{max} + \log A_{D_{max}}, \end{aligned}$$

where $A_{D_{max}} \triangleq (a_1 + a_2 + \cdots + a_{D_{max}})/D_{max}$ is the arithmetic mean of $a_1, a_2, \ldots, a_{D_{max}}$. Using a multi-level dictionary the average indexing codelength is

$$\begin{aligned} L^{multi} &= \frac{1}{D_{max}} \sum_{d=1}^{D_{max}} (\log D_{max} + \log a_d) \\ &= \log D_{max} + \log G_{D_{max}}, \end{aligned}$$

where $G_{D_{max}} \triangleq \left(\prod_{d=1}^{D_{max}} a_d\right)^{1/D_{max}}$ is the geometric mean of $a_1, a_2, \ldots, a_{D_{max}}$. Hence, the comparison between L^{uni} and L^{multi} comes down to comparing the arithmetic mean and the geometric mean of $a_1, a_2, \ldots, a_{D_{max}}$. Thus, $A_{D_{max}} \geq G_{D_{max}}$, which established the first part of the theorem. For the second part of the theorem, we use lower and upper bounds on $A_{D_{max}} - G_{D_{max}}$ derived in [69]

$$\begin{aligned} \frac{\left(\sqrt{a_{D_{max}}} - \sqrt{a_1}\right)^2}{D_{max}} &\leq A_{D_{max}} - G_{D_{max}} \leq \\ &\left[w a_1 + (1-w) a_{D_{max}} - a_1^w a_{D_{max}}^{1-w}\right], \end{aligned}$$

where $w = \frac{\ln[(a_{D_{max}}/(a_{D_{max}} - a_1)) \ln(a_{D_{max}}/a_1)]}{\ln(a_{D_{max}}/a_1)}$. Since $a_1 \leq G_{D_{max}} \leq a_{D_{max}}$ and $L^{uni} - L^{multi} = \log \frac{A_{D_{max}}}{G_{D_{max}}}$, the proof is complete. □

Theorem 1 shows that a multi-level dictionary gives shorter average indexing codelength than a uni-level dictionary. $\log D_{max} + \log a_d$ is the indexing codelength for patterns of depth d, where a_d is the total number of observed patterns of the depth d. In order to reduce the indexing codelength even further, the patterns of the same length in each set $\mathcal{S}_{\mathcal{D}}^{(d)}$ can be ordered according to their relative frequency (empirical probability) in the training sequence. This allows Huffman or Shannon–Fano–Elias source coding [4] to be used to assign prefix codes to patterns in each set $\mathcal{S}_{\mathcal{D}}^{(\mathrm{d})}$ separately. In this case, for any pattern $x_1^d \in \mathcal{S}_{\mathcal{D}}^{(\mathrm{d})}$, the indexing codelength becomes

$$L^{multi}\left(x_1^d\right) = \log D_{max} + L_{\mathcal{D}}^{(d)}\left(x_1^d\right), \tag{1}$$

where $L_{\mathcal{D}}^{(d)}\left(x_1^d\right)$ is the codelength assigned to the pattern x_1^d based on its empirical probability using a Huffman or Shannon–Fano–Elias encoder. If such encoders are used, the codelength (1) is optimal ([4] Theorem 5.8.1). Since the whole purpose of creating a pattern dictionary is to learn the patterns in the training data, assigning the shorter codelength to the more frequent patterns and assigning longer codelength to the less frequent patterns in any pattern set $\mathcal{S}_{\mathcal{D}}^{(\mathrm{d})}$ will improve the efficiency of the coded representation.

Example 2. *Suppose the alphabet is $\mathcal{X} = \{A, B, C, D\}$ and the training sequence is $x = ABACADABBACCADDABABACADAB$. Table 1 shows the dictionary with $D_{max} = 3$ created by the patterns inside the training sequence, and the codelength assigned for each pattern using Huffman coding.*

Table 1. Filling (training) the dictionary (of maximum depth $D_{max} = 3$) with the patterns in the training sequence $ABACADABBACCADDABABACADAB$.

Depth 1			Depth 2			Depth 3		
x_1^d	$\Pr(x_1^d)$	$L_{\mathcal{D}}^{(1)}(x_1^d)$	x_1^d	$\Pr(x_1^d)$	$L_{\mathcal{D}}^{(2)}(x_1^d)$	x_1^d	$\Pr(x_1^d)$	$L_{\mathcal{D}}^{(3)}(x_1^d)$
A	0.44	1	AB	0.2083	2	ABA	0.1304	3
B	0.24	2	BA	0.1667	3	BAC	0.1304	3
C	0.16	3	AC	0.1250	3	CAD	0.1304	3
D	0.16	3	CA	0.1250	3	DAB	0.1304	3
			AD	0.1250	3	ACA	0.0870	4
			DA	0.1250	3	ADA	0.0870	4
			BB	0.0417	4	ABB	0.0435	4
			CC	0.0417	5	BBA	0.0435	4
			DD	0.0417	5	ACC	0.0435	4
						CCA	0.0435	4
						ADD	0.0435	4
						DDA	0.0435	5
						BAB	0.0435	5

4.3. Pattern Dictionary for Detection (PDD)

Suppose we want to sequentially compress a test sequence $x_1^l = \{x_n, n = 1, \ldots, l\}$ using a trained pattern dictionary $\mathcal{D}$ with maximum depth $D_{max} < l$. The encoder parses the test sequence x_1^l into c phrases, $x_{v_1}^{v_2-1}, x_{v_2}^{v_3-1}, \ldots, x_{v_c}^l$ where v_i is the index of the start of the ith phrase, and each phrase $x_{v_i}^{v_{i+1}-1}$ is a pattern in the pattern dictionary $\mathcal{D}$. Let $\mathcal{S}_{\mathcal{D}}\left(x_1^l\right) = \left\{x_{v_1}^{v_2-1}, x_{v_2}^{v_3-1}, \ldots, x_{v_c}^l\right\}$ denote the set of the parsed phrases using pattern dictionary $\mathcal{D}$. The parsing process begins with setting $v_1 = 1$ and finding the largest $v_2 \le D_{max}$ and $v_2 \le l$ such that $x_{v_1}^{v_2-1} \in \mathcal{D}$ but $x_{v_1}^{v_2} \notin \mathcal{D}$. This results in the first phrase $x_1^{v_2-1}$. Similarly, the same procedure is performed in order to find the largest $v_3 \le D_{max}$ and $v_3 \le l$ such that $x_{v_2}^{v_3-1} \in \mathcal{D}$ but $x_{v_2}^{v_3} \notin \mathcal{D}$. This type of cross-parsing was first introduced in [48] in order to estimate an empirical relative entropy between two individual sequences that are independent realizations of two finite-order, finite-alphabet and stationary Markov processes. Here, we do not impose such an assumption on the sources generating the sequences. Algorithm 1 summarizes the procedure of the proposed pattern dictionary (PD) parser. After parsing the whole test sequence x_1^l into c phrases, $x_{v_1}^{v_2-1}, x_{v_2}^{v_3-1}, \ldots, x_{v_c}^{l}$, the codelength will be

$$L\left(x_1^l\right) = \sum_{i=1}^{c} L_{\mathcal{D}}\left(x_{v_i}^{v_{i+1}-1}\right) + c \log D_{max}. \quad (2)$$

Algorithm 1 Pattern Dictionary (PD) Parser

Require: Pattern Dictionary $\mathcal{D}$, Test Sequence x_1^l

Set $c = 1, v_c = 1, d = 1$
while $v_c + d - 1 < l$ **do**
    if $x_{v_c}^{v_c+d-1} \in \mathcal{S}_{\mathcal{D}}^{(\mathrm{d})}$ **then**
        if $d + 1 \le D_{max}$ **then**
            $d = d + 1$
        else
            $v_{c+1} = v_c + d$
            $c = c + 1$
            $d = 1$
    else
        $v_{c+1} = v_c + d - 1$
        $c = c + 1$
        $d = 1$
return $x_{v_1}^{v_2-1}, x_{v_2}^{v_3-1}, \ldots, x_{v_c}^{l}$

For detection purposes, on a test sequence x_1^l, either the number of parsed phrases or the codelength can be used as anomaly scores with respect to the trained pattern dictionary $\mathcal{D}$. In other words, for any test sequence x_1^l and given a pattern dictionary, if the number of parsed phrases $\left|\mathcal{S}_{\mathcal{D}}\left(x_1^l\right)\right|$ or the codelength $L\left(x_1^l\right)$ in Equation (2) are greater than a certain threshold, then x_1^l is declared to be anomalous. While the proposed pattern dictionary technique can be used as a stand-alone anomaly detection technique, below we show how it can be used for atypicality detection [2,3] as a training-based fixed source coder (data-dependent encoder).

5. Pattern Dictionary-Based Atypicality (PDA)

In [2,3], an *atypicality framework* was introduced as a data discovery and anomaly detection framework that is based on a central definition: "a sequence (or subsequence) is atypical if it can be described (coded) with fewer bits in itself rather than using the (optimum) code for typical sequences". In this framework, detection is based on the comparison of a lossless descriptive codelength between an optimum encoder (if the typical model is known) or a training-based fixed source coder (if the typical model is unknown, but training data are available) and a universal source coder in order to detect atypical subsequences in the data [2,3]. In this section, we apply our proposed pattern dictionary as a training-based fixed source coder (typical encoder) in an atypicality framework. We call it pattern dictionary-based atypicality (PDA) method.

The pattern dictionary-based source coder can be considered as a generalization of the Context Tree [70–72] based fixed source coder that was used in [2] for discrete data. The universal source coder (atypical encoder) used here is the Tree-Structured Lempel–Ziv (LZ78) [4,5]. The primary reason for choosing LZ78 as the universal encoder is that its sequential parsing procedure is similar to the proposed pattern dictionary described in Section 4, and it is (asymptotically) optimal [4,5]. One might ask why do we even need to compare descriptive codelengths of a training-based (or optimum) encoder with a universal encoder for data discovery purposes when, as alluded to in the end of last section, a training-based fixed source coder can be a stand-alone anomaly detector. The necessity of such concurrent comparison is articulated in [2]. In fact, such a codelength comparison enables the atypicality framework to go beyond the detection of anomalies and outliers, extending to the detection of *rare* parts of data that might have a data structure of interest to the practitioner.

We give an example to provide further intuition for why anomaly detection can benefit from our framework that compares the outputs of a typical encoder and an atypical encoder. Consider an i.i.d. binary sequence of length L with $P(X = 1) = p$ in which there is embedded an anomalous subsequence of length $l \ll L$ with $P(X = 1) = \widehat{p} \neq p$ that we would like to detect. If $p = \frac{1}{2}$ and $\widehat{p} = 1$, the typical encoder cannot catch the anomaly while the atypical encoder can. On the other hand, if $p = \frac{1}{3}$ and $\widehat{p} = \frac{2}{3}$, the typical encoder identifies the anomaly while an atypical encoder fails to do so (since the entropy for $p = \frac{1}{3}$ and $\widehat{p} = \frac{2}{3}$ is the same). Note that in both cases, our framework would catch the anomaly since it uses the difference between the descriptive codelengths of these two encoders.

Recall that in Section 4, we supposed that a test sequence x_1^l has been parsed using a trained pattern dictionary $\mathcal{D}$ with maximum depth $D_{max} < l$. This parsing results in $\left|\mathcal{S}_{\mathcal{D}}\left(x_1^l\right)\right|$ parsed phrases. Using Equation (2), the typical codelength of the sequence x_1^l is given by

$$L_T\left(x_1^l\right) = \sum_{y \in \mathcal{S}_{\mathcal{D}}\left(x_1^l\right)} L_{\mathcal{D}}(y) + \left|\mathcal{S}_{\mathcal{D}}\left(x_1^l\right)\right| \log D_{max}.$$

For the atypical encoder, the LZ78 algorithm results in a distinct parsing of the test sequence x_1^l. Let $\mathcal{S}_{LZ}\left(x_1^l\right)$ denote the set of parsed phrases in the LZ78 parsing of x_1^l. As such, the resulting atypical codelength is [4,5]

$$L_A\left(x_1^l\right) = \left|\mathcal{S}_{LZ}\left(x_1^l\right)\right| \left[\log\left|\mathcal{S}_{LZ}\left(x_1^l\right)\right| + 1\right].$$

Since $L\left(x_1^l\right)$ using both LZ78 and the pattern dictionary depends on the number of parsed phrases, we investigate the possible range and properties of $\left|\mathcal{S}_{\mathcal{D}}\left(x_1^l\right)\right| - \left|\mathcal{S}_{LZ}\left(x_1^l\right)\right|$. While the LZ78 encoder is a well-known compression method which is asymptotically optimal [4,5], its non-asymptotic behavior is not well understood. In the next section, we establish a novel non-asymptotic property of an LZ78 parser, and then compare it with the pattern dictionary parser.

5.1. Lempel–Ziv Parser

We start this section with a theorem that establishes the non-asymptotic lower and upper bounds on the number of distinct phrases in a sequence parsed by LZ78.

Theorem 2. *The number of distinct phrases $c(l)$ resulting from LZ78 parsing of an $|\mathcal{X}|$-ary sequence $x_1^l = \{x_n, n = 1, \ldots, l\}$ satisfies*

$$\frac{1}{2}\left(\sqrt{8l+1} - 1\right) \leq c(l) \leq \frac{l \ln|\mathcal{X}|}{\mathrm{W}\left(\frac{\beta}{\alpha}|\mathcal{X}|^{\frac{\alpha+1}{-\alpha}} \ln|\mathcal{X}|\right)},$$

where $\alpha = |\mathcal{X}| - 1$, $\beta = (|\mathcal{X}| - 1)^2 l - |\mathcal{X}|$, and $\mathrm{W}(.)$ is the Lambert W function [6].

Proof. First, we establish the upper bound. Note that the number of parsed distinct phrases $c(l)$ is maximized when all the phrases are as short as possible. Define $M \triangleq |\mathcal{X}|$ and let l_k be the sum of the lengths of all distinct strings of length less than or equal to k. Then,

$$l_k = \sum_{j=1}^{k} jM^j = \frac{1}{(M-1)^2}\left[\{(M-1)k - 1\}M^{k+1} + M\right].$$

Since $l = l_k$ occurs when all the phrases are of length $\leq k$,

$$c(l_k) \leq \sum_{j=1}^{k} M^j = \frac{M\left(M^k - 1\right)}{M-1} < \frac{M^{k+1}}{M-1} \leq \frac{l_k}{k - \frac{1}{M-1}}.$$

If $l_k \leq l < l_{k+1}$, we write $l = l_k + \triangle$ where

$$\begin{aligned}\triangle < l_{k+1} - l_k &= (Mk + M - 1 - k)\frac{M^{k+1}}{M-1} \\ &= (k+1)\frac{M^{k+1}}{M-1}.\end{aligned}$$

We conclude that the parsing ends up with $c(l_k)$ phrases of length $\leq k$ and $\frac{l-l_k}{k+1}$ phrases of length $k+1$. Therefore,

$$\begin{aligned}c(l) &\leq c(l_k) + \frac{l - l_k}{k+1} \leq \frac{l_k}{k - \frac{1}{M-1}} + \frac{\triangle}{k+1} \\ &\leq \frac{l_k + \triangle}{k - \frac{1}{M-1}} = \frac{l}{k - \frac{1}{M-1}}.\end{aligned} \quad (3)$$

We now bound the size of k for a given sequence of length l by setting $l = l_k$. Define $\alpha \triangleq M - 1$ and $\beta \triangleq (M-1)^2 l - M$. Then,

$$\begin{aligned}&\frac{1}{(M-1)^2}\left[((M-1)k - 1)M^{k+1} + M\right] = l \\ \Longleftrightarrow &((M-1)k-1)M^{k+1} = (M-1)^2 l - M \\ \Longleftrightarrow &(\alpha k - 1)M^{k+1} = \beta \\ \Longleftrightarrow &\widehat{k} M^{(\widehat{k}+1)/\alpha + 1} = \beta \\ \Longleftrightarrow &\widehat{k}\frac{\ln M}{\alpha}\exp\left(\widehat{k}\frac{\ln M}{\alpha}\right) = \frac{\beta}{\alpha} M^{-1-1/\alpha} \ln M.\end{aligned}$$

where $\widehat{k} = \alpha k - 1$. The last equation can be solved using the Lambert W function [6]. Since all the involved numbers are real and for $M > 1$ and $l \geq 2$, we have $\frac{\beta}{\alpha} M^{-1-1/\alpha} \ln M \geq 0 > -\frac{1}{e}$, it follows that

$$\begin{aligned}&\widehat{k}\frac{\ln M}{\alpha} = \mathrm{W}\left(\frac{\beta}{\alpha} M^{-1-1/\alpha} \ln M\right) \\ \Longleftrightarrow &k = \frac{\alpha \mathrm{W}\left(\frac{\beta}{\alpha} M^{-1-1/\alpha} \ln M\right) + \ln M}{\alpha \ln M},\end{aligned}$$

where W(.) is the Lambert W function. Using equation (3), we write

$$c(l) \leq \frac{l}{k - \frac{1}{\alpha}} = \frac{l \ln M}{\mathrm{W}\left(\frac{\beta}{\alpha} M^{-1-1/\alpha} \ln M\right)}.$$

To prove the lower bound, note that the number of parsed distinct phrases $c(l)$ is minimized when the sequence of length l consists of only one symbol that repeats. Let $\widetilde{l_k}$ be the sum of the lengths of all such distinct strings of length less than or equal to k. Then,

$$\widetilde{l_k} = \sum_{j=1}^{k} j = \frac{k(k+1)}{2}.$$

Thus, given a sequence of length l by enforcing $l = \frac{k(k+1)}{2}$, we obtain the lower bound. □

Figure 3 illustrates the lower and upper bounds established in Theorem 2 against the sequence length for various alphabet sizes. Note that the lower bound on the number of distinct phrases is independent of the alphabet size.

While numerical experiments are not a substitute for the mathematical proof of Theorem 2 provided above, the reader may find it useful to understand the theorem in terms of a simple example. In Figures 4–6, we compare the theoretical bound with numerical results of simulation for binary i.i.d. sequences. In these experiments, for each value of $P(X = 1)$, a thousand binary sequences are generated; then, the number of distinct phrases resulting from LZ78 parsing of each sequence is calculated, and hence, the average, minimum, and maximum of these counts are found and represented by error bars.

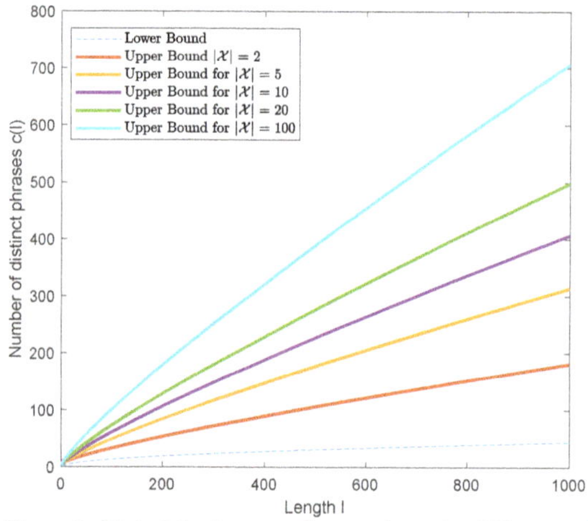

Figure 3. Plot of the lower and upper bounds of Theorem 2 on the number of distinct phrases resulting from LZ78-parsing of an $|\mathcal{X}|$-ary sequence of length l.

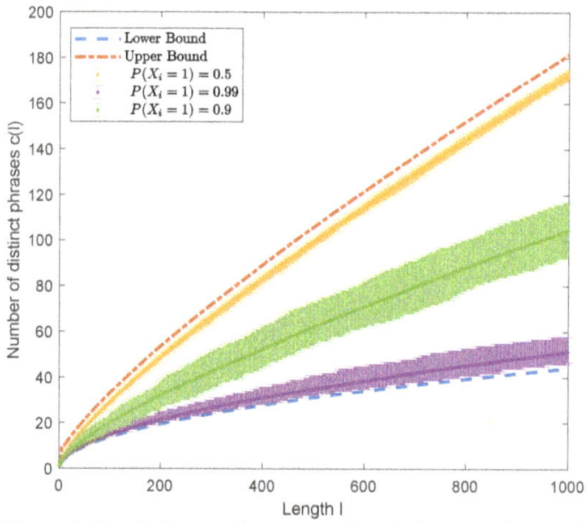

Figure 4. Simulation results compared to the lower and upper bounds of Theorem 2 on the number of distinct phrases resulting from LZ78-parsing of binary sequences of length l generated by sources with three different source probabilities $P(X = 1)$. For every $P(X = 1)$, one thousand binary sequences of length l are generated. Error bars represent the maximum, minimum, and average number of distinct phrases.

Next, we verify the convergence of the non-asymptotic upper bound achieved in Theorem 2 to the asymptotic upper bound of the LZ78 parser. Using a lower bound on Lambert W function $\ln x - \ln(\ln x) \leq W(x)$ [73], we write

$$\begin{aligned} W\left(\frac{\beta}{\alpha}\frac{\ln M}{M^{1+1/\alpha}}\right) &= W\left(\left((M-1)l - \frac{M}{M-1}\right)\frac{\ln M}{M^{\frac{M}{M-1}}}\right) \\ &\approx W(c_M l \ln M) \\ &\geq \ln \frac{c_M l \ln M}{\ln(c_M l \ln M)} \\ &= \ln \frac{c_M l}{\log(c_M l \ln M)}, \end{aligned}$$

where the logarithm is base $M = |\mathcal{X}|$ and $c_M = \frac{M-1}{M^{M/(M-1)}}$. Hence, we can further simplify the asymptotic upper bound of $c(l)$ as follows

$$\begin{aligned} c(l) &\leq \frac{l \ln M}{W\left(\frac{\beta}{\alpha} M^{-1-1/\alpha} \ln M\right)} \\ &\leq \frac{l \ln M}{\ln \frac{c_M l}{\log(c_M l \ln M)}} \\ &= \frac{l}{\log \frac{c_M l}{\log(c_M l \ln M)}} \\ &= \frac{l}{\log l + \log c_M - \log\log(c_M l \ln M)} \\ &= \frac{l}{\left(1 - \frac{\log\log l + \widehat{c_M}}{\log l}\right)\log l}, \end{aligned}$$

where $\widehat{c_M} = \log c_M - \log\log(c_M \ln M)$. Therefore, as $l \to \infty$, we have $c(l) \leq \frac{l}{\log l}$. This is consistent with the binary case $M = 2$ proved in ([4] Lemma 13.5.3) or [5]. The following Lemma extends the result of ([4] Lemma 13.5.3) to $|\mathcal{X}|$-ary case.

Lemma 1. *The number of distinct phrases $c(l)$ resulting from LZ78-parsing of an $|\mathcal{X}|$-ary sequence $x_1^l = \{x_n, n = 1, \ldots, l\}$ satisfies*

$$c(l) \leq \frac{l}{(1-\epsilon_l)\log l},$$

where the logarithm is base $|\mathcal{X}|$ and $\epsilon_l = \min\left\{1, \frac{\log\log l - \log(|\mathcal{X}|-1) + \frac{3|\mathcal{X}|-2}{|\mathcal{X}|-1}}{\log l}\right\} \to 0$ *as* $l \to \infty$.

Proof. The proof is similar to the proof in ([4] Lemma 13.5.3) or ([74] Theorem 2). Let $M \triangleq |\mathcal{X}|$. In Theorem 2, we defined l_k as the sum of the lengths of all distinct strings of length less than or equal to k, and we showed that for any given l such that $l_k \leq l < l_{k+1}$, we have $c(l) \leq c(l_k) + \frac{l-l_k}{k+1} \leq \frac{l}{k-\frac{1}{M-1}}$. Next, we bound the size of k. As such, we have $l \geq l_k \geq M^k$ or, equivalently, $k \leq \log l$ where the logarithm is base M. Additionally,

$$
\begin{aligned}
l \leq l_{k+1} &= \left(k+1-\frac{1}{M-1}\right)\frac{M^{k+2}}{M-1}+\frac{M}{(M-1)^2} \\
&= \left(\frac{k}{M-1}+\frac{M-2}{(M-1)^2}\right)M^{k+2}+\frac{M}{(M-1)^2} \\
&\leq \frac{k+2}{M-1}M^{k+2} \leq \frac{\log l+2}{M-1}M^{k+2},
\end{aligned}
$$

therefore, $k+2 \geq \log \frac{(M-1)l}{\log l+2}$. Equivalently, for $l \geq M^2$,

$$
\begin{aligned}
k-\frac{1}{M-1} &\geq \log l - \log(\log l+2) + \log(M-1) - 2 - \frac{1}{M-1} \\
&= \left(1-\frac{\log(\log l+2)-\log(M-1)+\frac{2M-1}{M-1}}{\log l}\right)\log l \\
&\geq \left(1-\frac{\log(2\log l)-\log(M-1)+\frac{2M-1}{M-1}}{\log l}\right)\log l \\
&= \left(1-\frac{\log\log l-\log(M-1)+\frac{3M-2}{M-1}}{\log l}\right)\log l \\
&= (1-\epsilon_l)\log l,
\end{aligned}
$$

where $\epsilon_l = \min\left\{1, \frac{\log\log l-\log(M-1)+\frac{3M-2}{M-1}}{\log l}\right\}$. □

Next, we analyze the properties of the number of distinct phrases $c(l)$ resulting from LZ78-parsing of an $|\mathcal{X}|$-ary sequence $x_1^l = \{x_n, n = 1, \ldots, l\}$ when l is fixed. The error bar representation in Figure 4 shows the variation of $c(l)$ when l is fixed. A possible explanation for such variations is that the statistical distribution of the pseudorandomly generated data are different from the theoretical distribution of the generating source. To elucidate this possibility, we enforce the exact matching of the source probability mass function and the empirical probability mass function of the generated data. Figure 5 represents the number of distinct phrases $c(l)$ resulting from LZ78-parsing of a binary sequence of fixed length where the characteristic of the generating source and the generated data matches. As seen, there is still some variation around the average value of $c(l)$. We can specify a distribution-dependent bound on $c(l)$ when both l and the distribution of the source are fixed.

In ([75] Theorem 1), for sequences generated from a memoryless source, $c(l)$ is assumed to be a random variable with the following mean and variance:

$$
\begin{aligned}
\mathrm{E}(c(l)) &\sim \frac{hl}{\log l}, \\
\mathrm{Var}(c(l)) &\sim \frac{(h_2-h^2)l}{\log^2 l},
\end{aligned}
\tag{4}
$$

where $h = -\sum_{a\in\mathcal{X}} p_a \log p_a$ is the entropy rate, and $h_2 = \sum_{a\in\mathcal{X}} p_a \log^2 p_a$ with p_a being the probability of symbol $a \in \mathcal{X}$. Note that the approximations (4) are asymptotic as $l \to \infty$. Below, we obtain a finite sample characterization of $c(l)$.

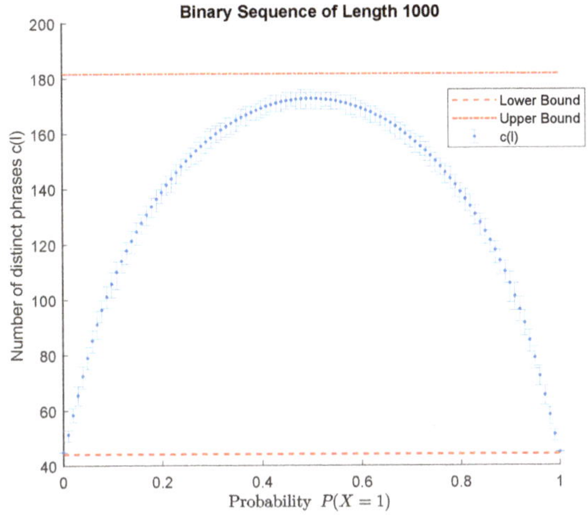

Figure 5. Similar to Figure 4, the number of distinct phrases resulting from LZ78-parsing of binary sequences of fixed length $l = 1000$ varies over the source probability parameter $P(X = 1)$. For every $P(X = 1)$, one thousand binary sequences of length l are generated. Error bars represent the maximum, minimum, and average number of distinct phrases.

Consider an $|\mathcal{X}|$-ary sequence $x_1^l = \{x_n, n = 1, \ldots, l\}$ with fixed length l generated from a source with the probability mass function $p(x)$. Here, the notations x_1^l and x^l are used interchangeably. Let $c(l, p)$ denote the number of distinct phrases resulting from LZ78-parsing of the sequence x_1^l of length l and the generating probability mass function is defined by $p(x)$. In order to find a distribution-dependent bound on the number of distinct phrases in LZ78-based parsing of x_1^l, we note that since the generating distribution is not necessarily uniform, all the strings x^n for $n < l \ll \infty$ do not necessarily appear as parsed phrases. For instance, consider the binary case with $P(X = 1) = 0.9$. Then, it is very unlikely to have a string with multiple consecutive zeros in any parsing of a realization of the finite sequence x^l. As such, using the Asymptotic Equipartition Properties (AEP) ([4] Chapter 3) or Non-asymptotic Equipartition Properties (NEP) [76], we define the *typical set* $\mathcal{A}_\epsilon^{(n)}$ with respect to $p(x)$ as the set of subsequences $x^n \in \mathcal{X}^n$ of x_1^l with the property

$$2^{-n(h+\epsilon)} \leq p(x^n) \leq 2^{-n(h-\epsilon)},$$

where h is the entropy. Then, we have

$$1 = \sum_{x^n \in \mathcal{X}^n} p(x^n) \geq \sum_{x^n \in \mathcal{A}_\epsilon^{(n)}} p(x^n) \geq \left|\mathcal{A}_\epsilon^{(n)}\right| 2^{-n(h+\epsilon)},$$

therefore, $\left|\mathcal{A}_\epsilon^{(n)}\right| \leq 2^{n(h+\epsilon)}$. Let l_k be the sum of the lengths of all the distinct strings x^n in the set $\left|\mathcal{A}_\epsilon^{(n)}\right|$ of length less than or equal to k. We write,

$$\begin{aligned} l_k &= \sum_{n=1}^{k} n \left|\mathcal{A}_\epsilon^{(n)}\right| \\ &\leq \sum_{n=1}^{k} n 2^{n(h+\epsilon)} \\ &= \frac{1}{(m-1)^2}\Big[((m-1)k - 1)m^{k+1} + m\Big], \end{aligned}$$

where $m \triangleq 2^{h+\epsilon}$. Therefore, $l = \frac{1}{(m-1)^2}\left[((m-1)k-1)m^{k+1}+m\right]$ can be solved for k which leads into an upper bound for $c(l,p)$ as follows

$$\begin{aligned} k &= \frac{\alpha \mathrm{W}\left(\frac{\beta}{\alpha} m^{-1-1/\alpha} \ln m\right) + \ln m}{\alpha \ln m} \\ c(l,p) &\leq \sum_{n=1}^{k} \left|\mathcal{A}_{\epsilon}^{(n)}\right| = \frac{m\left(m^k-1\right)}{m-1} \\ &= \frac{2^{k(h+\epsilon)}-1}{1-2^{-h-\epsilon}}, \end{aligned}$$

where $\alpha = m-1$ and $\beta = (m-1)^2 l - m$. Therefore, the dependency of the $c(l,p)$ upper bound on the distribution is only through the entropy. Figure 6 depicts the upper bound on $c(l,p)$ for $\epsilon = 0.1$.

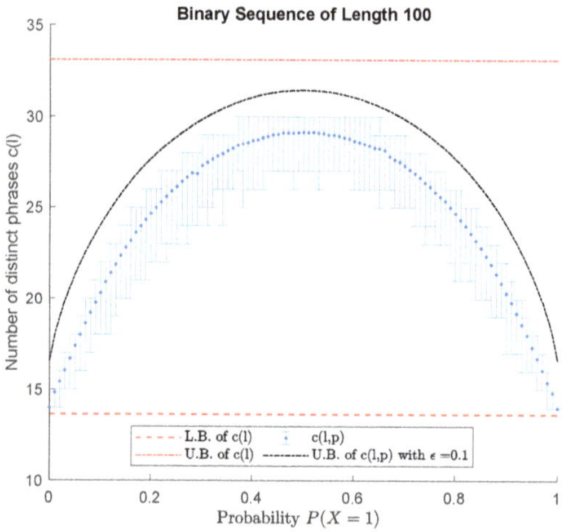

Figure 6. Simulation of the probability-dependent upper bound $c(l,p)$ for binary sequences of fixed length $l = 100$ with various probability parameters $P(X=1)$. For every $P(X=1)$, one thousand binary sequences of length l are generated. Error bars represent the maximum, minimum, and average number of distinct phrases.

5.2. *Pattern Dictionary Parser versus LZ78 Parser*

Given an $|\mathcal{X}|$-ary sequence $x_1^l = \{x_n, n = 1, \ldots, l\}$, let $c_T(l)$ be the number of parsed phrases of x_1^l when the typical encoder (pattern dictionary with D_{max}) is used, and $c_A(l)$ be the number of parsed phrases of x_1^l when the atypical encoder (LZ78) is used. Clearly, $\frac{l}{D_{max}} \leq c_T(l) \leq l$ where the lower bound is achieved when $\mathcal{S}_{\mathcal{D}}\left(x_1^l\right) = \left\{x_{v_1}^{v_2-1}, x_{v_2}^{v_3-1}, \ldots, x_{v_c}^{l}\right\}$, and each $x_{v_i}^{v_i-1} \in \mathcal{S}_{\mathcal{D}}^{(\mathrm{D_{max}})}$, namely $x_{v_i}^{v_i-1}$ is of length D_{max} and exists in the dictionary. The upper bound is achieved when $\mathcal{S}_{\mathcal{D}}\left(x_1^l\right) = \{x_1, x_2, \ldots, x_l\}$ where each $x_n \in \mathcal{S}_{\mathcal{D}}^{(1)}$. Using the result of Theorem 2 and a lower bound on the Lambert W function, $\ln x - \ln(\ln x) \leq \mathrm{W}(x)$ [73], we have

$$\begin{aligned} \frac{l}{D_{max}}\left(1 - \frac{D_{max}}{\log \frac{l}{\log(l \ln|\mathcal{X}|)}}\right) &\leq c_T(l) - c_A(l) \\ &\leq l\left(1 - \frac{\sqrt{8l+1}-1}{2l}\right). \end{aligned} \tag{5}$$

The above bounds have asymptotic and non-asymptotic implications. The asymptotic analysis of the bounds in (5) suggests that as $l \to \infty$, for a dictionary with fixed D_{max}, we have $\frac{l}{D_{max}} \leq c_T(l) - c_A(l) \leq l$. This inequality implies the asymptotic dominance of the parser using a typical encoder. This is to be expected due to the asymptotic optimality of LZ78. However, the above inequality also implies a more interesting result: if $D_{max} > \log \frac{l}{\log(l \ln|\mathcal{X}|)}$ as $l \to \infty$, then $c_T(l)$ can be smaller than $c_A(l)$. The non-asymptotic behavior of the bounds in (5) is more relevant to the anomaly detection problem. These bounds suggest that for a fixed l and $|\mathcal{X}|$, increasing D_{max} has a vanishing effect on the possible range of the anomaly score. Additionally, the achieved bounds on $c_T(l) - c_A(l)$ provide the range of values of the anomaly score. This facilitates the search for a data-dependent threshold for anomaly detection, as the search can be restricted to this range.

5.3. *Atypicality Criterion for Detection of Anomalous Subsequences*

Consider the problem of finding the atypical (anomalous) subsequences of a long sequence with respect to a trained pattern dictionary $\mathcal{D}$. Suppose we are looking for an infrequent anomalous subsequence $x_n^{n+l-1} = \{x_n, n = n, \ldots, n+l-1\}$ embedded in a test sequence $\{x_n, n = 1, \ldots, L\}$ from the finite alphabet $\mathcal{X}$. Using Equation (2), the typical codelength of the subsequence x_n^{n+l-1} is

$$L_T\left(x_n^{n+l-1}\right) = \sum_{y \in \mathcal{S}_\mathcal{D}\left(x_n^{n+l-1}\right)} L_\mathcal{D}(y) + \left|\mathcal{S}_\mathcal{D}\left(x_n^{n+l-1}\right)\right| \log D_{max},$$

while using LZ78, the atypical codelength of the subsequence x_n^{n+l-1} is

$$L_A\left(x_n^{n+l-1}\right) = \left|\mathcal{S}_{LZ}\left(x_n^{n+l-1}\right)\right|\left[\log\left|\mathcal{S}_{LZ}\left(x_n^{n+l-1}\right)\right| + 1\right] + \log^*(l) + \tau,$$

where $\log^*(l) + \tau$ is an additive penalty for not knowing in advance the start and end points of the anomalous sequence [2,3], and $\log^*(l) = \log l + \log\log l + \ldots$ where the sum continues as long as the argument to the outer log is positive. Let $L_A^{'} = L_A - \tau$. We propose the following atypicality criterion for detection of an anomalous subsequence:

$$\triangle L(n) = \max_{l}\left\{L_T\left(x_n^{n+l-1}\right) - L_A^{'}\left(x_n^{n+l-1}\right)\right\} > \tau, \tag{6}$$

where τ can be treated as an anomaly detection threshold. In practice, τ can be set to ensure a false positive constraint, e.g., using bootstrap estimation of the quantiles in the training data.

6. Experiment

In this section, we illustrate the proposed pattern dictionary anomaly detection on a synthetic time series, known as Mackey–Glass [77], as well as on a real-world time series of physiological signals. In both experiments, first, the real-valued samples are discretized using a uniform quantizer [78], and then, anomaly detection methods are applied.

6.1. *Anomaly Detection in Mackey–Glass Time Series*

In this section, we illustrate the proposed anomaly detection method for the case of a chaotic Mackey–Glass (MG) time series that has an anomalous segment grafted into the middle of the sequence. MG time series are generated from a nonlinear time delay differential equation. The MG model was originally introduced to represent the appearance of complex dynamic in physiological control systems [77]. The nonlinear differential equation is of the form $\frac{dx(t)}{dt} = -ax(t) + \frac{bx(t-\delta)}{1+x^{10}(t-\delta)}$, $t \geq 0$, where a, b and δ are constants. For the training data, we generated 3000 samples of the MG time series with $a = 0.2$, $b = 0.1$, and $\delta = 17$. For the test data, we normalized and embedded 500 samples of the

MG time series with $a = 0.4$, $b = 0.2$, and $\delta = 17$ inside 1000 samples of a MG time series generated from the same source as the training data, resulting in a test sequence of length 1500. Figure 7 shows a realization of the training data and the test data.

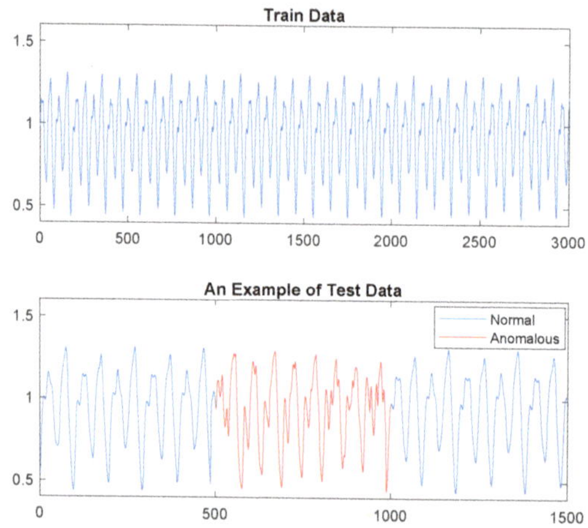

Figure 7. Mackey–Glass time series: the training data (**top**) and an example of the test data (**bottom**) in which samples in $[501, 1000]$ are anomalous (shown in red).

The anomaly detection performance of our proposed pattern dictionary is evaluated. To illustrate the effect of the model parameter, i.e., the maximum depth D_{max}, on the detection and compression performance of the pattern dictionary, we run two experiments. First, we use a 30-fold cross-validation on the training data (resulting in 30 sequences of length 100) and calculate the number of distinct parsed phrases against D_{max}. Second, we train a pattern dictionary with various D_{max} using the training data and then evaluate the sensitivity of detector of the anomalous subsequences in the test data using Equation (6) with $\tau = 0$. In this experiment, the detection sensitivity (true positive rate) is defined as the ratio of number of samples correctly identified as anomalous over the total number of anomalous samples. Figure 8 illustrates the result of both experiments. As seen, after some point, increasing D_{max} has diminishing effect on both detection sensitivity and the number of distinct parsed phrases. Note that this behavior is to be expected as it was suggested by the bounds in (5).

Next, we compare anomaly detection performance of our proposed pattern dictionary methods, PDD and PDA, with the nearest neighbors-based similarity (NNS) technique [7], the compression-based dissimilarity measure (CDM) method [12–14], Ziv–Merhav method (ZM) [48], and the threshold Sequence Time-Delay Embedding (t-STIDE) technique [8–11]. In this experiment, a window of length 100 is slid over the test data and each method measures the *anomaly score* (as described below) of the current subsequence with respect to the training data. The anomaly is detected when the score exceeds a threshold, determined to ensure a specified false positive rate. In the following, we compute AUC (area under the curve) of the ROC (receiver operating characteristic) and Precision-Recall curves as performance measures. In the following, we provide details of the implementation.

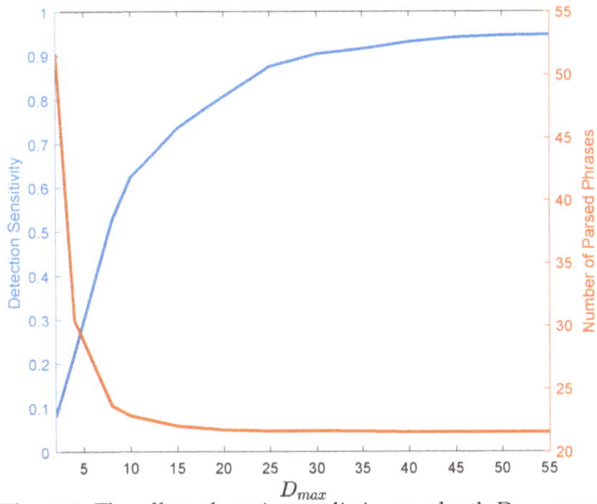

Figure 8. The effect of maximum dictionary depth D_{max} on parsing and detection sensitivity (true positive rate) of the Mackey–Glass time series presented in Figure 7.

Pattern Dictionary for Detection (PDD)

First, the training data are used to create a pattern dictionary with $D_{max} = 40$, as described in Section 4. Then, for each subsequence x^{100} (the sliding window of length 100) of the test data, the anomaly score is computed as the codelength $L(x^{100})$ of Equation (2) described in Section 4.3.

Pattern Dictionary Based Atypicality (PDA)

Similar to PDD, first the training data are used to create a pattern dictionary with $D_{max} = 40$, as described in Section 4. Then, for each subsequence x^{100} of the test data, the anomaly score is the atypicality measure described in Section 5, i.e., $L_T(x^{100}) - L_A(x^{100})$, the difference between the compression codelength of the test subsequence using typical encoder (pattern dictionary) and atypical encoder (LZ78).

Ziv–Merhav Method (ZM) [48]

In this method, a cross-parsing procedure is used in which for each subsequence x^{100} of the test data, the anomaly score is computed as the number of the distinct phrases of x^{100} with respect to the training data.

Nearest Neighbors-Based Similarity (NNS) [7]

In this method, a list $\mathcal{S}$ of all the subsequence of length 100 (the length of the sliding window) of the training data is created. Then, for each subsequence x^{100} of the test data, the distance between x^{100} and all the subsequences in the list $\mathcal{S}$ is calculated. Finally, the anomaly score of x^{100} is its distance to the nearest neighbor in the list $\mathcal{S}$.

Compression-Based Dissimilarity Measure (CDM) [12–14]

In this method, given the training data x_{train}, for each subsequence x^{100} of the test data the anomaly score is

$$CDM(x_{train}, x^{100}) = \frac{\mathcal{L}(C(x_{train}, x^{100}))}{\mathcal{L}(x_{train}) + \mathcal{L}(x^{100})},$$

where $C(y, x)$ represents concatenation of sequences y and z, and $\mathcal{L}(x)$ is the size of the compressed version of the sequence x using any standard compression algorithm. The CDM anomaly score is close to 1 if the two sequence are not related, and smaller than one if the sequences are related.

Threshold Sequence Time-Delay Embedding (t-STIDE) [8–11]

In this method, given $l < 100$, for each sub-subsequence x^l of the subsequence x^{100} of the test data, the likelihood score of x^l is the normalized frequency of its occurrence in the training data, and the anomaly score of x^{100} is one minus the average likelihood score of all its sub-subsequences of length l. In this experiment, various values of l are tested and the best performance is reported.

We compare the detection performance of the aforementioned methods by generating 200 test data sequences with different anomaly segments (the anomalous MG segments have different initializations in each test dataset). The detection results of comparisons are reported in Table 2. As seen, our proposed PDD and PDA methods outperform the rest, with ZM and CDM coming in third place. The effect of alphabet size of the quantized data (the resolution parameter of the uniform quantizer [78]) on anomaly detection performance is summarized in Table 3. Table 3 shows that our proposed PDD and PDA methods outperform in all three cases of data resolution.

Table 2. Comparison of anomaly detection methods ($\mu \pm \sigma$ representation is used where μ is the mean and σ is the standard deviation). The proposed PDA method attains overall best performance (bold entries of table).

	ROC AUC	PR AUC
PDA	**0.963 ± 0.009**	**0.909 ± 0.044**
PDD	0.959 ± 0.009	0.907 ± 0.044
ZM	0.959 ± 0.009	0.895 ± 0.049
CDM	0.957 ± 0.012	0.907 ± 0.057
NNS	0.920 ± 0.021	0.777 ± 0.091
t-STIDE	0.897 ± 0.013	0.857 ± 0.044

Since the parsing procedure of our proposed PD-based methods and the ZM method [48] are similar, it is of interest to compare the running time of these two methods. While the cross-parsing procedure of the ZM method was introduced as an on the fly process [48], we can also consider another implementation similar to our proposed PD by creating a codebook of all the subsequences of the training data prior to the parsing procedure. As such, in order to compare the running time of the dictionary/codebook creation and parsing procedure of our PD-based methods with the aforementioned two implementations of the ZM method, we use the same MG training data of length 3000, one test dataset of length 1500 while a sliding window of length 100 is slid over it for anomaly score calculation, and the PD-based method with $D_{max} = 40$. Note that since a sliding window of length 100 over the test data is considered, for the codebook-based implementation of ZM, all the subsequences of the training data up to length 100 are extracted which make its codebook creation process significantly faster. Table 4 summarizes the running time comparison. As it can be seen, our PD-based method is faster in both dictionary/codebook creation and parsing process.

Table 3. Comparison of anomaly detection methods for different cases of data resolutions: high resolution corresponds to an alphabet size of 90, medium resolution corresponds to an alphabet size of 45, and low resolution corresponds to an alphabet size of 10. In this table, $\mu \pm \sigma$ representation is used where μ is the mean and σ is the standard deviation. The proposed PDA method achieves overall best performance (bold entries of table).

	Resolution	PDA	PDD	ZM	CDM	NNS	t-STIDE
ROC AUC	Low	**0.948 ±0.011**	0.930 ±0.013	0.943 ±0.014	0.787 ±0.017	0.901 ±0.027	0.725 ±0.025
	Medium	**0.955 ±0.010**	0.943 ±0.011	0.954 ±0.011	0.940 ±0.014	0.918 ±0.022	0.881 ±0.017
	High	**0.963 ±0.009**	0.959 ±0.009	0.959 ±0.009	0.957 ±0.012	0.920 ±0.021	0.897 ±0.013
PR AUC	Low	**0.876 ±0.050**	0.871 ±0.052	0.826 ±0.071	0.669 ±0.067	0.719 ±0.098	0.678 ±0.067
	Medium	**0.885 ±0.046**	0.882 ±0.047	0.881 ±0.053	0.880 ±0.060	0.777 ±0.093	0.828 ±0.050
	High	**0.909 ±0.044**	0.907 ±0.044	0.895 ± 0.044	0.907 ±0.057	0.777 ±0.091	0.857 ±0.044

Table 4. Comparison of running time (in second) of PD-based method and two implementations of the ZM method for different cases of data resolutions: high resolution corresponds to an alphabet size of 90, medium resolution corresponds to an alphabet size of 45, and low resolution corresponds to an alphabet size of 10. This experiment is performed on a Hansung laptop with 2.60 GHz CPU, 500 GB of SSD, and 16 GB of RAM using MATLAB R2021a. The proposed PD-based method has fastest run time overall (bold entries in table).

	Resolution	PD-Based	ZM-Codebook	ZM
dictionary generation	Low	**6.80**	29.98	N/A
	Medium	**13.12**	39.01	N/A
	High	**15.46**	40.80	N/A
parsing procedure	Low	**6.07**	9.23	142.77
	Medium	**10.81**	11.10	433.55
	High	**14.83**	16.70	670.18

6.2. Infection Detection Using Physiological Signals

Finally, we apply the proposed pattern dictionary method to detect unusual patterns in physiological signals of two human subjects after exposure to a pathogen while only one of these subjects became symptomatically ill. The time series data were collected in a human viral challenge study that was performed in 2018 at the University of Virginia under a DARPA grant. Consented volunteers were recruited into this study following an IRB-approved protocol and the data was processed and analyzed at Duke University and the University of Michigan. The challenge study design and data collection protocols are described in [79]. Volunteers' skin temperature and heart rate were recorded by a wearable device (Empatica E4) over three consecutive days before and five consecutive days after exposure to a strain of human Rhinovirus (RV) pathogen. During this period, the wearable time series were continuously recorded while biospecimens (viral load) were collected daily. The infection status can be clinically detected by biospecimen samples, but in practice, the collection process of these types of biosamples can be invasive and costly. As such, here, we apply the proposed anomaly detection framework to the measured two-dimensional heart rate and temperature time series to detect unusual patterns after exposure with respect to the normal (healthy) baseline patterns.

In the preprocessing phase, we followed the wearable data preprocessing procedure described in [80]. Specifically, we first downsample the time series to one sample per minute by averaging. Then, we apply an outlier detection procedure to remove technical noise, e.g., sensor contact loss. After preprocessing, the two-dimensional space of temperature and heart rate time series is discretized using a two-dimensional uniform quantizer [78] with step size of 5 for heart rate and 0.5 for temperature, resulting in one-dimensional discrete sequence data. The first three days of data are used as the training data, and the PDA methods with maximum depth $D_{max} = 30$ are used to learn the patterns in the training data. In order to detect anomalous patterns of the test data (the last five days), we used the result of Section 5.3 and the atypicality criterion of Equation (6), which requires choosing the threshold τ. While this threshold can be chosen freely, we selected it using cross-validation on the training data. Leave-one-out cross-validation over the training data generates an empirical null distribution of the PDA anomaly score function $L_T - L_A$. The threshold τ was chosen as the upper 99% quantile of this distribution. Figure 9 illustrates the result of anomaly detection on one subject who became infected as measured by viral shedding as shown in Figure 9C. All the anomalous patterns occur when the subject was shedding the virus. Figure 10 also depicts the result of anomaly detection on one subject who had a mild infection with a low level of viral shedding, as shown in Figure 10C. Note that in this case, no anomalous patterns were detected.

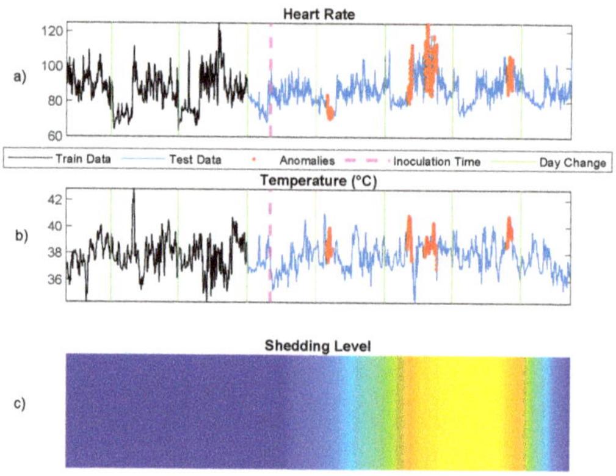

Figure 9. Anomaly detection using the proposed PDA method for a subject based on heart rate and temperature data collected from a wearable wrist sensor. Anomalies are shown in red in (**a**,**b**). (**c**) shows the subject's infection level.

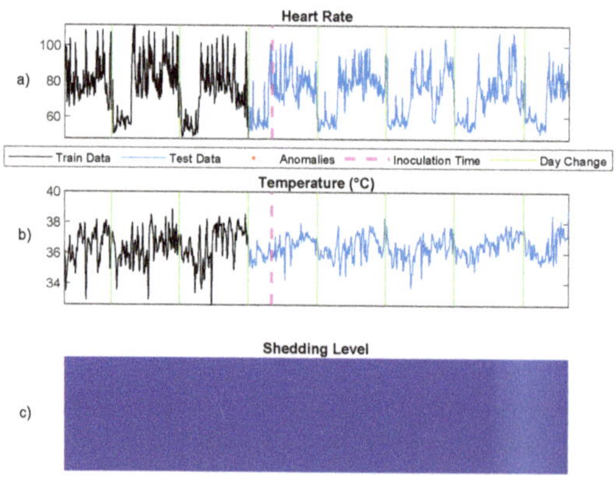

Figure 10. Anomaly detection using the proposed PDA method for a subject who had a mild infection with low level of viral shedding based on heart rate and temperature data collected from a wearable wrist sensor. Note that no anomaly has been detected: (**a**) heart rate, (**b**) temperature, and (**c**) infection level.

7. Conclusions

In this paper, we have developed a universal nonparametric model-free anomaly detection method for time series and sequence data using a pattern dictionary. We proved that using a multi-level dictionary that separates the patterns by their depth results in a shorter average indexing codelength in comparison to a uni-level dictionary that uses a uniform indexing approach. We illustrated that the proposed pattern dictionary method can be used as a stand-alone anomaly detector, or integrated with Tree-Structured Lempel–Ziv (LZ78) and incorporated into an atypicality framework. We developed novel non-asymptotic lower and upper bounds of the LZ78 parser and demonstrated that the non-asymptotic upper bound on the number of distinct phrases resulting from LZ78-parsing of an $|\mathcal{X}|$-ary sequence can be explicitly derived in terms of the Lambert W function, an important theoretical result that is not trivial. We showed that the achieved non-asymptotic bounds on LZ78 and pattern dictionary determine the range of the anomaly score and the anomaly detection threshold. We also presented an empirical study in which the pattern dictionary approach is used to detect anomalies in physiological time series. In the future work, we will investigate the generalization of the context tree weighting methods to the general discrete case, using the pattern dictionary since the pattern dictionary handles sparsity well and is computationally less expensive when the alphabet size is large.

Author Contributions: Data curation, E.S. and S.O.; Formal analysis, E.S.; Funding acquisition, A.O.H.; Methodology, E.S.; Project administration, A.O.H.; Software, E.S. and S.O.; Supervision, A.O.H.; Validation, E.S., P.X.K.S. and A.O.H.; Visualization, E.S. and S.O.; Writing—original draft, E.S.; Writing—review & editing, P.X.K.S. and A.O.H. All authors have read and agreed to the published version of the manuscript.

Funding: This work was supported in part by Michigan Institute for Data Science, and grants from the Army Research Office, grant W911NF-15-0479, the Defense Advanced Research Projects Agency, grant N66001-17-2-401, and the Department of Energy/National Nuclear Security Administration, grant DE-NA0003921.

Institutional Review Board Statement: University of Michigan ethical review and approval were waived since only de-identified data artifacts from a previously approved human challenge study were made available to the co-authors for the analysis reported in Section 6.2 Information concerning provenance of the data, i.e., the human rhinovirus (RV) challenge study experiment and its IRB approved protocol, was reported in [79].

Informed Consent Statement: The de-identified data used for our analysis in Section 6.2 came from a human rhinovirus (RV) challenge study in which informed consent was obtained from all subjects involved in the study. See the [79] for details.

Data Availability Statement: The experimental data used in Section 6.2 will be made available upon request.

Conflicts of Interest: The authors declare no conflict of interest.

References

1. Chandola, V.; Banerjee, A.; Kumar, V. Anomaly detection: A survey. *ACM Comput. Surv. (CSUR)* **2009**, *41*, 15. [CrossRef]
2. Høst-Madsen, A.; Sabeti, E.; Walton, C. Data discovery and anomaly detection using atypicality: Theory. *IEEE Trans. Inf. Theory* **2019**, *65*, 5302–5322. [CrossRef]
3. Sabeti, E.; Høst-Madsen, A. Data Discovery and Anomaly Detection Using Atypicality for Real-Valued Data. *Entropy* **2019**, *21*, 219. [CrossRef] [PubMed]
4. Cover, T.; Thomas, J. *Information Theory*, 2nd ed.; John Wiley: Hoboken, NJ, USA, 2006.
5. Ziv, J.; Lempel, A. Compression of individual sequences via variable-rate coding. *Inf. Theory IEEE Trans.* **1978**, *24*, 530–536. [CrossRef]
6. Corless, R.M.; Gonnet, G.H.; Hare, D.E.; Jeffrey, D.J.; Knuth, D.E. On the LambertW function. *Adv. Comput. Math.* **1996**, *5*, 329–359. [CrossRef]
7. Chandola, V.; Mithal, V.; Kumar, V. Comparative evaluation of anomaly detection techniques for sequence data. In Proceedings of the 2008 Eighth IEEE International Conference on Data Mining, Pisa, Italy, 15–19 December 2008; pp. 743–748.

8. Cabrera, J.B.; Lewis, L.; Mehra, R.K. Detection and classification of intrusions and faults using sequences of system calls. *ACM SIGMOD Rec.* **2001**, *30*, 25–34. [CrossRef]
9. Hofmeyr, S.A.; Forrest, S.; Somayaji, A. Intrusion detection using sequences of system calls. *J. Comput. Secur.* **1998**, *6*, 151–180. [CrossRef]
10. Lane, T.; Brodley, C.E. Temporal sequence learning and data reduction for anomaly detection. *ACM Trans. Inf. Syst. Secur. (TISSEC)* **1999**, *2*, 295–331. [CrossRef]
11. Warrender, C.; Forrest, S.; Pearlmutter, B. Detecting intrusions using system calls: Alternative data models. In Proceedings of the 1999 IEEE Symposium on Security and Privacy (Cat. No. 99CB36344), Oakland, CA, USA, 14 May 1999; pp. 133–145.
12. Keogh, E.; Lonardi, S.; Ratanamahatana, C.A. Towards parameter-free data mining. In Proceedings of the Tenth ACM SIGKDD International Conference on Knowledge Discovery and Data Mining, Seattle, WA, USA, 22–25 August 2004; pp. 206–215.
13. Keogh, E.; Lonardi, S.; Ratanamahatana, C.A.; Wei, L.; Lee, S.H.; Handley, J. Compression-based data mining of sequential data. *Data Min. Knowl. Discov.* **2007**, *14*, 99–129. [CrossRef]
14. Keogh, E.; Keogh, L.; Handley, J.C. Compression-based data mining. In *Encyclopedia of Data Warehousing and Mining*, 2nd ed.; IGI Global: Pennsylvania, PA, USA, 2009; pp. 278–285.
15. Keogh, E.; Lonardi, S.; Chiu, B.C. Finding surprising patterns in a time series database in linear time and space. In Proceedings of the Eighth ACM SIGKDD International Conference on Knowledge Discovery and Data Mining, Edmonton, AB, Canada, 23–26 July 2002; pp. 550–556.
16. Keogh, E.; Lin, J.; Lee, S.H.; Van Herle, H. Finding the most unusual time series subsequence: Algorithms and applications. *Knowl. Inf. Syst.* **2007**, *11*, 1–27. [CrossRef]
17. Ferguson, T.S. *Mathematical Statistics: A decision Theoretic Approach*; Academic Press: Cambridge, MA, USA, 2014; Volume 1.
18. Siegmund, D.; Venkatraman, E. Using the generalized likelihood ratio statistic for sequential detection of a change-point. *Ann. Stat.* **1995**, 23, 255–271. [CrossRef]
19. Hirai, S.; Yamanishi, K. Detecting changes of clustering structures using normalized maximum likelihood coding. In Proceedings of the 18th ACM SIGKDD International Conference on Knowledge Discovery and Data Mining, Beijing, China, 12–16 August 2012; pp. 343–351.
20. Yamanishi, K.; Miyaguchi, K. Detecting gradual changes from data stream using MDL-change statistics. In Proceedings of the 2016 IEEE International Conference on Big Data (Big Data), Washington, DC, USA, 5–8 December 2016; pp. 156–163.
21. Killick, R.; Fearnhead, P.; Eckley, I.A. Optimal detection of changepoints with a linear computational cost. *J. Am. Stat. Assoc.* **2012**, *107*, 1590–1598. [CrossRef]
22. Gama, J.; Žliobaitė, I.; Bifet, A.; Pechenizkiy, M.; Bouchachia, A. A survey on concept drift adaptation. *ACM Comput. Surv. (CSUR)* **2014**, *46*, 44. [CrossRef]
23. Chernoff, H. Sequential design of experiments. *Ann. Math. Stat.* **1959**, *30*, 755–770. [CrossRef]
24. Basseville, M.; Nikiforov, I.V. *Detection of Abrupt Changes: Theory and Application*; Prentice Hall Englewood Cliffs: Hoboken, NJ, USA, 1993; Volume 104.
25. Veeravalli, V.V.; Banerjee, T. Quickest change detection. *Acad. Press Libr. Signal Process. Array Stat. Signal Process.* **2013**, *3*, 209–256.
26. Han, C.; Willett, P.; Chen, B.; Abraham, D. A detection optimal min-max test for transient signals. *Inf. Theory IEEE Trans.* **1998**, *44*, 866–869. [CrossRef]
27. Wang, Z.; Willett, P. A performance study of some transient detectors. *Signal Process. IEEE Trans.* **2000**, *48*, 2682–2685. [CrossRef]
28. Wang, Z.; Willett, P.K. All-purpose and plug-in power-law detectors for transient signals. *Signal Process. IEEE Trans.* **2001**, *49*, 2454–2466. [CrossRef]
29. Wang, Z.J.; Willett, P. A variable threshold page procedure for detection of transient signals. *IEEE Trans. Signal Process.* **2005**, *53*, 4397–4402. [CrossRef]
30. Hero, A.O. Geometric entropy minimization (GEM) for anomaly detection and localization. *NIPS* **2006**, *19*, 585–592.
31. Sricharan, K.; Hero, A. Efficient anomaly detection using bipartite k-nn graphs. *Adv. Neural Inf. Process. Syst.* **2011**, *24*, 478–486.
32. Sen, P.K. *Theory and Applications of Sequential Nonparametrics*; SIAM: Philadelphia, PA, USA, 1985.
33. Balsubramani, A.; Ramdas, A. Sequential Nonparametric Testing with the Law of the Iterated Logarithm. In Proceedings of the Thirty-Second Conference on Uncertainty in Artificial Intelligence, Jersey City, NJ, USA, 25–29 June 2016; AUAI Press: Arlington, VA, USA, 2016; pp. 42–51.
34. Chandola, V.; Banerjee, A.; Kumar, V. Anomaly Detection for Discrete Sequences: A Survey. *Knowl. Data Eng. IEEE Trans.* **2012**, *24*, 823–839. [CrossRef]
35. Evans, S.; Barnett, B.; Bush, S.; Saulnier, G. Minimum description length principles for detection and classification of FTP exploits. In Proceedings of the Military Communications Conference, Monterey, CA, USA, 31 October–3 November 2004; Volume 1, pp. 473–479. [CrossRef]
36. Wang, N.; Han, J.; Fang, J. An Anomaly Detection Algorithm Based on Lossless Compression. In Proceedings of the 2012 IEEE 7th International Conference on Networking, Architecture and Storage (NAS), Xiamen, China, 28–30 June 2012; pp. 31–38. [CrossRef]
37. Lee, W.; Xiang, D. Information-theoretic measures for anomaly detection. In Proceedings of the 2001 IEEE Symposium on Security and Privacy, Oakland, CA, USA, 14–16 May 2001; pp. 130–143. [CrossRef]
38. Paschalidis, I.; Smaragdakis, G. Spatio-Temporal Network Anomaly Detection by Assessing Deviations of Empirical Measures. *Netw. IEEE/ACM Trans.* **2009**, *17*, 685–697. [CrossRef]

39. Han, C.K.; Choi, H.K. Effective discovery of attacks using entropy of packet dynamics. *Netw. IEEE* **2009**, *23*, 4–12. [CrossRef]
40. Baliga, P.; Lin, T. Kolmogorov complexity based automata modeling for intrusion detection. In Proceedings of the 2005 IEEE International Conference on Granular Computing, Beijing, China, 25–27 July 2005; Volume 2, pp. 387–392. [CrossRef]
41. Shahriar, H.; Zulkernine, M. Information-Theoretic Detection of SQL Injection Attacks. In Proceedings of the 2012 IEEE 14th International Symposium on High-Assurance Systems Engineering (HASE), Omaha, NE, USA, 25–27 October 2012; pp. 40–47. [CrossRef]
42. Xiang, Y.; Li, K.; Zhou, W. Low-Rate DDoS Attacks Detection and Traceback by Using New Information Metrics. *Inf. Forensics Secur. IEEE Trans.* **2011**, *6*, 426–437. [CrossRef]
43. Pan, F.; Wang, W. Anomaly detection based-on the regularity of normal behaviors. In Proceedings of the 1st International Symposium on Systems and Control in Aerospace and Astronautics, Harbin, China, 19–21 January 2006. [CrossRef]
44. Eiland, E.; Liebrock, L. An application of information theory to intrusion detection. In Proceedings of the Fourth IEEE International Workshop on Information Assurance, London, UK , 13–14 April 2006. [CrossRef]
45. Li, M.; Chen, X.; Li, X.; Ma, B.; Vitanyi, P. The similarity metric. *Inf. Theory IEEE Trans.* **2004**, *50*, 3250–3264. [CrossRef]
46. Li, Y.; Nitinawarat, S.; Veeravalli, V.V. Universal outlier hypothesis testing. *IEEE Trans. Inf. Theory* **2014**, *60*, 4066–4082. [CrossRef]
47. Li, Y.; Nitinawarat, S.; Veeravalli, V.V. Universal outlier detection. In Proceedings of the Information Theory and Applications Workshop (ITA), San Diego, CA, USA, 10–15 February 2013; pp. 1–5.
48. Ziv, J.; Merhav, N. A measure of relative entropy between individual sequences with application to universal classification. *IEEE Trans. Inf. Theory* **1993**, *39*, 1270–1279. [CrossRef]
49. Chandola, V. Anomaly Detection for Symbolic Sequences and Time Series Data. Ph.D. Thesis, University of Minnesota, Minneapolis, MN, USA, 2009.
50. Rousseeuw, P.J.; Leroy, A.M. *Robust Regression and Outlier Detection*; John Wiley & Sons: Hoboken, NJ, USA, 2005; Volume 589.
51. Wu, Q.; Shao, Z. Network anomaly detection using time series analysis. In Proceedings of the Joint International Conference on Autonomic and Autonomous Systems and International Conference on Networking And Services-(icas-isns' 05), Papeete, France, 23–28 October 2005; p. 42.
52. Pincombe, B. Anomaly detection in time series of graphs using arma processes. *Asor Bull.* **2005**, *24*, 2.
53. Moayedi, H.Z.; Masnadi-Shirazi, M. Arima model for network traffic prediction and anomaly detection. In Proceedings of the 2008 International Symposium on Information Technology, Kuala Lumpur, Malaysia, 26–28 August 2008; Volume 4, pp. 1–6.
54. Ma, J.; Perkins, S. Online novelty detection on temporal sequences. In Proceedings of the Ninth ACM SIGKDD International Conference on Knowledge Discovery and Data Mining, Washington, DC, USA, 24–27 August 2003; pp. 613–618.
55. Knorn, F.; Leith, D.J. Adaptive kalman filtering for anomaly detection in software appliances. In Proceedings of the IEEE INFOCOM Workshops, Phoenix, AZ, USA, 13–18 April 2008; pp. 1–6.
56. Gusfield, D. Algorithms on stings, trees, and sequences: Computer science and computational biology. *Acm Sigact News* **1997**, *28*, 41–60. [CrossRef]
57. Thottan, M.; Ji, C. Anomaly detection in IP networks. *Signal Process. IEEE Trans.* **2003**, *51*, 2191–2204. [CrossRef]
58. Chakrabarti, S.; Sarawagi, S.; Dom, B. Mining surprising patterns using temporal description length. In Proceedings of the VLDB'98, 24rd International Conference on Very Large Data Bases, New York, NY, USA, 24–27 August 1998; pp. 606–617.
59. Akoglu, L.; Tong, H.; Koutra, D. Graph based anomaly detection and description: A survey. *Data Min. Knowl. Discov.* **2015**, *29*, 626–688. [CrossRef]
60. Ranshous, S.; Shen, S.; Koutra, D.; Harenberg, S.; Faloutsos, C.; Samatova, N.F. Anomaly detection in dynamic networks: A survey. *Wiley Interdiscip. Rev. Comput. Stat.* **2015**, *7*, 223–247. [CrossRef]
61. Yu, R.; Qiu, H.; Wen, Z.; Lin, C.; Liu, Y. A survey on social media anomaly detection. *ACM SIGKDD Explor. Newsl.* **2016**, *18*, 1–14. [CrossRef]
62. Aggarwal, C.C.; Philip, S.Y. An effective and efficient algorithm for high-dimensional outlier detection. *VLDB J.* **2005**, *14*, 211–221. [CrossRef]
63. Goldstein, M.; Dengel, A. Histogram-based outlier score (hbos): A fast unsupervised anomaly detection algorithm. In Proceedings of the KI-2012: Poster and Demo Track, Saarbrücken, Germany, 24–27 September 2012; 2012; pp. 59–63.
64. Foorthuis, R. SECODA: Segmentation-and combination-based detection of anomalies. In Proceedings of the 2017 IEEE International Conference on Data Science and Advanced Analytics (DSAA), Tokyo, Japan, 19–21 October 2017; pp. 755–764.
65. Foorthuis, R. The Impact of Discretization Method on the Detection of Six Types of Anomalies in Datasets. *arXiv* **2020**, arXiv:2008.12330.
66. Holland, J.H. Genetic algorithms. *Sci. Am.* **1992**, *267*, 66–73. [CrossRef]
67. Yang, Y.; Newsam, S. Bag-of-visual-words and spatial extensions for land-use classification. In Proceedings of the 18th SIGSPATIAL International Conference on Advances in Geographic Information Systems, San Jose, CA, USA, 2–5 November 2010; pp. 270–279.
68. Bradski, G.; Kaehler, A. *Learning OpenCV: Computer vision with the OpenCV Library*; O'Reilly Media, Inc.: Sebastopol, CA, USA, 2008.
69. Tung, S. On lower and upper bounds of the difference between the arithmetic and the geometric mean. *Math. Comput.* **1975**, *29*, 834–836. [CrossRef]
70. Willems, F. The context-tree weighting method: Extensions. *Inf. Theory IEEE Trans.* **1998**, *44*, 792–798. [CrossRef]

71. Willems, F.M.J.; Shtarkov, Y.; Tjalkens, T. The context-tree weighting method: Basic properties. *Inf. Theory IEEE Trans.* **1995**, *41*, 653–664. [CrossRef]
72. Willems, F.; Shtarkov, Y.; Tjalkens, T. Reflections on "The Context Tree Weighting Method: Basic properties". *Newsl. IEEE Inf. Theory Soc.* **1997**, *47*.
73. Hoorfar, A.; Hassani, M. Inequalities on the Lambert W function and hyperpower function. *J. Inequal. Pure Appl. Math* **2008**, *9*, 5–9.
74. Lempel, A.; Ziv, J. On the Complexity of Finite Sequences. *Inf. Theory IEEE Trans.* **1976**, *22*, 75–81. [CrossRef]
75. Jacquet, P.; Szpankowski, W. Limiting Distribution of Lempel Ziv'78 Redundancy. In Proceedings of the 2011 IEEE International Symposium on Information Theory Proceedings (ISIT), St. Petersburg, Russia, 31 July–5 August 2011; pp. 1509–1513.
76. Yang, E.H.; Meng, J. Non-asymptotic equipartition properties for independent and identically distributed sources. In Proceedings of the 2012 Information Theory and Applications Workshop, San Diego, CA, USA, 5–10 February 2012; pp. 39–46.
77. Mackey, M.C.; Glass, L. Oscillation and chaos in physiological control systems. *Science* **1977**, *197*, 287–289. [CrossRef] [PubMed]
78. Gersho, A.; Gray, R.M. *Vector Quantization and Signal Compression*; Springer Science & Business Media: Berlin/Heidelberg, Germany, 2012; Volume 159.
79. Grzesiak, E.; Bent, B.; McClain, M.T.; Woods, C.W.; Tsalik, E.L.; Nicholson, B.P.; Veldman, T.; Burke, T.W.; Gardener, Z.; Bergstrom, E.; et al. Assessment of the Feasibility of Using Noninvasive Wearable Biometric Monitoring Sensors to Detect Influenza and the Common Cold Before Symptom Onset. *JAMA Netw. Open* **2021**, 4, e2128534. [CrossRef] [PubMed]
80. She, X.; Zhai, Y.; Henao, R.; Woods, C.; Chiu, C.; Ginsburg, G.S.; Song, P.X.; Hero, A.O. Adaptive multi-channel event segmentation and feature extraction for monitoring health outcomes. *IEEE Trans. Biomed. Eng.* **2020**, 68, 2377–2388. [CrossRef]

entropy MDPI

Article

A Maximal Correlation Framework for Fair Machine Learning

Joshua Lee [1,†,‡], Yuheng Bu [1,*,‡], Prasanna Sattigeri [2], Rameswar Panda [2], Gregory W. Wornell [1], Leonid Karlinsky [2] and Rogerio Schmidt Feris [2]

1 Department of Electrical Engineering and Computer Science, Massachusetts Institute of Technology, Cambridge, MA 02139, USA; jk_lee@mit.edu (J.L.); gww@mit.edu (G.W.W.)

2 MIT-IBM Watson AI Lab, IBM Research, Cambridge, MA 02139, USA; psattig@us.ibm.com (P.S.); rpanda@ibm.com (R.P.); leonidka@il.ibm.com (L.K.); rsferis@us.ibm.com (R.S.F.)

* Correspondence: buyuheng@mit.edu

† Current address: Snap Inc., Santa Monica, CA 90405, USA.

‡ These authors contributed equally to this work.

Abstract: As machine learning algorithms grow in popularity and diversify to many industries, ethical and legal concerns regarding their fairness have become increasingly relevant. We explore the problem of algorithmic fairness, taking an information–theoretic view. The maximal correlation framework is introduced for expressing fairness constraints and is shown to be capable of being used to derive regularizers that enforce independence and separation-based fairness criteria, which admit optimization algorithms for both discrete and continuous variables that are more computationally efficient than existing algorithms. We show that these algorithms provide smooth performance–fairness tradeoff curves and perform competitively with state-of-the-art methods on both discrete datasets (COMPAS, Adult) and continuous datasets (Communities and Crimes).

Keywords: fairness; HGR maximal correlation; independence criterion; separation criterion

Citation: Lee, J.; Bu, Y.; Sattigeri, P.; Panda, R.; Wornell, G.W.; Karlinsky, L.; Schmidt Feris, R. A Maximal Correlation Framework for Fair Machine Learning. *Entropy* **2022**, *24*, 461. https://doi.org/10.3390/e24040461

Academic Editor: Friedhelm Schwenker

Received: 15 February 2022
Accepted: 24 March 2022
Published: 26 March 2022

1. Introduction

The use of machine learning in many industries has raised many ethical and legal concerns, especially that of fairness and bias in predictions, e.g., [1,2]. As systems are trusted to aid or make decisions regarding loan applications, criminal sentencing, and even health care, it is vital that unfair biases do not influence them.

However, mitigating these biases is complicated by ever-changing perspectives on fairness, and a good system for enforcing fairness must be adaptable to new settings. In particular, there are often competing notions on fairness. Two of these popular notions are independence and separation (a third condition, sufficiency, is beyond the scope of this paper), as discussed in [3]. Independence ensures that predictions are independent from membership in a protected class, so that one achieves equal favorable outcome rates across all groups, and it arises in applications such as affirmative action [4]. Separation is designed to achieve equal type I/II error rates across all groups by enforcing independence between predictions and membership in a protected class conditional on the class label. This criterion is used to measure fairness in recidivism predictions and bank loan applications. A significant body of work, including [3,5–7], has gone into explaining that independence and separation are inherently incompatible for non-trivial cases, and their applicability needs to be determined by the application and the stakeholders. This motivates us to construct a framework that is flexible enough to handle different fairness criteria and to do it with different modalities of data (discrete vs. continuous data, for example).

This bias mitigation must also be balanced out with the system's usefulness, and often, one must tune the tradeoff between the fairness (as measured in the particular context) and performance according to a current situation, which can be a difficult process if the tradeoff curve is not smooth. Generating the frontier of possible values can be computationally infeasible or impossible if the algorithm does not have a regularization parameter to adjust (see, [8,9]), thus making it difficult to achieve this balance, which makes the fast generation of fair classifiers even more important.

Different contexts also require different points of intervention during the learning process to ensure fairness. *Pre-processing* approaches ([8,10–14]) modify the data to eliminate bias, whereas *post-processing* approaches ([15–18]) modify learned features/predictions from existing models to be more fair. We focus on the *in-processing* approach [9,19–21], where the fairness criteria are directly incorporated in the training objective to produce fairer learned features. Motivated by few-shot applications where only a pre-trained network and few samples labeled with the sensitive attribute are available, we also seek a method that is applicable in a post-processing manner when we have access to only a small number of samples labeled with the sensitive attribute that we wish to be fair about, which would arise in settings where collecting this information can be very difficult.

In this paper, we frame the ideas of independence and separation in a way that allows a relevant regularizer or penalty term to be derived in addition to a measure of fairness, which is useful in enforcing fairness while also tractable, admitting an optimization algorithm (e.g., if used as an objective for a neural net trained using gradient descent, it must be differentiable), and easily computed. Existing approaches can struggle with efficiency, can fail to provide good control over the performance–fairness tradeoff, and/or can only deal with either discrete or continuous data.

We make the following contributions in this paper:

- We present a universal framework justified by an information–theoretic view that can inherently handle the popular fairness criteria, namely independence and separation, while seamlessly adopting both discrete and continuous cases, which uses the maximal correlation to construct measures of fairness associated with different criteria; then, we use these measures to further develop fair learning algorithms in a fast, efficient, and effective manner.
- We show empirically that these algorithms can provide the desired smooth tradeoff curve between the performance and the measures of fairness on several standard datasets (COMPAS, Adult, and Communities and Crimes), so that a desired level of fairness can be achieved.
- Finally, we perform experiments to illustrate that our algorithms can be used to impose fairness on a model originally trained without any fairness constraint in the few-shot regime, which further demonstrates the versatility of our algorithms in a post-processing setup.

2. Background

2.1. Fairness Objectives in Machine Learning

Consider the standard supervised learning scenario where we predict the value of a target variable $Y \in \mathcal{Y}$ using a set of decision or predictive variables $X \in \mathcal{X}$ with training samples $\{(x_1, y_1), \ldots, (x_n, y_n)\}$. For example, X may be information about an individual's credit history, and Y is whether the individual will pay back a certain loan. In general, we wish to find features $f(x)$, which are predictive of Y, so that we can construct a good predictor $\hat{y} = T(f(x))$ of y under some loss criteria $L(\hat{y}, y)$.

Now, suppose we have some sensitive attributes $D \in \mathcal{D}$ we wish to be "fair" about (e.g., race, gender), and training samples $\{(x_1, y_1, d_1), \ldots, (x_n, y_n, d_n)\}$. For example, in the criminal justice system, predictions about the chance of recidivism of a convicted criminal (Y) given factors such as the nature of the crime and the number of prior arrests (X) should not be determined by race (D). This is a known issue with the COMPAS recidivism score, which, despite not using race as an input to make decisions, still leads to systematic bias toward members of certain races in the output score as in [22,23].

The two most popular criteria for fairness are independence and separation. Independence states that for a feature to be fair, it must satisfy the independence property $\hat{Y} \perp D$ or $f(x) \perp D$. The intuition is simple: if the prediction/feature is independent of the sensitive attribute, then no information about the sensitive attribute is used to predict Y. This criterion has been studied under the lens of *demographic parity* and *disparate impact* in [3], and it admits a class of fairness measures based on the degree of dependence between $f(X)$ and D. For example, independence is satisfied if and only if the mutual information $I(f(X);D)$ is zero. When D is binary, another popular class of measures used by the US Equal Employment Opportunity Commission [4] is the disparate impact, which is defined as $\mathbb{D}(\mathbb{P}(Y|D=1);\mathbb{P}(Y|D=0)) = \frac{\mathbb{P}(\hat{Y}=1|D=0)}{\mathbb{P}(\hat{Y}=1|D=1)}$.

Separation requires the conditional independence property $(\hat{Y} \perp D)|Y$ or $(f(X) \perp D)|Y$. This criterion allows for a violation of demographic parity to the extent that it is justified by the target variable. In the general case, this criterion suggests a fairness measure based on the conditional dependence between $\hat{Y}$ and D conditioned on Y. In the case where D is binary, we obtain the *equalized opportunities* (EO) measures in [3], which are given by the differences in error rates for the two groups (e.g., the difference between the false positive rates for $D = 0, 1$). For a more complete discussion of the advantages and disadvantages of these two criteria, please refer to [3].

2.2. Maximal Correlation

Since these fairness criteria are expressed as enforcing independencies with respect to joint distributions, we look for constraints that reduce the dependency between variables. In particular, the right formulation of correlation between learned features and sensitive attributes can provide a framework for measuring and optimizing for fairness. One effective measure applicable to both continuous and discrete data is the Hirschfeld–Gebelein–Renyi (HGR) maximal correlation, which is a measure of nonlinear correlation that originated in [24] and is further developed in [25,26]. The HGR maximal correlation between two random variables is equal to zero if and only if the two variables are independent, and it increases in value the more correlated they are (i.e., the more biased/unfair).

Definition 1. *For two jointly distributed random variables $X \in \mathcal{X}$ and $Y \in \mathcal{Y}$, given $1 \le k \le K-1$ with $K = \min\{|\mathcal{X}|, |\mathcal{Y}|\}$, the HGR maximal correlation problem is*

$$(\mathbf{f}^*, \mathbf{g}^*) \triangleq \underset{\mathbf{f}:\, \mathcal{X}\to\mathbb{R}^k,\ \mathbf{g}:\, \mathcal{Y}\to\mathbb{R}^k}{\arg\max}\ \mathbb{E}\left[\mathbf{f}^{\mathrm{T}}(X)\,\mathbf{g}(Y)\right], \tag{1}$$

with constraints

$$\mathbb{E}[\mathbf{f}(X)] = \mathbb{E}[\mathbf{g}(Y)] = \mathbf{0}, \quad \mathbb{E}\left[\mathbf{f}(X)\mathbf{f}^{\mathrm{T}}(X)\right] = \mathbb{E}\left[\mathbf{g}(Y)\mathbf{g}^{\mathrm{T}}(Y)\right] = \mathbf{I}, \tag{2}$$

and expectations taken over $P_{X,Y}$. We refer to $\mathbf{f}^*$ and $\mathbf{g}^*$ as maximal correlation functions, with $\mathbf{f}^* = (f_1^*, \ldots, f_k^*)^{\mathrm{T}}$ and $\mathbf{g}^* = (g_1^*, \ldots, g_k^*)^{\mathrm{T}}$, and the associated maximal correlations are

$$\sigma(f_i^*\, g_i^*) \triangleq \mathbb{E}[f_i^*(X)\, g_i^*(Y)],\ \text{for}\ i = 1, \ldots, k, \tag{3}$$

and the HGR maximal correlation is

$$\mathrm{HGR}_k(X,Y) \triangleq \mathbb{E}\left[\mathbf{f}^{*\mathrm{T}}(X)\,\mathbf{g}^*(Y)\right] = \sum_{i=1}^{k} \sigma(f_i^*\, g_i^*). \tag{4}$$

Note that the original definition of HGR maximal correlation is the special case of our definition when $k = 1$ (see, [27]). This generalization of maximal correlation analysis enables us to produce more than one feature mapping by solving the maximal correlation problem, and these feature mappings can be used in other applications, including ensemble learning, multi-task learning, and transfer learning [28,29].

2.3. Related Work

Independence and separation have been studied in many works. Most existing approaches fail to provide an efficient solution in both discrete/continuous settings. Ref. [11] develops an optimizer using absolute difference in odds $|\mathbb{P}(\hat{Y} = 1|D = 1) - \mathbb{P}(\hat{Y} = 1|D = 0)|$ as a regularizer, which requires discrete Y and D and was only applied to Naïve Bayes and Logistic Regression to enforce the independence criterion. In [16], a post-processing method is provided using a probabilistic combination of classifiers to achieve the desired ROC curves, which only applies when D is discrete. Alternatively, Ref. [8] proposes pre-processing the data beforehand to enforce fairness before learning, based on randomized mappings of the data subject to a fairness constraint defined by $J = \max(|\frac{\mathbb{P}(\hat{Y}=1|D=1)}{\mathbb{P}(\hat{Y}=1|D=0)} - 1|, |\frac{\mathbb{P}(\hat{Y}=1|D=0)}{\mathbb{P}(\hat{Y}=1|D=1)} - 1|)$. Again, this method is only designed for independence with discrete Y and D, and it requires processing the entire dataset, which is computationally complex. Ref. [30] propose the use of a robust log-loss predictor for fairness, but in practice, it requires that Y be discrete.

Other methods can also be limited in their ability to handle all dependencies between variables. Ref. [31] uses a covariance-based constraint to enforce fairness, so it likely would not do well on other metrics. Furthermore, it is strictly a linear penalty rather than our non-linear formulation and penalizes the predictions of the system rather than the features learned. This limits the relationships between variables it can capture. An adversarial method is proposed in [20] to enforce independence or separation, but it requires the training of an adversary to predict the sensitive attribute, which can introduce issues of convergence and bias.

Recently, Ref. [9] propose the use of the HGR maximal correlation as a regularizer for either the independence or the separation constraint. In contrast to our approach dealing with the maximal correlation directly, they use a χ^2 divergence computed over a mesh grid to upper bound the HGR maximal correlation during the optimization of the classifier (either a linear regressor or a Deep Neural Net (DNN)). This method applies to cases where X is continuous and Y and D are either continuous or discrete variables, but it scales poorly with the bandwidth and dimensionality of D, and it treats the discrete case in the same way as the continuous case, resulting in slow performance on discrete datasets.

There are other works that use either an HGR-based or mutual information-based formulation of fairness but do not generalize to more than one setting. Refs. [32,33] use correlation-based regularizers but can only be used in the independence case. Furthermore, Ref. [33] only works with discrete targets, and only uses a single mode of the HGR maximal correlation (as opposed to multiple modes, which our method makes use of) for regularization, which limits the information it can encapsulate, and it is also not designed for continuous sensitive attributes. Ref. [34] also develops a method that can only be used for independence, and it requires training an additional network in order to evaluate a bound for the mutual information which can be used to as a fairness penalty, thus increasing the complexity and required runtime. Finally, Ref. [35] approximates the mutual information with a variational formulation, but it does not include a formulation for continuous labels.

3. Maximal Correlation for Fairness

Equipped with the HGR maximal correlation as a measure of dependence, we explore its use as a fairness penalty. Depending on the data modality (discrete/continuous) and the fairness criteria (independence/separation), the resulting fair learning algorithm takes different specifically tailored forms. In this section, we demonstrate how to derive these regularizers and algorithms to ensure the aforementioned fairness objectives for both discrete and continuous cases.

3.1. Maximal Correlation for Discrete Learning

In this subsection, the decision variable X, target variable Y, and sensitive attribute D are discrete random variables defined on alphabets $\mathcal{X}$, $\mathcal{Y}$, and $\mathcal{D}$, respectively.

We first describe how to solve the discrete maximal correlation problem using a divergence transfer matrix (DTM)-based approach. As it is shown later, it is more convenient to work with their equivalent representation via DTM instead of the joint distribution $P_{X,Y}$.

Definition 2. *The divergence transfer matrix (DTM)* $\mathbf{B}_{Y,X} \in \mathbb{R}^{|\mathcal{Y}|\times|\mathcal{X}|}$ *associated with joint distribution* $P_{X,Y}$ *is given by*

$$\mathbf{B}_{X,Y}(x,y) \triangleq \frac{P_{X,Y}(x,y)}{\sqrt{P_X(x)}\sqrt{P_Y(y)}}. \tag{5}$$

The following useful result expresses that the maximal correlation problem can be solved by simply computing the singular value decomposition (SVD) of the DTM **B** in the discrete case.

Theorem 1 ([27]). *Assume that the SVD of DTM* $\mathbf{B}_{Y,X}$ *takes the form*

$$\mathbf{B}_{Y,X} = \sum_{i=0}^{K-1} \sigma_i \psi_i^Y (\psi_i^X)^{\mathrm{T}}, \tag{6}$$

with singular values $\sigma_0 \geq \sigma_1 \geq \cdots \geq \sigma_{K-1}$, *singular vectors* ψ_i^Y, ψ_i^X, *and* $K = \min\{|\mathcal{X}|, |\mathcal{Y}|\}$. *Then, we have*

$$\sigma_0 = 1, \quad \psi_0^X(x) = \sqrt{P_X(x)}, \quad \psi_0^Y(y) = \sqrt{P_Y(y)}, \tag{7}$$

and the maximal correlation functions are related to the singular vectors in the SVD:

$$f_i^*(x) = \frac{\psi_i^X(x)}{\sqrt{P_X(x)}}, \quad g_i^*(x) = \frac{\psi_i^Y(y)}{\sqrt{P_Y(y)}}, \tag{8}$$

with associated maximal correlations $\sigma(f_i^*\, g_i^*) = \sigma_i$, *for* $i = 1, \cdots, K-1$. *Thus, the conditional distribution* $P_{Y|X}$ *has the following decomposition:*

$$P_{Y|X}(y|x) = P_Y(y)\left[1 + \sum_{i=1}^{K-1} \sigma_i f_i^*(x) g_i^*(y)\right]. \tag{9}$$

As we can see from this theorem, the singular values σ_i (since the associated maximal correlations is equal to the corresponding singular values of DTM, we abuse the notation a little bit and use σ to denote both of them) of the matrix $\mathbf{B}_{Y,X}$ essentially characterize the dependence between two discrete random variables, and the singular vectors $\Phi^X = [\psi_1^X, \cdots, \psi_k^X]$ and $\Phi^Y = [\psi_1^Y, \cdots, \psi_k^Y]$ are equivalent to the maximal correlation functions **f** and **g**.

Since our goal is to construct feature mappings $\mathbf{f}(x)$ under fairness constraints, our algorithms in the discrete case are built on the following variational characterization of an SVD, which does not involve $\mathbf{g}(y)$:

Lemma 1 ([36]). *For any* $k \leq K-1$ *and* $\Phi_X \in \mathbb{R}^{|\mathcal{X}|\times(k+1)}$,

$$\max_{\Phi_X^{\mathrm{T}}\Phi_X = \mathbf{I}} \|\mathbf{B}\Phi_X\|_{\mathrm{F}}^2 = \sum_{i=0}^{k} \sigma_i^2, \tag{10}$$

where $\|A\|_{\mathrm{F}} \triangleq \sqrt{\mathrm{tr}(A^{\mathrm{T}}A)}$ *denotes the Frobenius norm.*

3.1.1. Independence

To ensure sufficient independence, we must construct feature mappings $\mathbf{f}: \mathcal{X} \to \mathbb{R}^k$ so that the maximal correlations between $\mathbf{f}(X)$ and Y are large, while the ones between

$\mathbf{f}(X)$ and D are small. Motivated by Lemma 1 and Theorem 1, we propose the following DTM-based approach to construct $\mathbf{f}$:

$$\max_{\Phi\in\mathbb{R}^{|\mathcal{X}|\times(k+1)}:\Phi^{\mathrm{T}}\Phi=\mathbf{I}} \|\mathbf{B}_{Y,X}\Phi\|_{\mathrm{F}}^2-\lambda\|\mathbf{B}_{D,X}\Phi\|_{\mathrm{F}}^2, \tag{11}$$

where $\mathbf{B}_{Y,X}$ and $\mathbf{B}_{D,X}$ denote the DTMs of distribution $P_{Y,X}$ and $P_{D,X}$, respectively, and λ is the regularization coefficient that controls the penalty of the maximal correlations between $\mathbf{f}(X)$ and D. $\Phi^*=[\phi_0^*,\phi_1^*,\cdots,\phi_k^*]$ is the solution of the optimization problem (11). As shown in Theorem 1, $\mathbf{B}_{Y,X}$ and $\mathbf{B}_{D,X}$ have a shared right singular vector $\sqrt{P_X(x)}$, and we can let $\phi_0^*=\sqrt{P_X(x)}$. Then, the feature mappings for independence can be obtained by normalizing other column vectors in Φ^*

$$f_i(x)=\phi_i^*(x)/\sqrt{P_X(x)},\ i=1,\cdots,k. \tag{12}$$

We have the following remarks:

(1) The optimization problem in (11) can be written as $\max\operatorname{tr}(\Phi^{\mathrm{T}}(\mathbf{B}_{Y,X}^{\mathrm{T}}\mathbf{B}_{Y,X}-\lambda\mathbf{B}_{D,X}^{\mathrm{T}}\mathbf{B}_{D,X})\Phi)$, and it can be solved exactly by computing the eigen decomposition of $\mathbf{B}_{Y,X}^{\mathrm{T}}\mathbf{B}_{Y,X}-\lambda\mathbf{B}_{D,X}^{\mathrm{T}}\mathbf{B}_{D,X}$.

(2) Lemma 1 states that the Frobenius norm squared $\|\mathbf{B}_{Y,X}F\|_{\mathrm{F}}^2$ corresponds to the squared sum of the singular values. Actually, the following lemma shows that $\|\mathbf{B}_{Y,X}F\|_{\mathrm{F}}^2$ can be further related to the mutual information $I(X;Y)$ when the dependence between X and Y is weak.

Lemma 2 ([27]). *Let $X\in\mathcal{X}$ and $Y\in\mathcal{Y}$ be ϵ-dependent random variables; i.e., the χ^2-divergence is bounded $D_{\chi^2}(P_{X,Y}\|P_XP_Y)\le\epsilon$, then*

$$I(X;Y)=\frac{1}{2}\sum_{i=1}^{K-1}\sigma_i^2+o(\epsilon^2). \tag{13}$$

(3) As suggested by Lemma 2, the optimization problem in (11) can also be interpreted as maximizing the mutual information between $\mathbf{f}(X)$ and Y while penalizing the mutual information $I(\mathbf{f}(X);D)$.

Once we solve (11) and obtain the feature mappings $\mathbf{f}(x)$, we can obtain the corresponding maximal correlation function $\mathbf{g}(y)$ for the target variable Y via one step of the alternating conditional expectations algorithm by [37]:

$$g_i(y)\propto\mathbb{E}_{p_{X|Y}(\cdot|y)}[f_i(X)],\ i=1,\ldots,k. \tag{14}$$

In turn, $\mathbf{g}(y)$ can be computed by further normalizing the conditional expectations of $\mathbf{f}(X)$, so that the condition $\mathbb{E}[\mathbf{g}(Y)\mathbf{g}^{\mathrm{T}}(Y)]=\mathbf{I}$ is satisfied. Finally, the predictions $\hat{Y}$ can be made following the Maximum A Posteriori (MAP) rule, where the posteriori distribution $P_{Y|X}(y|x)$ can be approximately computed by plugging the learned feature mappings $\mathbf{f}(X)$ and $\mathbf{g}(Y)$ into (9), i.e.,

$$\hat{Y}=\operatorname*{arg\,max}_{y\in\mathcal{Y}}P_Y(y)\left[1+\sum_{i=1}^{k}\sigma_if_i(x)g_i(y)\right]. \tag{15}$$

3.1.2. Separation

For the separation criterion, we want to ensure sufficient conditional independence $(f(X)\perp D)|Y$. Here, we cannot simply replace the $\mathbf{B}_{D,X}$ in (11) with a conditional DTM, as it involves three random variables and thus cannot be usefully expressed as a matrix. Since maximal correlation is related to mutual information as shown in Lemma 2, we consider the following formulation:

$$
\begin{aligned}
&\max_{\mathbf{f}} I(\mathbf{f}(X);Y) - \lambda I(\mathbf{f}(X);D,Y) \\
&= \max_{\mathbf{f}} I(\mathbf{f}(X);Y) - \lambda\big(I(\mathbf{f}(X);Y) + I(\mathbf{f}(X);D|Y)\big) \\
&= \max_{\mathbf{f}} (1-\lambda) I(\mathbf{f}(X);Y) - \lambda I(\mathbf{f}(X);D|Y),
\end{aligned} \tag{16}
$$

where the first equality follows from the chain rule of mutual information and $\lambda \in (0,1)$. Thus, we can control the conditional mutual information $I(\mathbf{f}(X);D|Y)$ by adding the joint mutual information $I(\mathbf{f}(X);D,Y)$ as a regularizer in the training process.

Note that Lemma 1 and Lemma 2 imply that mutual information can be approximated using DTM, as shown in (11) in an independence case. Accordingly, we approximate (16) using the following optimization problem to ensure the separation criterion for discrete data:

$$
\max_{\Phi \in \mathbb{R}^{|\mathcal{X}|\times(k+1)}:\Phi^{\mathrm{T}}\Phi=\mathbf{I}} \|B_{Y,X}\Phi\|_{\mathrm{F}}^2 - \lambda \|B_{D\otimes Y,X}\Phi\|_{\mathrm{F}}^2, \tag{17}
$$

where $D \otimes Y$ is the Cartesian product of D and Y, and $B_{D\otimes Y,X}$ denotes the DTM of distribution $P_{D\otimes Y,X}$. Once we obtained the solution Φ^*, we could follow similar steps as in the independence case to get $\mathbf{f}(x)$ and $\mathbf{g}(y)$ and make predictions for the test samples.

3.2. *Maximal Correlation for Continuous Learning*

When X, Y, and D are all continuous and real-valued, computing the HGR maximal correlation becomes much more difficult, since the space of functions over real numbers is not tractable. Thus, we turn to approximations and begin by limiting our scope of learning algorithms to those that train models (e.g., neural nets) via gradient descent (or SGD) using samples, which encompasses most of the commonly used methods. Then, it follows that any approximation of the HGR maximal correlation used must be differentiable to calculate the gradient. Thus, we restrict the space of maximal correlation functions to be the family of functions that can be learned by neural nets, allowing us to compute the gradient while still providing a rich set of functions to search over.

3.2.1. Independence

To ensure sufficient independence, we want to minimize the loss function $L(\hat{Y},Y)$ and the maximal correlation between $\mathbf{f}(X)$ and D. Then, our optimization (for a given λ) becomes:

$$
\min_{\substack{\mathbf{f}:\,\mathcal{X}\to\mathbb{R}^m\\ T:\,\mathbb{R}^m\to\mathcal{Y}}} L(T(\mathbf{f}(X)),Y) + \lambda \mathrm{HGR}_k(\mathbf{f}(X),D), \tag{18}
$$

where $\mathrm{HGR}_k(\mathbf{f}(X),D) = \max_{\mathbf{g},\,\mathbf{h}} \mathbb{E}\big[\mathbf{g}^{\mathrm{T}}(\mathbf{f}(X))\,\mathbf{h}(D)\big]$, with $\mathbb{E}[\mathbf{g}(\mathbf{f}(X))] = \mathbb{E}[\mathbf{h}(D)] = \mathbf{0}$, and $\mathbb{E}\big[\mathbf{g}(\mathbf{f}(X))\mathbf{g}^{\mathrm{T}}(\mathbf{f}(X))\big] = \mathbb{E}\big[\mathbf{h}(D)\mathbf{h}^{\mathrm{T}}(D)\big] = \mathbf{I}$. m is the dimension of the features $\mathbf{f}(X)$, k is the number of maximal correlation functions, and $\mathbf{g}\colon \mathbb{R}^m \to \mathbb{R}^k$, $\mathbf{h}\colon \mathcal{D} \to \mathbb{R}^k$ are the maximal correlation functions relating $\mathbf{f}(X)$ with D. Given the difficulty of enforcing the orthogonalization constraint, we use a variational characterization of the HGR maximal correlation called Soft-HGR proposed in [29], which relaxes the orthogonal constraint:

$$
\mathrm{HGR}_{\mathrm{soft}}(X,Y) \triangleq \max_{\substack{\mathbb{E}[\mathbf{g}(X)]=\mathbf{0}\\ \mathbb{E}[\mathbf{h}(Y)]=\mathbf{0}}} \mathbb{E}\Big[\mathbf{g}^{\mathrm{T}}(X)\,\mathbf{h}(Y)\Big] - \frac{1}{2}\mathrm{tr}(\mathrm{cov}[\mathbf{g}(X)]\,\mathrm{cov}[\mathbf{h}(Y)]), \tag{19}
$$

where $\text{cov}[X]$ is the covariance matrix of X. [29] shows that this Soft-HGR formulation can be viewed as a low-rank approximation of the original HGR maximal correlation problem in the discrete case. Then, our learning objective becomes:

$$\min_{\substack{\mathbf{f}:\, \mathcal{X}\to\mathbb{R}^m \\ T:\, \mathbb{R}^m\to\mathcal{Y}}} \quad \max_{\substack{\mathbf{g}:\, \mathbb{R}^m\to\mathbb{R}^k,\ \mathbf{h}:\, \mathcal{D}\to\mathbb{R}^k \\ \mathbb{E}[\mathbf{g}(\mathbf{f}(X))]=\mathbb{E}[\mathbf{h}(D)]=\mathbf{0}}} C, \tag{20}$$

where

$$C = L(T(\mathbf{f}(X)), Y) + \lambda\mathbb{E}\Big[\mathbf{g}^{\mathrm{T}}(\mathbf{f}(X))\,\mathbf{h}(D)\Big] - \frac{\lambda}{2}\,\mathrm{tr}\left(\mathrm{cov}[\mathbf{g}(\mathbf{f}(X))]\,\mathrm{cov}[\mathbf{h}(D)]\right).$$

We solve this optimization by alternating between optimizing $\mathbf{f}, T$ and optimizing $\mathbf{g}, \mathbf{h}$. In practice, we implement this by alternating between one step of gradient descent for $\mathbf{f}$ and T and five steps of gradient descent on $\mathbf{g}$ and $\mathbf{h}$ to allow the maximal correlation functions to adapt to the changing of features $\mathbf{f}$.

3.2.2. Separation

For separation, we use a similar argument as in the discrete case to ensure the conditional independence. Specifically, we solve the following optimization problem:

$$\min_{\substack{\mathbf{f}:\, \mathcal{X}\to\mathbb{R}^m \\ T:\, \mathbb{R}^m\to\mathcal{Y}}} L(T(\mathbf{f}(X)), Y) + \lambda\big(\mathrm{HGR}_{\mathrm{soft}}(f(X), D \otimes Y) - \mathrm{HGR}_{\mathrm{soft}}(f(X), Y)\big). \tag{21}$$

Note that for the first Soft-HGR term, we use $\mathbf{g}, \mathbf{h}$ to denote the maximal correlation functions and $\mathbf{g}', \mathbf{h}'$ to denote the functions for the second term. Similar to the discrete case, the difference term allows us to approximate the conditional mutual information using two unconditional terms. Once again, we solve this optimization by alternating between optimizing $\mathbf{f}$, T and optimizing $\mathbf{g}, \mathbf{h}, \mathbf{g}', \mathbf{h}'$.

3.2.3. Few-Shot Learning

In the continuous case, our learning objective can also be applied a posteriori in a few-shot setting with a clasifier that has already been trained in a fairness-unaware manner on a large number of samples without the sensitive attribute label. In this case, we can formulate our objective as before and use the few samples containing the sensitive attribute to further train the network and force it to learn fairer features that are still predictive of the desired labels.

4. Experimental Results

In order to illustrate the effectiveness of our algorithms, we run experiments using the proposed algorithms on discrete (Adult and COMPAS) and continuous (Communities and Crimes) datasets.

4.1. Discrete Case

We test the proposed DTM-based approach on the ProPublica's COMPAS recidivism dataset (https://github.com/propublica/compas-analysis (accessed on 14 February 2022)) and the UCI Adult dataset (https://archive.ics.uci.edu/ml/datasets/adult (accessed on 14 February 2022)), which were chosen as they contain categorical features and are used in prior works. More experiments for the discrete case can be found in the Appendix A.

For the COMPAS dataset, the goal is to predict whether the individual recidivated (re-offended) (Y) using the severity of charge, number of prior crimes, and age category as the decision variables (X). As discussed in [8], COMPAS scores are biased against African-Americans, so race is set to be the sensitive attribute (D) and filtered to contain only Caucasian and African-American individuals. As for the Adult dataset, the goal is to predict the binary indicator (Y) of whether the income of the individual is more than

50K or not based on the following decision variables (X): age (quantized to decades) and education (in years), and the sensitive attribute (D) is the gender of the individual.

For both datasets, we randomly split all data into 80%/20% training/test samples. We first construct an estimate of DTM $\hat{\mathbf{B}}$ with the empirical distribution of the training set; then, we solve the proposed optimization in (11) and (17) using $\hat{\mathbf{B}}$ to obtain fair feature mappings $\hat{\mathbf{f}}(x), \hat{\mathbf{g}}(y)$. The predictions $\hat{Y}$ of the test samples X' are given by plugging the learned feature mappings $\hat{\mathbf{f}}(x'), \hat{\mathbf{g}}(y)$ into the MAP rule (15), where P_Y can be estimated from the empirical distribution $\hat{P}_Y$ on the training set.

For the independence case, we compare the tradeoff between the performance and the discrimination achieved by our method with that of the optimized pre-processing methods proposed in [8]. Note that we adopt the same settings as the experiments in [8] to do a fair comparison, and the reported results for their method are from their work. We plot the area under the ROC curve (AUC) of $\hat{P}_{Y|X'}(y|x')$ compared to the true test labels Y' against the following standard discrimination measure derived from legal proceedings [4]:

$$J = \max_{d,d' \in \mathcal{D}} \left| \mathbb{P}_{\hat{Y}|D}(1|d) / \mathbb{P}_{\hat{Y}|D}(1|d') - 1 \right|. \tag{22}$$

Figures 1 and 2 (Top) show the results. For both datasets, it can be seen that simply dropping the sensitive attribute D and applying logistic regression (LR) and random forest (RF) algorithms cannot ensure independence between $\hat{Y}$ and D. However, the proposed DTM-based algorithm provides a tradeoff between performance and discrimination by varying the value of the regularizer λ in the optimization (11), which outperforms the optimized pre-processing methods in [8] on the Adult dataset and achieves similar performance on the COMPAS dataset. More importantly, the DTM-based algorithm provides a smooth tradeoff curve between the performance and discrimination, so that a desired level of fairness can be achieved by setting λ in practice. In addition, since our method only requires us to perform eigen-decomposition, it runs significantly faster than the optimized pre-processing method, which needs to solve a much more complex optimization problem. Empirically, we find at least a tenfold speed up in runtime compared to the existing methods.

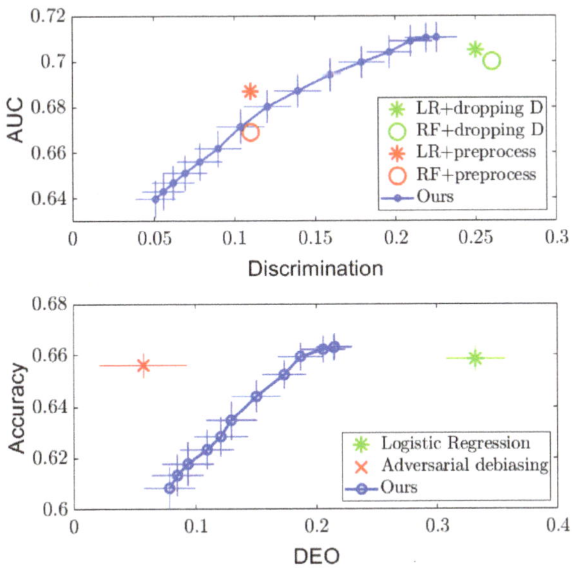

Figure 1. Regularization results on the COMPAS dataset, with AUC plotted against discrimination measure for independence (**Top**), and accuracy plotted against DEO for separation (**Bottom**), respectively.

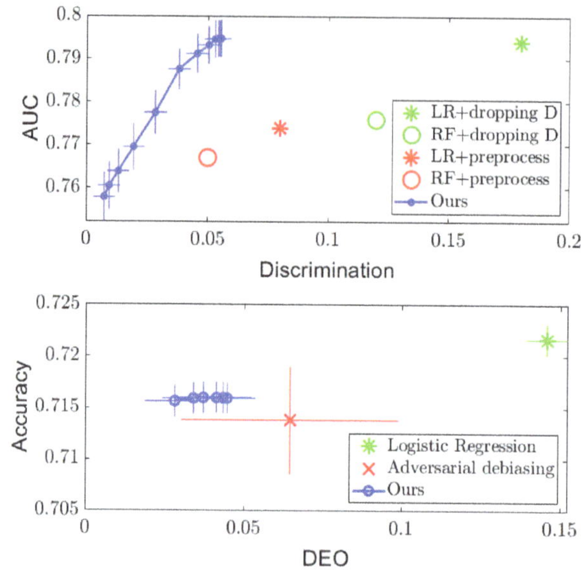

Figure 2. Regularization results on Adult dataset, with AUC plotted against discrimination measure for independence (**Top**), and accuracy plotted against DEO for separation (**Bottom**), respectively.

For the separation criterion, we compare the balanced accuracy achieved by our algorithm with that of the adversarial debiasing method in [20] (implementation given in [2]) against the difference in equalized opportunities (DEO), which is another standard measure used commonly in the literature:

$$\text{DEO} = \left|\mathbb{P}(\hat{Y}{=}1|D{=}1, Y{=}1) - \mathbb{P}(\hat{Y}{=}1|D{=}0, Y{=}1)\right|. \tag{23}$$

The results on the COMPAS and Adult datasets are presented in Figures 1 and 2 (Bottom). Compared to the naïve logistic regression, the proposed DTM-based algorithm dramatically decreases the DEO while maintaining similar accuracy performance on both datasets, which outperforms the adversarial debiasing method in [20] on the Adult dataset. We note that the accuracy and DEO curve achieved by the proposed algorithm in the separation setting has a smaller range compared to that in the independence setting. This is because the value of the regularizer λ is restricted in the separation optimization problem (17) to $\lambda \in [0, 1)$, but only to $\lambda > 0$ for the optimization in (11). More details about the influence of the regularizer λ can be found in Appendix A.

4.2. Continuous Case

In the continuous case, we experiment on the Communities and Crimes (C&C) dataset (http://archive.ics.uci.edu/ml/datasets/communities+and+crime (accessed on 14 February 2022)). The goal is to predict the crime rate Y of a community given a set of 121 statistics X (distributions of income, age, urban/rural, etc.). The 122-th statistic (percentage of black people in the community) is used as the sensitive variable D. All variables in this dataset are real-valued. The dataset was split into 1794 training and 200 test samples. Following [9], we use a Neural Net with a 50-node hidden layer (which we denote as $f(x)$) and train a predictor $\hat{y} = T(f(x))$ with the mean squared error (MSE) loss and the Soft-HGR penalty, varying λ. For Soft-HGR, we use two two-layer NNs with scalar outputs as the two maximal correlation functions **g** and **h**, and then, we trained them according to (20) (independence) or (21) (separation). Then, we computed the test MSE and test "discrimination" in each case.

For independence, our metric was $I(\hat{Y}; D)$, which was approximated using a standard kNN-based mutual information estimator [38]. For separation, we computed $I(\hat{Y}; D|Y)$ using the same estimator. We report the results of our experiment as well as that of the χ^2 method of [9] with the same architecture. The results of the experiments are presented in Figure 3.

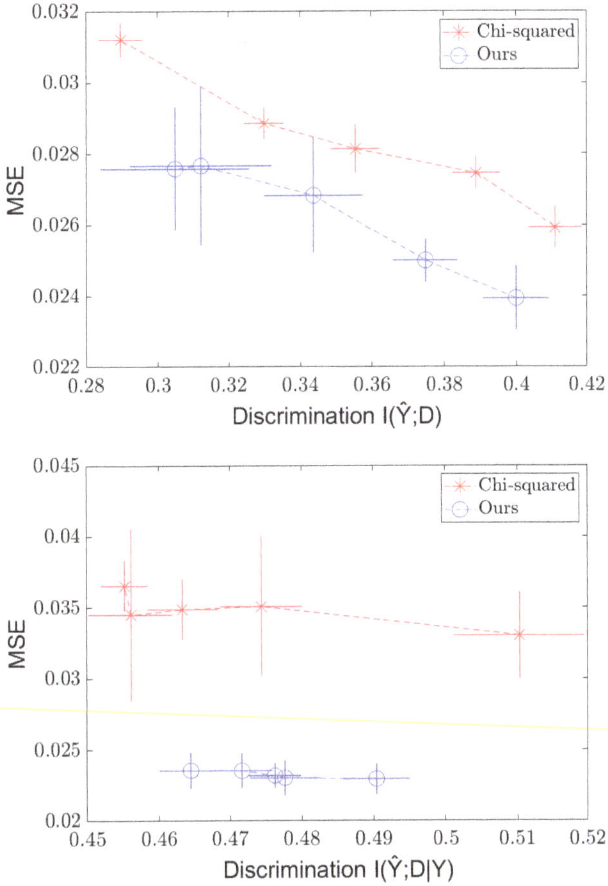

Figure 3. Independence (**top**) and Separation (**bottom**) regularization on the C&C dataset, with MSE plotted against $I(\hat{Y}; D|Y)$.

As expected, we see a tradeoff between the MSE and discrimination, creating a frontier of possible values. We also see that the Soft-HGR penalty provides modest gains compared to the χ^2 method for both independence and separation.

Moreover, our method runs significantly faster than the χ^2 method (on the order of seconds per iteration for our method versus just under a minute per iteration for the comparison method), as the χ^2 method requires computation over a mesh grid of a Gaussian KDE, which scales with the product of the number of "bins" (mesh points) and the number of training samples, while our method only scales with the number of samples ($O(n)$), since it only requires passing over all the training samples a constant number of times per iteration. For large bandwidths, d can become quite large. KDE methods also scale poorly with dimensionality (see, [39]) in an exponential manner, and thus, if d is high-dimensional, the χ^2 method would run much slower than our method, which can take in an arbitrarily-sized input and scale linearly with the dimensionality of the input multiplied by the number of samples. Empirically, we find that our method runs around five times faster.

We also run experiments to illustrate how our method's simplicity allows it to adapt to the few-shot, few-epoch regime faster than that of the χ^2 method. We take 10 "few-shot" samples from the training set; then, we train a network to predict Y from X without any fairness regularizer using the full training set. Then, we run five more iterations of gradient descent on the trained model using the fairness-regularized objective and the 10 few-shot samples, and we compare the separation results between the Soft-HGR and χ^2 regularizer. We choose to compare to the χ^2 regularizer as it is one of the few methods designed to handle continuous D. The results are shown in Figure 4. Once again, we see the tradeoff curve, and we see that our method outperform the χ^2 method, and that it appears to be competitive with the standard case in just a few iterations, while the χ^2 method is still far from achieving the original MSE. We also vastly outperform the baseline (before fairness regularization) model in reducing discrimination, at the cost of only a small increase in error. Thus, in situations where, due to ethical/legal issues, only a few samples labeled with the sensitive attribute can be collected, fairness can still be enforced.

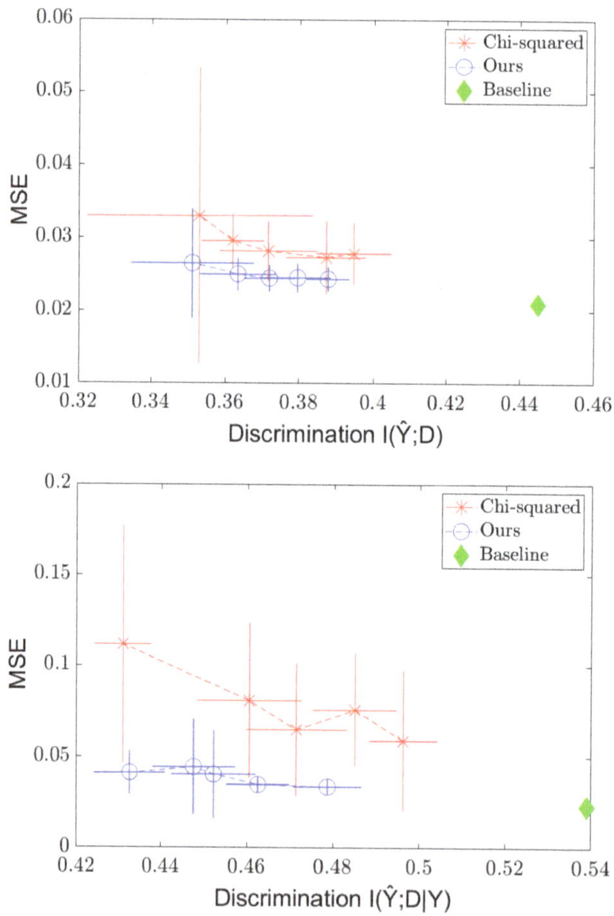

Figure 4. Independence (**top**) and Separation (**bottom**) regularization on the C&C dataset in the *few-shot* settings, with MSE plotted against $I(\hat{Y}; D|Y)$.

5. Conclusions

As machine learning algorithms gain more relevance, more focus will be placed upon ensuring their fairness. We have presented a framework using the HGR maximal correlation, which provides effective and computationally efficient methods for enforcing independence and separation constraints, and derived algorithms for fair learning on discrete and continuous data, which provide competitive tradeoff curves. In addition, we have also shown promising results in the few-shot setting and suggested a method for rapidly adapting a classifier to improve fairness. In the future, it would be beneficial to extend this framework to other criteria (e.g., sufficiency) and to to determine how to use this framework to enforce fairness in a transfer learning setup coupled with the few-shot setting, to determine how to fairly adapt a classifier to a new task.

However, this method requires knowledge of the sensitive attribute for all samples during the training time, which can be impractical in some cases. Further extension into developing these regularizers with a limited number of such samples would be very useful.

Author Contributions: J.L.: software, continuous algorithm design, writing—original draft. Y.B.: software, discrete algorithm design, writing—original draft. P.S. and R.P.: conceptualization, writing—review and editing. G.W.W.: supervision, funding acquisition, and writing—review and editing. L.K. and R.S.F.: writing—review and editing. All authors have read and agreed to the published version of the manuscript.

Funding: This work was supported, in part, by the MIT-IBM Watson AI Lab under Agreement No. W1771646, and NSF under Grant No. CCF-1717610.

Institutional Review Board Statement: Not applicable.

Informed Consent Statement: Not applicable.

Data Availability Statement: Data and code can be found in Section 4.

Conflicts of Interest: The authors declare no conflict of interest.

Appendix A. Effect of the Regularizer in Discrete Case

In this section, we provide additional experiment results to demonstrate how the performance of classification and fairness measures change with different values of the regularizer.

We use the same setup described in Section 4.1 and present the results in Figures A1–A4. In Figures A1 and A2, we plot the achieved AUC and Discrimination (measured with J in (21)) versus the value of λ for both COMPAS and Adult data using independence criterion. In Figures A3 and A4, we plot the accuracy of the classifier and DEO versus λ for both datasets using separation criterion. As shown by all the figures, the performance of classification and fairness measures are all decreasing as we increase λ, and the proposed DTM-based algorithm is able to provide a smooth tradeoff curve between the performance and fairness measures.

Note that the value of the regularizer λ is restricted in the separation optimization problem to $\lambda \in [0, 1)$; therefore, the range of the achieved performance in Figures A3 and A4 is smaller than that in Figures A1 and A2.

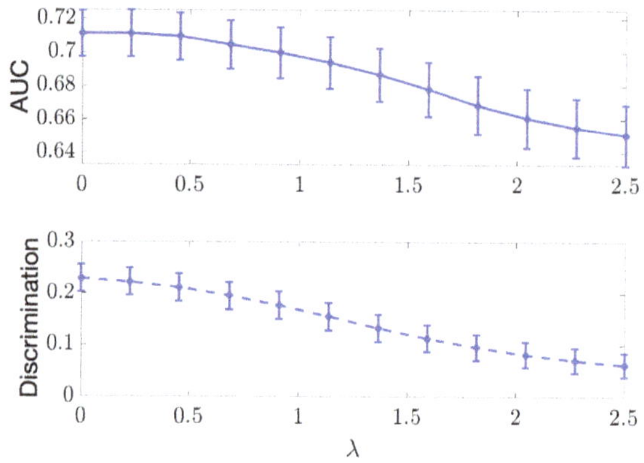

Figure A1. Results for independence regularization on the discrete COMPAS dataset, AUC results (**Top**) and discrimination measure J (**Bottom**) are plotted with respect to different values of λ.

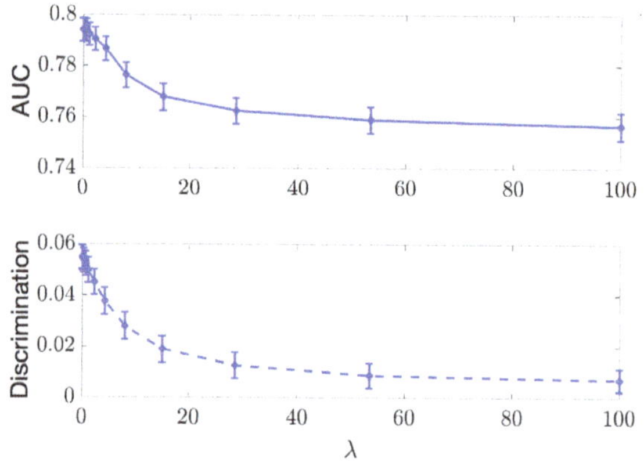

Figure A2. Results for independence regularization on the discrete Adult dataset, AUC results (**Top**) and discrimination measure J (**Bottom**) are plotted with respect to different values of λ.

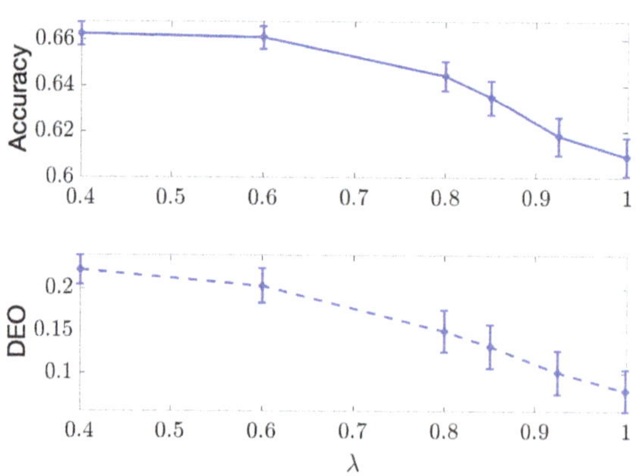

Figure A3. Results for separation regularization on the discrete COMPAS dataset, accuracy (**Top**) and DEO (**Bottom**) are plotted with respect to different values of λ.

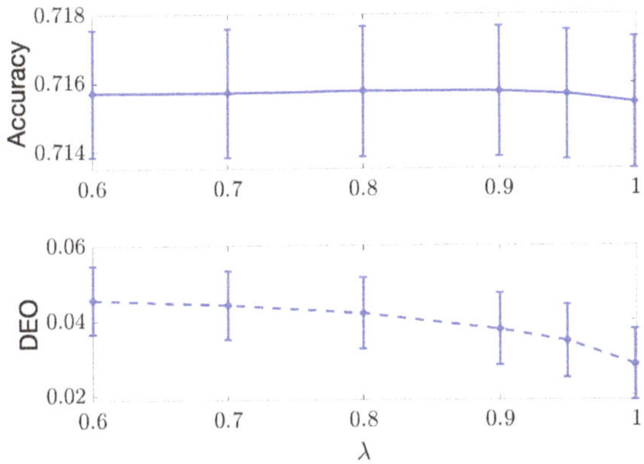

Figure A4. Results for separation regularization on the discrete Adult dataset, accuracy (**Top**) and DEO (**Bottom**) are plotted with respect to different values of λ.

References

1. Selbst, A.D.; Boyd, D.; Friedler, S.A.; Venkatasubramanian, S.; Vertesi, J. Fairness and abstraction in sociotechnical systems. In Proceedings of the Conference on Fairness, Accountability, and Transparency, Atlanta, GA, USA, 29–31 January 2019; pp. 59–68.
2. Bellamy, R.K.; Dey, K.; Hind, M.; Hoffman, S.C.; Houde, S.; Kannan, K.; Lohia, P.; Martino, J.; Mehta, S.; Mojsilovic, A.; et al. AI Fairness 360: An extensible toolkit for detecting, understanding, and mitigating unwanted algorithmic bias. *arXiv* **2018**, arXiv:1810.01943.
3. Barocas, S.; Hardt, M.; Narayanan, A. Fairness and Machine Learning. 2019. Available online: http://www.fairmlbook.org (accessed on 14 February 2022).
4. EEOC. *Department of Labor, & Department of Justice. Uniform Guidelines on Employee Selection Procedures;* Federal Register: Washington, DC, USA, 1978.
5. Locatello, F.; Abbati, G.; Rainforth, T.; Bauer, S.; Schölkopf, B.; Bachem, O. On the fairness of disentangled representations. In Proceedings of the Advances in Neural Information Processing Systems, Vancouver, BC, Canada, 8–14 December 2019; pp. 14584–14597.
6. Gölz, P.; Kahng, A.; Procaccia, A.D. Paradoxes in Fair Machine Learning. In Proceedings of the Advances in Neural Information Processing Systems, Vancouver, BC, Canada, 8–14 December 2019; pp. 8340–8350.
7. Corbett-Davies, S.; Goel, S. The measure and mismeasure of fairness: A critical review of fair machine learning. *arXiv* **2018**, arXiv:1808.00023.
8. Calmon, F.; Wei, D.; Vinzamuri, B.; Ramamurthy, K.N.; Varshney, K.R. Optimized pre-processing for discrimination prevention. In Proceedings of the Advances in Neural Information Processing Systems, Long Beach, CA, USA, 4–9 December 2017; pp. 3992–4001.
9. Mary, J.; Calauzenes, C.; El Karoui, N. Fairness-aware learning for continuous attributes and treatments. In Proceedings of the International Conference on Machine Learning, Long Beach, CA, USA, 9–15 June 2019; pp. 4382–4391.
10. Kamiran, F.; Calders, T. Data preprocessing techniques for classification without discrimination. *Knowl. Inf. Syst.* **2012**, *33*, 1–33. [CrossRef]
11. Zemel, R.; Wu, Y.; Swersky, K.; Pitassi, T.; Dwork, C. Learning fair representations. In Proceedings of the International Conference on Machine Learning, Atlanta, GA, USA, 16–21 June 2013; pp. 325–333.
12. Feldman, M.; Friedler, S.A.; Moeller, J.; Scheidegger, C.; Venkatasubramanian, S. Certifying and Removing Disparate Impact. In Proceedings of the ACM SIGKDD International Conference on Knowledge Discovery and Data Mining, Sydney, Australia, 10–13 August 2015; pp. 259–268.
13. Sattigeri, P.; Hoffman, S.C.; Chenthamarakshan, V.; Varshney, K.R. Fairness gan. *arXiv* **2018**, arXiv:1805.09910.
14. Xu, D.; Yuan, S.; Zhang, L.; Wu, X. Fairgan: Fairness-aware generative adversarial networks. In Proceedings of the 2018 IEEE International Conference on Big Data (Big Data), Seattle, WA, USA, 10–13 December 2018; pp. 570–575.
15. Kamiran, F.; Karim, A.; Zhang, X. Decision Theory for Discrimination-Aware Classification. In Proceedings of the IEEE International Conference on Data Mining, Brussels, Belgium, 10–13 December 2012; pp. 924–929. [CrossRef]
16. Hardt, M.; Price, E.; Srebro, N. Equality of Opportunity in Supervised Learning. In Proceedings of the Advances in Neural Information Processing Systems 29, Barcelona, Spain, 5–10 December 2016; pp. 3315–3323.
17. Pleiss, G.; Raghavan, M.; Wu, F.; Kleinberg, J.; Weinberger, K.Q. On fairness and calibration. In Proceedings of the Advances in Neural Information Processing Systems, Long Beach, CA, USA, 4–9 December 2017; pp. 5680–5689.

18. Wei, D.; Ramamurthy, K.N.; Du Pin Calmon, F. Optimized Score Transformation for Fair Classification. *arXiv* **2019**, arXiv:1906.00066.
19. Kamishima, T.; Akaho, S.; Asoh, H.; Sakuma, J. Fairness-Aware classifier with Prejudice Remover Regularizer. In *Machine Learning and Knowledge Discovery in Databases;* Springer: Berlin/Heidelberg, Germany, 2012; pp. 35–50.
20. Zhang, B.H.; Lemoine, B.; Mitchell, M. Mitigating unwanted biases with adversarial learning. In Proceedings of the 2018 AAAI/ACM Conference on AI, Ethics, and Society, New Orleans, LA, USA, 2–3 February 2018; pp. 335–340.
21. Celis, L.E.; Huang, L.; Keswani, V.; Vishnoi, N.K. Classification with fairness constraints: A meta-algorithm with provable guarantees. In Proceedings of the Conference on Fairness, Accountability, and Transparency, Atlanta, GA, USA, 29–31 January 2019; pp. 319–328.
22. Angwin, J.; Larson, J.; Mattu, S.; Kirchner, L. *Machine Bias: There's Software Used across the Country to Predict Future Criminals. And It's Biased against Blacks;* ProPublica: New York, NY, USA, 2016.
23. Chouldechova, A. Fair prediction with disparate impact: A study of bias in recidivism prediction instruments. *Big Data* **2017**, *5*, 153–163. [CrossRef] [PubMed]
24. Hirschfeld, H.O. A connection between correlation and contingency. *Proc. Camb. Phil. Soc.* **1935**, *31*, 520–524. [CrossRef]
25. Gebelein, H. Das statistische Problem der Korrelation als Variations-und Eigenwertproblem und sein Zusammenhang mit der Ausgleichsrechnung. *Z. Angew. Math. Mech.* **1941**, *21*, 364–379. [CrossRef]
26. Rényi, A. On Measures of Dependence. *Acta Math. Acad. Sci. Hung.* **1959**, *10*, 441–451. [CrossRef]
27. Huang, S.L.; Makur, A.; Wornell, G.W.; Zheng, L. On Universal Features for High-Dimensional Learning and Inference. Preprint, 2019. Available online: http://allegro.mit.edu/~gww/unifeatures (accessed on 14 February 2022).
28. Lee, J.; Sattigeri, P.; Wornell, G. Learning New Tricks From Old Dogs: Multi-Source Transfer Learning From Pre-Trained Networks. In Proceedings of the Advances in Neural Information Processing Systems, Vancouver, BC, Canada, 8–14 December 2019; pp. 4372–4382.
29. Wang, L.; Wu, J.; Huang, S.L.; Zheng, L.; Xu, X.; Zhang, L.; Huang, J. An efficient approach to informative feature extraction from multimodal data. In Proceedings of the AAAI Conference on Artificial Intelligence, Honolulu, HI, USA, 27 January–1 February 2019; Volume 33, pp. 5281–5288.
30. Rezaei, A.; Fathony, R.; Memarrast, O.; Ziebart, B. Fairness for robust log loss classification. In Proceedings of the AAAI Conference on Artificial Intelligence, New York, NY, USA, 7–12 February 2020; Volume 34, pp. 5511–5518.
31. Zafar, M.B.; Valera, I.; Rogriguez, M.G.; Gummadi, K.P. Fairness constraints: Mechanisms for fair classification. In Proceedings of theArtificial Intelligence and Statistics, PMLR, Fort Lauderdale, FL, USA, 20–22 April 2017; pp. 962–970.
32. Grari, V.; Ruf, B.; Lamprier, S.; Detyniecki, M. Fairness-Aware Neural Réyni Minimization for Continuous Features. *arXiv* **2019**, arXiv:1911.04929.
33. Baharlouei, S.; Nouiehed, M.; Beirami, A.; Razaviyayn, M. Rènyi Fair Inference. *arXiv* **2019**, arXiv:1906.12005.
34. Moyer, D.; Gao, S.; Brekelmans, R.; Galstyan, A.; Ver Steeg, G. Invariant representations without adversarial training. In Proceedings of the Advances in Neural Information Processing Systems, Montreal, QC, Canada, 3–8 December 2018; pp. 9084–9093.
35. Cho, J.; Hwang, G.; Suh, C. A fair classifier using mutual information. In Proceedings of the 2020 IEEE International Symposium on Information Theory (ISIT), Los Angeles, CA, USA, 21–26 June 2020; pp. 2521–2526.
36. Horn, R.A.; Johnson, C.R. *Matrix Analysis;* Cambridge University Press: Cambridge, UK, 2012.
37. Breiman, L.; Friedman, J.H. Estimating Optimal Transformations for Multiple Regression and Correlation. *J. Am. Stat. Assoc.* **1985**, *80*, 580–598. [CrossRef]
38. Gao, W.; Oh, S.; Viswanath, P. Demystifying Fixed *k*-Nearest Neighbor Information Estimators. *IEEE Trans. Inf. Theory* **2018**, *64*, 5629–5661. [CrossRef]
39. Wang, Z.; Scott, D.W. Nonparametric density estimation for high-dimensional data—Algorithms and applications. *Wiley Interdiscip. Rev. Comput. Stat.* **2019**, *11*, e1461. [CrossRef]

entropy

MDPI

Article

CTRL: Closed-Loop Transcription to an LDR via Minimaxing Rate Reduction

Xili Dai [1,2,†], Shengbang Tong [1,†], Mingyang Li [3,†], Ziyang Wu [4,†], Michael Psenka [1], Kwan Ho Ryan Chan [5], Pengyuan Zhai [6], Yaodong Yu [1], Xiaojun Yuan [2], Heung-Yeung Shum [4] and Yi Ma [1,3,*]

1 Department of EECS, University of California Berkeley, Berkeley, CA 94720, USA; daixili_cs@163.com (X.D.); tsb@berkeley.edu (S.T.); psenka@berkeley.edu (M.P.); yaodong_yu@berkeley.edu (Y.Y.)
2 School of Computer Science and Engineering, University of Electronic Science and Technology of China, Chengdu 610056, China; xjyuan@uestc.edu.cn
3 Tsinghua-Berkeley Shenzhen Institute, Shenzhen 518055, China; lmy17@mails.tsinghua.edu.cn
4 International Digital Economy Academy, Shenzhen 518048, China; robinwuzy@gmail.com (Z.W.); msraharry@hotmail.com (H.-Y.S.)
5 Mathematical Institute for Data Science, Johns Hopkins University, Baltimore, MD 21218, USA; kchan49@jhu.edu
6 Institute for Applied Computational Science, Harvard University, Cambridge, MA 02138, USA; pzhai@g.harvard.edu
* Correspondence: yima@eecs.berkeley.edu
† These authors contributed equally to this work.

Abstract: This work proposes a new computational framework for learning a structured generative model for real-world datasets. In particular, we propose to learn *a Closed-loop Transcription*between a multi-class, multi-dimensional data distribution and a *Linear discriminative representation* (*CTRL*) in the feature space that consists of multiple independent multi-dimensional linear subspaces. In particular, we argue that the optimal encoding and decoding mappings sought can be formulated as a *two-player minimax game between the encoder and decoder*for the learned representation. A natural utility function for this game is the so-called *rate reduction*, a simple information-theoretic measure for distances between mixtures of subspace-like Gaussians in the feature space. Our formulation draws inspiration from closed-loop error feedback from control systems and avoids expensive evaluating and minimizing of approximated distances between arbitrary distributions in either the data space or the feature space. To a large extent, this new formulation unifies the concepts and benefits of Auto-Encoding and GAN and naturally extends them to the settings of learning a *both discriminative and generative* representation for multi-class and multi-dimensional real-world data. Our extensive experiments on many benchmark imagery datasets demonstrate tremendous potential of this new closed-loop formulation: under fair comparison, visual quality of the learned decoder and classification performance of the encoder is competitive and arguably better than existing methods based on GAN, VAE, or a combination of both. Unlike existing generative models, the so-learned features of the multiple classes are structured instead of hidden: different classes are explicitly mapped onto corresponding *independent principal subspaces* in the feature space, and diverse visual attributes within each class are modeled by the *independent principal components* within each subspace.

Keywords: closed-loop transcription; linear discriminative representation; rate reduction; minimax game

Citation: Dai, X.; Tong, S.; Li, M.; Wu, Z.; Psenka, M.; Chan, K.H.R.; Zhai, P.; Yu, Y.; Yuan, X.; Shum, H.-Y.; et al. CTRL: Closed-Loop Transcription to an LDR via Minimaxing Rate Reduction. *Entropy* **2022**, *24*, 456. https://doi.org/10.3390/e24040456

Academic Editors: Lizhong Zheng and Chao Tian

Received: 10 February 2022
Accepted: 17 March 2022
Published: 25 March 2022

1. Introduction

One of the most fundamental tasks in modern data science and machine learning is to learn and model complex distributions (or structures) of real-world data, such as images or texts, from a set of observed samples. By "to learn and model", one typically means that we want to establish a (parametric) mapping between the distribution of the real data, say $x \in \mathbb{R}^D$, and a more compact random variable, say $z \in \mathbb{R}^d$:

$$f(\cdot, \theta) : x \in \mathbb{R}^D \mapsto z \in \mathbb{R}^d \quad \text{or the inverse} \quad g(\cdot, \eta) : z \in \mathbb{R}^d \mapsto x \in \mathbb{R}^D, \tag{1}$$

where z has a certain standard structure or distribution (e.g., normal distributions). The so-learned representation or feature z would be much easier to use for either generative (e.g., decoding or replaying) or discriminative (e.g., classification) purposes, or both.

Data embedding versus data transcription. *Be aware* that the support of the distribution of x (and that of z) is typically *extremely low-dimensional* compared to that of the ambient space (for instance, the well-known CIFAR-10 datasets consist of RGB images with a resolution of 32×32. Despite the images being in a space of $\mathbb{R}^{3072}$, our experiments will show that the intrinsic dimension of each class is less than a dozen, even after they are mapped into a feature space of $\mathbb{R}^{128}$) hence the above mapping(s) may not be uniquely defined based on the support in the space $\mathbb{R}^D$ (or $\mathbb{R}^d$). In addition, the data x may contain multiple components (e.g., modes, classes), and the intrinsic dimensions of these components are not necessarily the same. Hence, without loss of generality, we may assume the data x to be distributed over a union of low-dimensional nonlinear submanifolds $\cup_{j=1}^{k} \mathcal{M}_j \subset \mathbb{R}^D$, where each submanifold $\mathcal{M}_j$ is of dimension $d_j \ll D$. Regardless, we hope the learned mappings f and g are (locally dimension-preserving) *embedding* maps [1], when restricted to each of the components $\mathcal{M}_j$. In general, the dimension of the feature space d needs to be significantly higher than all of these intrinsic dimensions of the data: $d > d_j$. In fact, it should preferably be higher than the sum of all the intrinsic dimensions: $d \geq d_1 + \cdots + d_k$, since we normally expect that the features of different components/classes can be made fully independent or orthogonal in $\mathbb{R}^d$. Hence, without any explicit control of the mapping process, the actual features associated with images of the data under the embedding could still lie on some arbitrary nonlinear low-dimensional submanifolds inside the feature space $\mathbb{R}^d$. The distribution of the learned features remains "latent" or "hidden" in the feature space.

So, for features of the learned mappings (1) to be truly convenient to use for purposes such as data classification and generation, the goals of learning such mappings should not only simply reduce the dimension of the data x from D to d but also determine explicitly and precisely how the mapped feature $z = f(x)$ is distributed within the feature space $\mathbb{R}^d$, in terms of both its support and density. Moreover, we want to establish an explicit map $g(\cdot)$ from this distribution of feature z back to the data space such that the distribution of its image $\hat{x} = g(z)$ (closely) matches that of x. To differentiate from finding arbitrary feature embeddings (as most existing methods do), we call embeddings of data onto an explicit family of models (structures or distributions) in the feature space as *data transcription*.

Paper Outline. This work is to show how such transcription can be achieved for real-world visual data with one important family of models: the linear discriminative representation (LDR) introduced by [2]. Before we formally introduce our approach in Section 2, for the remainder of this section, we first discuss two existing approaches, namely autoencoding and GAN, that are closely related to ours. As these approaches are rather popular and known to the readers, we will mainly point out some of their main conceptual and practical limitations that have motivated this work. Although our objective and framework will be mathematically formulated, the main purpose of this work is to verify the effectiveness of this new approach empirically through extensive experimentation, organized and presented in Section 3 and Appendix A. Our work presents compelling evidence that the closed-loop data transcription problem and our rate-reduction-based formulation deserve serious attention from the information-theoretical and mathematical communities. This has raised many exciting and open theoretical problems or hypotheses about learning, representing, and generating distributions or manifolds of high-dimensional real-world data. We discuss some open problems in Section 4 and new directions in Section 5. Source code can be found at https://github.com/Delay-Xili/LDR (accessed on 9 February 2022).

1.1. Learning Generative Models via Auto-Encoding or GAN

Auto-Encoding and its variants. In the machine-learning literature, roughly speaking,there have been two representative approaches to such a distribution-learning task. One is the classic "Auto Encoding" (AE) approach [3,4] that aims to simultaneously learn an encoding mapping f from x to z and an (inverse) decoding mapping g from z back to x:

$$\boldsymbol{X} \xrightarrow{\ f(x,\theta)\ } \boldsymbol{Z} \xrightarrow{\ g(z,\eta)\ } \hat{\boldsymbol{X}}. \tag{2}$$

Here, we use bold capital letters to indicate a matrix of finite samples $\boldsymbol{X} = [x^1, \ldots, x^n] \in \mathbb{R}^{D\times n}$ of x and their mapped features $\boldsymbol{Z} = [z^1, \ldots, z^n] \subset \mathbb{R}^{d\times n}$, respectively. Typically, one wishes for two properties: firstly, the decoded samples $\hat{\boldsymbol{X}}$ are "similar" or close to the original $\boldsymbol{X}$, say in terms of maximum likelihood $p(\boldsymbol{X})$; and secondly, the (empirical) distribution of the mapped samples $\boldsymbol{Z}$, denoted as $\hat{p}(z|\boldsymbol{X})$, is close to certain desired prior distribution $p(z)$, say some much lower-dimensional multivariate Gaussian (The classical PCA can be viewed as a special case of this task. In fact, the original auto-encoding is precisely cast as *nonlinear* PCA [3], assuming the data lie on only one nonlinear submanifold $\mathcal{M}$).

However it is typically very difficult, often computationally intractable to maximize the likelihood function $p(\boldsymbol{X})$ or to minimize certain "distance", say the *KL-divergence* $\mathcal{D}_{KL}(\hat{p}, p)$, between $\hat{p}(z|\boldsymbol{X})$ and $p(z)$. Except for simple distributions such as Gaussian, the KL divergence usually does not have a closed-form, even for a mixture of Gaussians. The likelihood and the KL-divergence become ill-conditioned when the supports of the distributions are low-dimensional (i.e., degenerate) and not overlapping (which is almost always the case in practice when dealing with distributions of high-dimensional data in high-dimensional spaces). So in practice, one typically chooses to minimize instead certain approximate bounds or surrogates derived with various simplifying assumptions on the distributions involved, as is the case in variational auto-encoding (VAE) [5,6]. As a result, even after learning, the precise posterior distribution of $\hat{p}(z|\boldsymbol{X})$ remains unclear or hidden inside the feature space.

In this work, we will show that if we impose specific requirements on the (distribution of) learned feature z to be a mixture of subspace-like Gaussians, a natural closed-form distance can be introduced for such distributions based on rate distortion from the information theory. In addition, the optimal solution to the feature representation within this family can be learned directly from the data *without specifying any target $p(z)$ in advance*, which is particularly difficult in practice when the distribution of a mixed dataset is multi-modal and each component may have a different dimension.

GAN and its variants. Compared to measuring distribution distance in the (often controlled) feature space z, a much more challenging issue with the above auto-encoding approach is how to effectively measure the distance between the decoded samples $\hat{\boldsymbol{X}}$ and the original $\boldsymbol{X}$ in the data space x. For instance, for visual data such as images, their distributions $p(\boldsymbol{X})$ or generative models $p(\boldsymbol{X}|z)$ are often not known. Despite extensive studies in the computer vision and image processing literature [7], it remains elusive to find a good measure for similarity of real images that is both efficient to compute and effective in capturing visual quality and semantic information of the images equally well. Precisely due to such difficulties, it has been suggested early on by [8] that one may have to take a discriminative approach to learn the distribution or a generative model for visual data. More recently, *Generative Adversarial Nets (GAN)* [9] offers an ingenious idea to alleviate this difficulty by utilizing a powerful discriminator d, usually modeled and learned by a deep network, to discern differences between the generated samples $\hat{\boldsymbol{X}}$ and the real ones $\boldsymbol{X}$:

$$\boldsymbol{Z} \xrightarrow{\ g(z,\eta)\ } \hat{\boldsymbol{X}}, \boldsymbol{X} \xrightarrow{\ d(x,\theta)\ } \mathbf{0}, \mathbf{1}. \tag{3}$$

To a large extent, such a discriminator plays the role of minimizing certain distributional distance, e.g., the *Jensen–Shannon divergence*, between the data $\boldsymbol{X}$ and $\hat{\boldsymbol{X}}$. Compared to the KL-divergence, the JS-divergence is well-defined even if the supports of the two distributions are non-overlapping. (However, JS-divergence does not have a closed-form expression even between two Gaussians, whereas KL-divergence does). However, as shown in [10], since the data distributions are low-dimensional, the JS-divergence can be highly ill-conditioned to optimize. (This may explain why many additional heuristics are typically used in many subsequent variants of GAN). So, instead, one may choose to replace JS-divergence with the earth mover's distance or the Wasserstein distance. However both JS-divergence and W-distance can only be approximately computed between two general distributions. (For instance, the W-distance requires one to compute the maximal difference between expectations of the two distributions over all 1-Lipschitz functions). Furthermore, neither the JS-divergence nor the W-distance have closed-form formulae, even for the Gaussian distributions. (The (ℓ^1-norm) W-distance can be bounded by the (ℓ^2-norm) W2-distance which has a closed-form [11]. However, as is well-known in high-dimensional geometry, ℓ^1-norm and ℓ^2 norm deviate significantly in terms of their geometric and statistical properties as the dimension becomes high [12]. The bound can become very loose). However, from a data representation perspective, *subspace-like Gaussians (e.g., PCA) or a mixture of them are the most desirable family of distributions that we wish our features to become.* This would make all subsequent tasks (generative or discriminative) much easier. In this work, we will show how to achieve this with a different fundamental metric, known as the rate reduction, introduced by [13].

The original GAN aims to directly learn a mapping $g(\cdot)$, called a generator, from a standard distribution (say, a low-dimensional Gaussian random field) to the real (visual) data distribution in a high-dimensional space. However, distributions of real-world data can be rather sophisticated and often contain *multiple* classes and *multiple* factors in each class [14]. This makes learning the mapping g rather challenging in practice, suffering difficulties such as *mode-collapse* [15]. As a result, many variants of GAN have been subsequently developed in order to improve the stability and performance in learning multiple modes and disentangling different factors in the data distribution, such as *Conditional GAN* [16–20], *InfoGAN* [21,22], or *Implicit Maximum Likelihood Estimation (IMLE)* [23,24]. In particular, to learn a generator for multi-class data, prevalent conditional GAN literature requires label information as conditional inputs [16,25–27]. Recently, [28,29] has proposed training a k-class GAN by generalizing the two-class cross entropy to a $(k+1)$-class cross entropy. In this work, *we will introduce a more refined* $2k$*-class measure* for the k real and k generated classes. In addition, to avoid features for each class collapsing to a singleton [30], instead of cross entropy, *we will use the so-called rate-reduction measure that promotes multi-mode and multi-dimension in the learned features* [13]. One may view the rate reduction as a metric distance that has closed-form formulae for a mixture of (subspace-like) Gaussians, whereas neither JS-divergence nor W-distance can be computed in closed form (even between two Gaussians).

Another line of research is about how to stabilize the training of GAN. SN-GAN [31] has shown that spectral normalization on the discriminator is rather effective, which we will adopt in our work, although our formulation is not so sensitive to such choice designed for GAN (see ablation study in Appendix A.9). PacGAN [32] shows that the training stability can be significantly improved by packing a pair of real and generated images together for the discriminator. Inspired by this work, *we show how to generalize such an idea to discriminating an arbitrary number of pairs of real and decoded samples without concatenating the samples.* Our results in this work will even suggest that the larger the batch size discriminated, the merrier (see ablation study in Appendix A.10). In addition, ref. [29] has shown that optimizing the latent features leads to state-of-the-art visual quality. Their method is based on the deep compressed sensing GAN [28]. Hence, there are strong reasons to believe that their method essentially utilizes the *compressed sensing* principle [12] to implicitly exploit the low-dimensionality of the feature distribution. Our framework *will explicitly expose and exploit such low-dimensional structures on the learned feature distribution.*

Combination of AE and GAN. Although AE (VAE) and GAN originated with somewhat different motivations, they have evolved into popular and effective frameworks for learning and modeling complex distributions of many real-world data such as images. (In fact, in some idealistic settings, it can be shown that AE and GAN are actually equivalent: for instance, in the LOG settings, authors in [33] have shown that GAN coincides with the classic PCA, which is precisely the solution to auto-encoding in the linear case). Many recent efforts tend to combine both auto-encoding and GAN to generate more powerful generative frameworks for more diverse data sets, such as [15,34–42]. As we will see, in our framework, AE and GAN can be naturally interpreted as two different segments of a closed-loop data transcription process. However, unlike GAN or AE (VAE), the "origin" or "target" distribution of the feature z will no longer be specified *a priori*, and is instead learned from the data x. In addition, *this intrinsically low-dimensional distribution of z (with all of its low-dimensional supports) is explicitly modeled as a mixture of orthogonal subspaces (or independent Gaussians) within the feature space $\mathbb{R}^d$*, sometimes known as the principal subspaces.

Universality of Representations. Note that GANs (and most VAEs) are typically designed without explicit modeling assumptions on the distribution of the data nor on the features. Many even believe that it is this "universal" distribution learning capability (assuming minimizing distances between arbitrary distributions in high-dimensional space can be solved efficiently, which unfortunately has many caveats and often is impractical) that is attributed to their empirical success in learning distributions of complicated data such as images. In this work, we will provide empirical evidence that such an "arbitrary distribution learning machine" might not be necessary. (In fact, it may be computationally intractable in general). A *controlled and deformed* family of low-dimensional linear subspaces (Gaussians) can be more than powerful, and expressive enough to model real-world visual data. (In fact, a Gaussian mixture model is already a universal approximator of almost arbitrary densities [43]. Hence, we do not loose any generality at all). As we will also see, once we can place a proper and precise metric on such models, the associated learning problems can become much better conditioned and more amenable to rigorous analysis and performance guarantees in the future.

1.2. Learning Linear Discriminative Representation via Rate Reduction

Recently, the authors in [2] proposed a new objective for deep learning that aims to learn a *linear discriminative representation* (LDR) for multi-class data. The basic idea is to map distributions of real data, potentially on *multiple* nonlinear submanifolds $\cup_{j=1}^{k}\mathcal{M}_j \subset \mathbb{R}^D$ (in classical statistical settings, such nonlinear structures of the data were also referred to as principal curves or surfaces [44,45]. There has been a long quest of trying to extend PCA to handle potential nonlinear low-dimensional structures in data distribution (see [46] for a thorough survey) to a family of canonical models consisting of multiple independent (or orthogonal) linear subspaces, denoted as $\cup_{j=1}^{k}\mathcal{S}_j \subset \mathbb{R}^d$. To some extent, this generalizes the classic nonlinear PCA [3] to more general/realistic settings where we simultaneously apply *multiple nonlinear PCAs* to data on multiple nonlinear submanifolds. Or equivalently, the problem can also be viewed as a nonlinear extension to the classic *Generalized PCA* (GPCA) [46]. (Conventionally, "generalized PCA" refers to generalizing the setting of PCA to multiple *linear* subspaces. Here, we need to further generalize multiple *nonlinear* submanifolds. Unlike conventional discriminative methods that only aim to predict class labels as one-hot vectors, the LDR aims to learn the likely multi-dimensional distribution of the data, hence it is suitable for both discriminative and generative purposes. It has been shown that this can be achieved via maximizing the so-called "rate reduction" objective based on the rate distortion of subspace-like Gaussians [47].

LDR via MCR^2. More precisely, consider a set of data samples $\boldsymbol{X} = [\boldsymbol{x}^1, \ldots, \boldsymbol{x}^n] \in \mathbb{R}^{D\times n}$ from k different classes. That is, we have $\boldsymbol{X} = \cup_{j=1}^{k}\boldsymbol{X}_j$ with each subset of samples $\boldsymbol{X}_j$ belonging to one of the low-dimensional submanifolds: $\boldsymbol{X}_j \subset \mathcal{M}_j, j = 1, \ldots, k$. Following the notation in [2], we use a matrix $\boldsymbol{\Pi}^j(i,i) = 1$ to denote the membership of sample i

belonging to class j (and $\Pi^j = 0$ otherwise). One seeks a continuous mapping $f(\cdot,\theta) : x \mapsto z$ from X to an optimal representation $Z = [z^1, \ldots, z^n] \subset \mathbb{R}^{d\times n}$:

$$X \xrightarrow{\ f(x,\theta)\ } Z, \tag{4}$$

which maximizes the following coding rate-reduction objective, known as *the MCR^2 principle* [13]:

$$\max_{Z}\ \Delta R(Z\,|\,\Pi,\epsilon) \doteq \underbrace{\frac{1}{2}\log\det\left(I+\alpha ZZ^*\right)}_{R(Z\,|\,\epsilon)} - \underbrace{\sum_{j=1}^{k}\frac{\gamma_j}{2}\log\det\left(I+\alpha_j Z\Pi^j Z^*\right)}_{R_c(Z\,|\,\Pi,\epsilon)}, \tag{5}$$

where $\alpha = \frac{d}{n\epsilon^2}$, $\alpha_j = \frac{d}{\mathrm{tr}(\Pi^j)\epsilon^2}$, $\gamma_j = \frac{\mathrm{tr}(\Pi^j)}{n}$ for $j = 1, \ldots, k$. In this paper, for simplicity we denote $\Delta R(Z\,|\Pi,\epsilon)$ as $\Delta R(Z)$ assuming Π, ϵ are known and fixed. The first term $R(Z\,|\epsilon)$, or $R(Z)$ for short, is the coding rate of the whole feature set Z (coded as a Gaussian source) with a prescribed precision ϵ; the second term $R_c(Z\,|\Pi,\epsilon)$, or simply $R_c(Z)$, is the average coding rate of the k subsets of features $Z_j = f(X_j)$ (each coded as a Gaussian).

As has been shown by [13], maximizing the difference between the two terms will expand the whole feature set while compressing and linearizing features of each of the k classes. If the mapping f maximizes the rate reduction, it maps the features of different classes into independent (orthogonal) subspaces in $\mathbb{R}^d$. Figure 1 illustrates a simple example of data with $k = 2$ classes (on two submanifolds) mapped to two incoherent subspaces (solid black lines). Notice that, compared to AE (2) and GAN (3), the above mapping (4) is only one-sided: from the data X to the feature Z. In this work, we will see how to use the rate-reduction metric to establish inverse mapping from the feature Z back to the data X, while still preserving the subspace structures in the feature space.

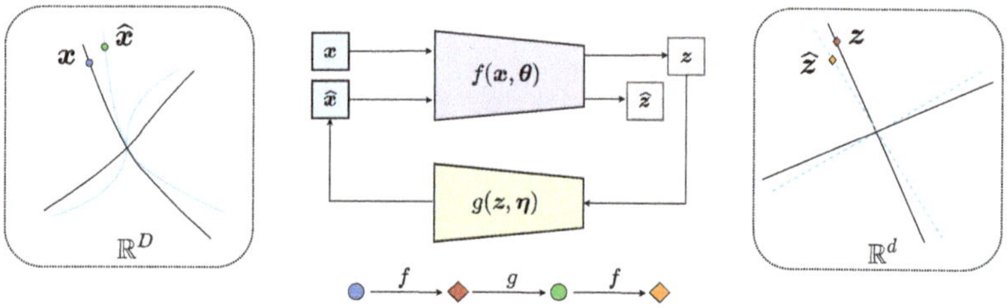

Figure 1. CTRL: A Closed-loop Transcription to an LDR. The encoder f has dual roles: it learns an LDR z for the data x via maximizing the rate reduction of z and it is also a "feedback sensor" for any discrepancy between the data x and the decoded $\hat{x}$. The decoder g also has dual roles: it is a "controller" that corrects the discrepancy between x and $\hat{x}$ and it also aims to minimize the overall coding rate for the learned LDR.

2. Data Transcription via Rate Reduction

2.1. Closed-Loop Transcription to an LDR (CTRL)

One issue with this one-sided LDR learning (4) is that maximizing the above objective (5) tends to expand the dimension of the learned subspace for features in each class (if the dimension of the feature space d is too high, maximizing the rate reduction may over-estimate the dimension of each class. Hence, to learn a good representation, one needs to pre-select a proper dimension for the feature space, as achieved in the experiments in [13]. In fact the same "model selection" problem persists even in the simplest single-subspace case, which is the classic PCA [48]. Selecting the correct number of principal components in a heterogeneous noisy situation remains an active research topic [49]). To verify whether the learned features are neither over-estimating nor under-estimating the data structure, we may consider learning a decoder $g(\cdot,\eta) : z \mapsto x$ from the representation $Z = f(X,\theta)$ back to the data space x: $\hat{X} = g(Z,\eta)$, and check how close X and $\hat{X}$ are or how close their features Z

and $\hat{\boldsymbol{Z}} = f(\hat{\boldsymbol{X}}, \theta)$ are. In principle, the decoder g should examine if all the learned features by the encoder f are both necessary and sufficient for achieving this task. The overall pipeline can be illustrated by the following "closed-loop" diagram:

$$\boldsymbol{X} \xrightarrow{\ f(\boldsymbol{x},\theta)\ } \boldsymbol{Z} \xrightarrow{\ g(\boldsymbol{z},\eta)\ } \hat{\boldsymbol{X}} \xrightarrow{\ f(\boldsymbol{x},\theta)\ } \hat{\boldsymbol{Z}}, \tag{6}$$

where the overall model has parameters: $\Theta = \{\theta, \eta\}$.

Notice that in the above process, the segment from $\boldsymbol{X}$ to $\hat{\boldsymbol{X}}$ resembles a typical *Auto-Encoding* process; although, as we will soon see, our MCR^2-based encoder f plays an additional role as a discriminator. The segment from $\boldsymbol{Z}$ to $\hat{\boldsymbol{Z}}$ draws resemblance to the typical GAN process; although, in our context, the distribution of the latent variable $\boldsymbol{z}$ will be learned from the data $\boldsymbol{x}$. Despite these connections, as we will soon see, this new closed-loop formulation will allow us to utilize the *error feedback* mechanism (widely practiced in control systems) and directly enforce loop consistency between encoding and decoding (networks) *without* using any additional discriminator(s) that are typically needed in existing VAE/GAN architectures.

Here, in the specific context of rate reduction, we name this special auto-encoding process "*Transcription to an LDR*" since the maximal rate-reduction principle explicitly transcribes the data $\boldsymbol{X}$, via f, to features $\boldsymbol{Z}$ on a linear discriminative representation (LDR) (through our extensive experiments on diverse real-world visual datasets, one does not lose any generality or expressiveness by restricting to this special but rich class of models. On the contrary, the restriction significantly simplifies and improves the learning process), which can be subsequently decoded back to the data space $\hat{\boldsymbol{X}}$, via g. Hence, the encoding and decoding maps f and g together form a "closed-loop" process, as illustrated in Figure 1. We hope that this closed-loop transcription to an LDR (CTRL) has the following good properties:

- **Injectivity:** the generated $\hat{\boldsymbol{x}} = g(f(\boldsymbol{x}, \theta), \eta) \in \hat{\boldsymbol{X}}$ should be as close to (ideally the same as) the original data $\boldsymbol{x} \in \boldsymbol{X}$, in terms of certain measures of similarity or distance.
- **Surjectivity:** for all mapped images $\boldsymbol{z} = f(\boldsymbol{x}) \in \boldsymbol{Z}$ of the training data $\boldsymbol{x} \in \boldsymbol{X}$, there are decoded samples $\hat{\boldsymbol{z}} = f(g(\boldsymbol{z}, \eta), \theta) \in \hat{\boldsymbol{Z}}$ close to (ideally the same as) $\boldsymbol{z}$.

Mathematically, we seek an *embedding* of the data $\boldsymbol{x}$ supported on certain nonlinear submanifolds $\cup_{j=1}^{k} \mathcal{M}_j$ in the space $\mathbb{R}^D$ to feature $\boldsymbol{z}$ on a set of (discriminative) linear subspaces $\cup_{j=1}^{k} \mathcal{S}_j$ in the feature space $\mathbb{R}^d$. Ideally, both f and g should be embeddings [1], when restricted on the support of the data distribution or that of the features. (That is, we hope $f\,|_{\mathcal{M}_j}$ and $g\,|_{\mathcal{S}_j}$ are all embeddings for all $j = 1, \ldots, k$.) In addition, more ideally, we hope f and g are mutually inverse embeddings: $g \circ f = \mathrm{Id}$ (when restricted on the submanifolds). Nevertheless, if we are only interested in learning the distribution, embeddings of the support would often suffice the purposes (e.g., classification or generative purposes). Notice that the above goals are similar to many VAE+GAN-related methods in the machine-learning literature, such as BiGAN [38] and ALI [39]. We will discuss the differences of our approach from these existing methods in Section 2.3 (as well as providing some experimental comparisons in the Appendix A).

At first sight, this is a rather daunting task, since we are trying to learn over a (seemingly infinite-dimensional) functional space of all embeddings and distributions from finite samples. In this work, we will take a more pragmatic approach and show how one can learn a good encoding, decoding, and representation tuple: $(f, g, \boldsymbol{z})$ from $\boldsymbol{X}$ via tractable computational means. In particular, we will convert the above goals to certain feasible programs that optimize a sensible measure of goodness for the learned representations $\boldsymbol{Z}$.

2.2. Measuring Distances in the Feature Space and Data Space

Contractive measure for the decoder. For the *second* item in the above wishlist, as the representations in the feature space $\boldsymbol{z}$ are by design linear subspaces or (degenerate) Gaussians, we have geometrically or statistically meaningful metrics for both samples and distributions in the feature space z. For example, we care about the distance between distributions between the features of the original data $\mathbf{Z}$ and the transcribed $\hat{\mathbf{Z}}$. Since the features of each class, $\mathbf{Z}_j$ and $\hat{\mathbf{Z}}_j$, are similar to subspaces/Gaussians, their "distance" can be measured by the rate reduction, with (5) restricted to two sets of equal size:

$$\Delta R(\mathbf{Z}_j, \hat{\mathbf{Z}}_j) \doteq R(\mathbf{Z}_j \cup \hat{\mathbf{Z}}_j) - \frac{1}{2}(R(\mathbf{Z}_j) + R(\hat{\mathbf{Z}}_j)). \tag{7}$$

According to the interpretation of the rate reduction given in [13], the above quantity precisely measures the volume of the space between $\mathbf{Z}_j$ and $\hat{\mathbf{Z}}_j$, illustrated as a pair of black and blue lines in Figure 1. Then, for the "distance" of all, say k, classes, we simply sum the rate reduction for all pairs:

$$d(\mathbf{Z}, \hat{\mathbf{Z}}) \doteq \min_{\eta} \sum_{j=1}^{k} \Delta R(\mathbf{Z}_j, \hat{\mathbf{Z}}_j) = \min_{\eta} \sum_{j=1}^{k} \Delta R(\mathbf{Z}_j, f(g(\mathbf{Z}_j, \eta), \theta)), \tag{8}$$

where $\mathbf{Z}_j = f(\mathbf{X}_j, \theta)$ and $\hat{\mathbf{Z}}_j = f(\hat{\mathbf{X}}_j, \theta)$. Obviously, a main goal of the learned decoder $g(\cdot, \eta)$ is to *minimize* the distance between these distributions. Notice that if the encoder f preserves (i.e., injective for) the intrinsic structures of the original data $\mathbf{X}$, (this is typically the case for MCR^2-based feature representation [13]) this criterion essentially aims to ensure there will be some decoded sample $\hat{x}$ close to every data sample x—hence the decoder g should be "surjective". According to the ideas of IMLE [23], such a requirement could effectively help to avoid mode-collapsing or mode-dropping.

Contrastive measure for the encoder. For the *first* item in our wishlist, however, we normally do not have a natural metric or "distance" for similarity of samples or distributions in the original data space x for data such as images. As mentioned before, finding proper metrics or distance functions on natural images has always been an elusive and challenging task [7]. To alleviate this difficulty, we can measure the similarity or difference between $\hat{\mathbf{X}}$ and $\mathbf{X}$ through their mapped features $\hat{\mathbf{Z}}$ and $\mathbf{Z}$ in the feature space (again assuming f is structure-preserving). If we are interested in discerning *any* differences in the distributions of the original and transcribed samples, we may view the MCR^2 feature encoder $f(\cdot, \theta)$ as a "discriminator" to *magnify* any difference between all pairs of $\mathbf{X}_j$ and $\hat{\mathbf{X}}_j$, by simply maximizing, instead of minimizing, the *same quantity* in (8):

$$d(\mathbf{X}, \hat{\mathbf{X}}) \doteq \max_{\theta} \sum_{j=1}^{k} \Delta R(\mathbf{Z}_j, \hat{\mathbf{Z}}_j) = \max_{\theta} \sum_{j=1}^{k} \Delta R(f(\mathbf{X}_j, \theta), f(\hat{\mathbf{X}}_j, \theta)). \tag{9}$$

That is, a "distance" between $\mathbf{X}$ and $\hat{\mathbf{X}}$ can be measured as the maximally achievable rate reduction between all pairs of classes in these two sets. In a way, this measures how well or badly the decoded $\hat{\mathbf{X}}$ aligns with the original data $\mathbf{X}$—hence measuring the goodness of "injectivity" of the encoder f. Notice that such a discriminative measure is consistent with the idea of GAN [9] that tries to separate $\mathbf{X}$ and $\hat{\mathbf{X}}$ into two classes, measured by the cross-entropy. Nevertheless, here the MCR^2-based discriminator f naturally generalizes to cases when the data distributions are multi-class and multi-modal, and the discriminativeness is measured with a more refined measure—the rate reduction—instead of the typical two-class loss (e.g., cross entropy) used in GANs. See Appendix A.8 for comparisons with some ablation studies.

One may wonder why we need the mapping $f(\cdot, \theta)$ to function as a discriminator between $\mathbf{X}$ and $\hat{\mathbf{X}}$ by maximizing $\max_{\theta} \Delta R(f(\mathbf{X}, \theta), f(\hat{\mathbf{X}}, \theta))$. Figure 2 gives a simple illustration: there might be many decoders g such that $f \circ g$ is an identity (Id) mapping. Here, we use the notion of "identity mapping" in a loose sense: depending on the context, it could simply mean an embedding from S_z to S_z. $f \circ g(z) = z$ for all z in the subspace S_z in the feature space. However, $g \circ f$ is not necessarily an auto-encoding map for x in the original distribution S_x (here for simplicity drawn as a subspace). That is, $g \circ f(S_x) \not\subset S_x$, let alone $g \circ f(S_x) = S_x$ or $g \circ f(x) = x$. One should expect, without careful control of the image of g, with high probability, this would be the case, especially when the support of the distribution of x is extremely low-dimensional in the original high-dimensional data space. For example, as we will see in the experiments, the intrinsic dimension of the submanifold associated with each image category is about a dozen, whereas images are embedded in a (pixel) space of thousands or tens of thousands of dimensions.

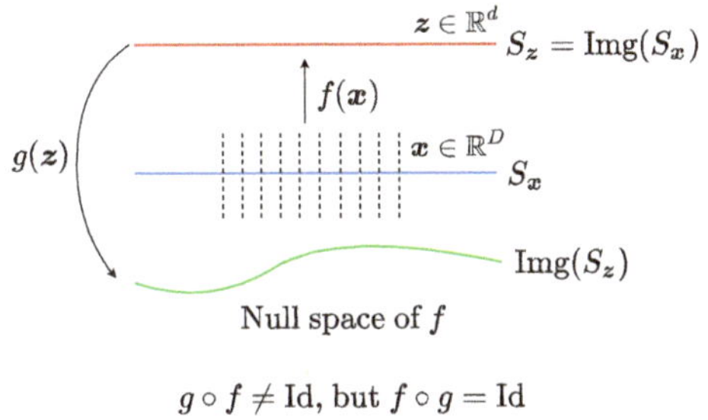

Figure 2. Embeddings of Low-Dimensional Submanifolds in High-Dimensional Spaces. S_x (blue) is the submanifold for the original data x; S_z (red) is the image of S_x under the mapping f, representing the learned feature z; and the green curve is the image of the feature z under the decoding mapping g.

Remark: representing the encoding and decoding mappings. Some practical questions arise immediately: how rich should the families of functions be that we should consider to use for the encoder f and decoder g that can optimize the above rate-reduction-type objectives? In fact, similar questions exist for the formulation of GAN, regarding the realizability of the data distribution by the generator, see [50]. Conceptually, here we know that the encoder f needs to be rich enough to discriminate (small) deviations from the true data support $\mathcal{M}_j$, while the decoder g needs to be expressive enough to generate the data distribution from the learned mixture of subspace-Gaussians. How should we represent or parameterize them, hence making our objectives computable and optimizable? For the most general cases, these remain widely open and challenging mathematical and computational problems. As we mentioned earlier, in this work, we will take a more pragmatic approach by simply representing these mappings with popular neural networks that have empirically proven to be good at approximating distributions of practical (visual) datasets or for achieving the maximum of the rate-reduction-type objectives [13]. Nevertheless, our experiments indicate that our formulation and objectives are *not so sensitive* to particular choices in network structures or many of the tricks used to train them. In addition, in the special cases when the real data distribution is benignly deformed from an LDR, the work of [2] has shown that one can explicitly construct these mappings from the rate-reduction objectives in the form of a deep network known as ReduNet. However, it remains unclear how such constructions could be generalized to closed-loop settings. Regardless, answers to these questions are beyond the scope of this work, as our purposes here are mainly to empirically verify the validity of the proposed closed-loop data transcription framework.

2.3. Encoding and Decoding as a Two-Player MiniMax Game

Comparing the contractive and contrastive nature of (8) and (9) on the same utility, we see the roles of the encoder $f(\cdot, \theta)$ and the decoder $g(\cdot, \eta)$ naturally as "**a two-player game**": *while the encoder f tries to magnify the difference between the original data and their transcribed data, the decoder g aims to minimize the difference.* Now for convenience, let us define the "closed-loop encoding" function:

$$h(x, \theta, \eta) \doteq f(g(f(x, \theta), \eta), \theta) : x \mapsto z. \tag{10}$$

Ideally, we want this function to be very close to $f(x, \theta)$ or at least the distributions of their images should be close. With this notation, combining (8) and (9), a closed-loop notion of "distance" between $\boldsymbol{X}$ and $\hat{\boldsymbol{X}}$ can be computed as *an equilibrium point* to the following

Min-Max (or Max-Min) program for the same utility in terms of rate reduction (theoretically, there might be significant difference in formulating and seeking the desired solution as the equilibrium point to a min-max or max-min game. In practice, we do not see major differences as we optimize the program by simply alternating between minimization and maximization. We leave a more careful investigation to future work):

$$\mathcal{D}(\boldsymbol{X}, \hat{\boldsymbol{X}}) \doteq \min_{\eta} \max_{\theta} \sum_{j=1}^{k} \Delta R\big(f(\boldsymbol{X}_j, \theta), h(\boldsymbol{X}_j, \theta, \eta)\big). \tag{11}$$

Notice that this only measures the difference between (features of) the original data and its transcribed version. It does not measure how good the representation $\boldsymbol{Z}$ (or $\hat{\boldsymbol{Z}}$) is for the multiple classes within $\boldsymbol{X}$ (or $\hat{\boldsymbol{X}}$). To this end, we may combine the above distance with the original MCR^2-type objectives (5): namely, the rate reduction $\Delta R(\boldsymbol{Z})$ and $\Delta R(\hat{\boldsymbol{Z}})$ for the learned LDR $\boldsymbol{Z}$ for $\boldsymbol{X}$ and $\hat{\boldsymbol{Z}}$ for the decoded $\hat{\boldsymbol{X}}$. Notice that although the encoder f tries to *maximize* the multi-class rate reduction of the features $\boldsymbol{Z}$ of the data $\boldsymbol{X}$, the decoder g should *minimize* the rate reduction of the multi-class features $\hat{\boldsymbol{Z}}$ of the decoded $\hat{\boldsymbol{X}}$. That is, the decoder g tries to use a minimal coding rate needed to achieve a good decoding quality.

Hence, the overall "multi-class" Min-Max program for learning the Closed-loop Transcription to an LDR, named CTRL-Multi, is subject to certain constraints (upper or lower bounds) on the first term and the second term. In this work, we only consider the simple case by adding these rate-reduction quantities together. Of course, in the future, one may consider other more delicate formulations. For instance, we may consider a Min-Max game on the third term (11). Such constrained minimax games have also started to draw attention lately [51].

$$\begin{aligned}\min_{\eta} \max_{\theta} \mathcal{T}_{\boldsymbol{X}}(\theta, \eta) \;&\doteq\; \underbrace{\Delta R\big(f(\boldsymbol{X}, \theta)\big)}_{\text{Expansive encode}} + \underbrace{\Delta R\big(h(\boldsymbol{X}, \theta, \eta)\big)}_{\text{Compressive decode}} + \sum_{j=1}^{k} \underbrace{\Delta R\big(f(\boldsymbol{X}_j, \theta), h(\boldsymbol{X}_j, \theta, \eta)\big)}_{\text{Contrastive encode \& Contractive decode}} \\ &= \;\Delta R\big(\boldsymbol{Z}(\theta)\big) + \Delta R\big(\hat{\boldsymbol{Z}}(\theta, \eta)\big) + \sum_{j=1}^{k} \Delta R\big(\boldsymbol{Z}_j(\theta), \hat{\boldsymbol{Z}}_j(\theta, \eta)\big).\end{aligned} \tag{12}$$

Empirically, we have evaluated the necessity of these terms in an ablation study (see Appendix A.8.3). Notice that, without the terms associated with the generative part h or with all such terms fixed as constant, the above objective is precisely the original MCR^2 objective proposed by [13]. In an unsupervised setting, if we view each sample (and its augmentations) as its own class, the above formulation remains exactly the same. The number of classes k is simply the number of independent samples. In addition, notice that the minimax objective function depends only on (features of) the data $\boldsymbol{X}$, hence one can learn the encoder and decoder (parameters) without the need for sampling or matching any additional distribution (as typically needed in GANs or VAEs).

As a special case, if $\boldsymbol{X}$ only has one class, the above Min-Max program reduces (as the first two rate reduction terms automatically become zero) to a special "two-class" or "binary" form, named CTRL-Binary, between $\boldsymbol{X}$ and the decoded $\hat{\boldsymbol{X}}$ by viewing $\boldsymbol{X}$ and $\hat{\boldsymbol{X}}$ as two classes $\{\boldsymbol{0}, \boldsymbol{1}\}$. Notice that this binary case resembles formulation of the original GAN (3). Nevertheless, instead of using cross entropy, our formulation adopts a more refined rate-reduction measure, which has been shown to promote diversity in the learned representation [13]).

$$\text{CTRL-Binary:} \quad \min_{\eta} \max_{\theta} \mathcal{T}^{b}_{\boldsymbol{X}}(\theta, \eta) \doteq \Delta R\big(f(\boldsymbol{X}, \theta), h(\boldsymbol{X}, \theta, \eta)\big) = \Delta R\big(\boldsymbol{Z}(\theta), \hat{\boldsymbol{Z}}(\theta, \eta)\big). \tag{13}$$

Sometimes, even when $\boldsymbol{X}$ contains multiple classes/modes, one could still view all classes together as one class. Then, the above binary objective is to align the union distribution of all classes with their decoded $\hat{\boldsymbol{X}}$. This is typically a simpler task to achieve than the multi-class one (12), since it does not require learning of a more refined multi-class CTRL for the data, as we will later see in experiments. Notice that one good characteristic of the above formulation is that *all quantities in the objectives are measured in terms of rate reduction for the learned features* (assuming features eventually become subspace Gaussians).

In all of our subsequent experiments, we solve the above minimax programs using the most basic gradient descent–ascent (GDA) algorithm [52] that alternates between the minimization and maximization, with the same learning rate and without any timescale separation (as typically needed for training GANs [53]). Although more refined optimization schemes can likely further improve the efficiency and performance, we leave these for future investigations.

Remark: closed-loop error correction. One may notice that our framework (see Figure 1) draws inspiration from closed-loop error correction widely practiced in feedback control systems. In the machine-learning and deep-learning literature, the idea of closed-loop error correction and closed-loop fixed point has been explored before to interpret the recursive error-correcting mechanism and explain stability in a forward (predictive) deep neural network, for example the *deep equilibrium networks* [54] and the *deep implicit networks* [55], again drawing inspiration from feedback control. Here, in our framework, the closed-loop mechanism is not used to interpret the encoding or decoding (forward) networks f and g. Instead, it is used to form an overall feedback system between the two encoding and decoding networks for correcting the "error" in the distributions between the data x and the decoded $\hat{x}$. Using terminology from control theory, one may view the encoding network f as a "sensor" for error feedback while the decoding network g as a "controller" for error correction. However, notice that here the "target" for control is not a scalar nor a finite dimensional vector, but a continuous mapping—in order for the distribution of $\hat{x}$ to match that of the data x. This is in general a control problem in an infinite dimensional space. The space of diffeomorphisms of submanifolds is infinite-dimensional [1]. Ideally, we hope when the sensor f and the controller g are optimal, the distribution of x becomes a "fixed point" for the closed loop while the distribution of z reaches a compact LDR. Hence, the minimax programs (12) and (13) can also be interpreted as games between an error-feedback sensor and an error-reducing controller.

Remark: relation to bi-directional or cycle consistency. The notion of "bi-directional" and "cycle" consistency between encoding and decoding has been exploited in the works of BiGAN [38] and ALI [39] for mappings between the data and features and in the work of CycleGAN [56] for mappings between two different data distributions. In our context, it is similar in order to promote $g \circ f$ and $f \circ g$ to be close to identity mappings (either for the distributions or for the samples). Interestingly, our new closed-loop formulation actually "decouples" the data $\boldsymbol{X}$, say, observed from the external world, from their internally represented features $\boldsymbol{Z}$. The objectives (12) and (13) are functions of *only* the internal features $\boldsymbol{Z}(\theta)$ and $\hat{\boldsymbol{Z}}(\theta, \eta)$, which can be learned and optimized by adjusting the neural networks $f(\cdot, \theta)$ and $g(\cdot, \eta)$ alone. There is no need for any additional external metrics or heuristics to promote how "close" the decoded images $\hat{\boldsymbol{X}}$ are to $\boldsymbol{X}$. This is very different from most VAE/GAN-type methods such as BiGAN and ALI that require additional discriminators (networks) for the images and the features. Some experimental comparison are given in the Appendix A.2. In addition, in Appendix A.8.1, we provide some ablation study to illustrate the importance and benefit of a closed loop for enforcing the consistency between the encoder and decoder.

Remark: transparent versus hidden distribution of the learned features. Notice that in our framework, there is no need to explicitly specify a prior distribution either as a target distribution to map to for AE (2) or as an initial distribution to sample from for GAN (3). The common practice in AEs or GANs is to specify the prior distribution as a generic Gaussian. This is however particularly problematic when the data distribution is multi-modal and has multiple low-dimensional structures, which is commonplace for multi-class data. In this case, the common practice in AEs or GANs is to train a conditional GAN for different classes or different attributes. However, here we only need to assume the desired target distribution belonging to the family of LDRs. The specific optimal distribution of the features within this family is then learned from the data directly, and then can be represented *explicitly* as a mixture of independent subspace Gaussians (or equivalently, a mixture of PCAs on independent subspaces). We will give more details in the experimental Section 3 as well as more examples in Appendices A.2–A.4. Although many GAN + VAE-type methods can learn bidirectional encoding and decoding mappings, the distribution of the learned features inside the feature space remains *hidden* or even *entangled*. This makes it difficult to sample the feature space for generative purposes or to use the features for discriminative tasks. (For instance, typically one can only use so-learned features for nearest-neighbor-type classifiers [38], instead of nearest subspace as in this work, see Section 3.3).

3. Empirical Verification on Real-World Imagery Datasets

This experiment section serves three purposes: First, we empirically justify the proposed formulation for data transcription by demonstrating good properties of the learned encoder, decoder, and representation tuple (f, g, z) from $\boldsymbol{X}$. Second, we compare our method with several representative methods from the GAN family and VAE family. The purpose of the comparison is *not* to compete for any state-of-the-art performance. Instead, we want to convincingly verify the validity of the proposed framework and its potential in going beyond. Finally, we evaluate the so-learned CTRL through both generative tasks (controlled visualization) and discriminative (classification) tasks. More extensive experimental results, evaluations, and ablation studies can be found in the Appendix A.

Datasets. We provide extensive qualitative and quantitative experimental results on the following datasets: MNIST [57], CIFAR-10 [58], STL-10 [59], CelebA [60], LSUN bedroom [61], and ImageNet ILSVRC 2012 [62]. The network architectures and implementation details can be found in Appendix A.1 and corresponding Appendix A for each dataset.

3.1. Empirical Justification of CTRL Transcription

To empirically validate our new framework, we conduct experiments from a small low-variety dataset (MNIST), to a small dataset of diverse real-world objects (CIFAR-10), to higher resolution images (STL-10, CelebA, LSUN-bedroom), to a large-scale diverse image set (ImageNet). The results are evaluated both quantitatively and qualitatively. Implementation details, more experimental results, and ablation studies are given in Appendix A.

Comparison (IS and FID) with other formulations. First, we conduct five experiments to fairly compare our formulation with GAN [63] and VAE(-GAN) [64] on MNIST and CIFAR-10. Except for the objective function, everything else is exactly the same for all methods (e.g., networks, training data, optimization method). These experiments are: (1). GAN; (2). GAN with its objective replaced by that of the CTRL-Binary (13); (3). VAE-GAN ; (4). Binary CTRL (13); and (5). Multi-class CTRL (12). Some visual comparison is given in Figure 3. IS [65] and FID [66] scores are summarized in Table 1. Here, for simplicity, we have chosen a uniform feature dimension $d = 128$ for all datasets. If we choose a higher feature dimension, say $d = 512$, for the more complex CIFAR-10 dataset, the visual quality can be further improved, see Table A14 in Appendix A.11.

(**a**) MNIST (**b**) CIFAR-10 (**c**) ImageNet

Figure 3. Qualitative comparison on (**a**) MNIST, (**b**) CIFAR-10 and (**c**) ImageNet. First row: original $\boldsymbol{X}$; other rows: reconstructed $\hat{\boldsymbol{X}}$ for different methods.

Table 1. Quantitative comparison on MNIST and CIFAR-10. Average Inception scores (IS) [65] and FID scores [66]. ↑ means higher is better. ↓ means lower is better.

Method		GAN	GAN (CTRL-Binary)	VAE-GAN	CTRL-Binary	CTRL-Multi
MNIST	IS ↑	2.08	1.95	**2.21**	2.02	2.07
	FID ↓	24.78	20.15	33.65	**16.43**	16.47
CIFAR-10	IS ↑	7.32	7.23	7.11	**8.11**	7.13
	FID ↓	26.06	22.16	43.25	**19.63**	23.91

As we see from Table 1, replacing cross-entropy with the Equation (13) can improve the generative quality. The two CTRL formulations are clearly on par with the others in terms of IS and significantly better in FID. Finally, with the same training datasets, the quality of CTRL-Multi is lower than that of CTRL-Binary. This is expected, as the multi-class task is more challenging. Nevertheless, as we will see soon, images decoded by CTRL-Multi align much better with their classes than Binary.

Visualizing correlation of features $\mathbf{Z}$ and decoded features $\hat{\mathbf{Z}}$. We visualize the cosine similarity between $\mathbf{Z}$ and $\hat{\mathbf{Z}}$ learned from the multi-class objective (12) on MNIST, CIFAR-10 and ImageNet (10 classes), which indicates how close $\hat{z} = f \circ g(z)$ is from z. Results in Figure 4 show that $\mathbf{Z}$ and $\hat{\mathbf{Z}}$ are aligned very well within each class. The block-diagonal patterns for MNIST are sharper than those for CIFAR-10 and ImageNet, as images in CIFAR-10 and ImageNet have more diverse visual appearances.

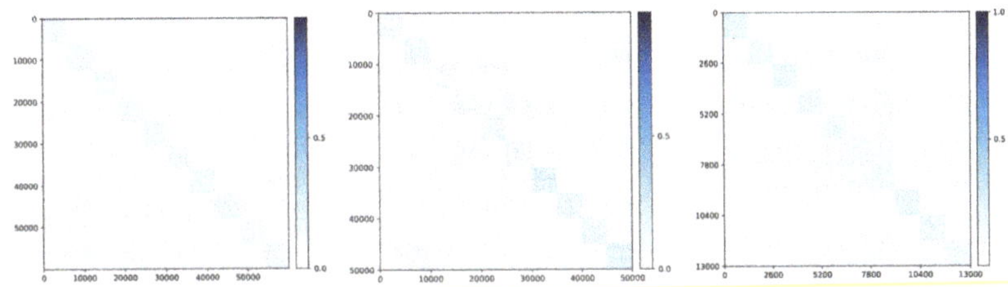

(a) MNIST **(b)** CIFAR10 **(c)** ImageNet

Figure 4. Visualizing the alignment between $\mathbf{Z}$ and $\hat{\mathbf{Z}}$: $|\mathbf{Z}^\top \hat{\mathbf{Z}}|$ and in the feature space for (**a**) MNIST, (**b**) CIFAR-10, and (**c**) ImageNet-10-Class.

Visualizing auto-encoding of the data X and the decoded $\hat{X}$. We compare some representative X and $\hat{X}$ on MNIST, CIFAR-10 and ImageNet (10 classes) to verify how close $\hat{x} = g \circ f(x)$ is to x. The results are shown in Figure 5, and visualizations are created from training samples. Visually, the auto-encoded $\hat{x}$ faithfully captures major visual features from its respective training sample x, especially the pose, shape, and layout. For the simpler dataset such as MNIST, auto-encoded images are almost identical to the original. The visual quality is clearly better than other GAN+VAE-type methods, such as VAE-GAN [34] and BiGAN [38]. We refer the reader to Appendices A.2, A.4 and A.7 for more visualization of results on these datasets, including similar results on transformed MNIST digits. More visualization results for learned models on real-life image datasets such as STL-10, CelebA, and LSUN can be found in the Appendices A.5 and A.6.

(**a**) MNIST X (**b**) MNIST $\hat{X}$ (**c**) CIFAR-10 X (**d**) CIFAR-10 $\hat{X}$ (**e**) ImageNet X (**f**) ImageNet $\hat{X}$

Figure 5. Visualizing the auto-encoding property of the learned closed-loop transcription ($x \approx \hat{x} = g \circ f(x)$) on MNIST, CIFAR-10, and ImageNet (zoom in for better visualization).

3.2. Comparison to Existing Generative Methods

Table 2 gives a quantitative comparison of visual quality of our method with others on CIFAR-10, STL-10, and ImageNet. In general, there is a large difference in terms of FID and IS scores between the GAN family and the VAE family of models. SNGAN [31] are commonly used methods in most generative applications, while LOGAN [29] is the state-of-the-art method on ImageNet in terms of FID and IS. More comparisons with existing methods, including results on on the higher-resolution ImageNet dataset, can be found in Table A10 of the Appendix A.7.

As we see, even if the rate-reduction objectives (12) and (13) are not specifically designed nor engineered for visual quality and the networks and hyper-parameters adopted in our experiments are rather basic compared to many of the state-of-the-art generative methods, our method is still rather competitive in terms of these metrics. In our current implementation, the original objectives are used without any other heuristics or regularization. The simplicity of our framework and formulation suggests that there is significant room for further improvement. For instance, in all experiments on all datasets, we have chosen a feature dimension of $d = 128$ for simplicity and uniformity. In the last Appendix A.11, we have conducted an ablation study on using a higher feature dimension $d = 512$. The visual quality of the learned model can be significantly improved (as shown in Figure A22 and Table A14 of Appendix A.11).

In fact, compared to these methods, our method has learned not just any generative model. It has learned a *structured* generative model that has many additional beneficial properties that we now present.

Table 2. Comparison of CIFAR-10 and STL-10. Comparison with more existing methods and on ImageNet can be found in Table A10 in the Appendix A. ↑ means higher is better. ↓ means lower is better.

Method		GAN Based Methods			VAE/GAN-Based Methods				
		SNGAN	CSGAN	LOGAN	VAE-GAN	NVAE	DC-VAE	CTRL-Binary	CTRL-Multi
CIFAR-10	IS ↑	7.4	8.1	**8.7**	7.4	-	**8.2**	**8.1**	7.1
	FID ↓	29.3	19.6	**17.7**	39.8	50.8	**17.9**	**19.6**	23.9
STL-10	IS ↑	**9.1**	-	-	-	-	8.1	8.4	7.7
	FID ↓	40.1	-	-	-	-	41.9	**38.6**	45.7

3.3. Benefits of the Learned LDR Transcription Model

As we have argued before, the learned LDR transcription model (including the feature z, the encoder f, and the decoder g) can be used for both generative and discriminative purposes. In particular, unlike almost all existing generative methods, the internal structures or distribution of the learned z are no longer "hidden" as they have clear subspace structures. Hence, we can easily derive an explicit (parametrizable) model for the distribution of the learned features as a mixture of independent subspace-like Gaussians. This gives us full control in sampling the learned distribution for generative purposes.

Principal subspaces and principal components for the feature. To be more specific, given the learned k-class features $\cup_{j=1}^{k} \mathbf{Z}_j$ for the training data, we have observed that the leading singular subspaces for different classes are all approximately orthogonal to each other: $\mathbf{Z}_i \perp \mathbf{Z}_j$ (see Figure 4). This corroborates with our above discussion about the theoretical properties of the rate-reduction objective. They essentially span k independent principal subspaces. We can further calculate the mean $\bar{z}_j$ and the singular vectors $\{\boldsymbol{v}_j^i\}_{i=1}^{r_j}$ (or principal components) of the learned features $\mathbf{Z}_j$ for each class. Although we conceptually view the support of each class is a subspace, the actual support of the features is close to being on the sphere due to feature (scale) normalization. Hence, it is more precise to find its mean and its support centered around the mean. Here, r_j is a rank we may choose to model the dimension of each principal subspace (say, based on a common threshold on the singular values). Hence, we obtain an explicit model for how the feature $\boldsymbol{z}$ is distributed in each of the k principal subspaces in the feature space $\mathbb{R}^d$:

$$z_j \sim \bar{z}_j + \sum_{l=1}^{r_j} n_l^j \sigma_j^l \boldsymbol{v}_j^l, \quad \text{where} \quad n_l^j \sim \mathcal{N}(0,1), \;\; j = 1, \ldots, k. \tag{14}$$

Hence, this essentially gives an explicit mixture of a subspace-like Gaussians model for the learned features: statistical differences between different classes are modeled as k independent principal subspaces; statistical differences within each class j are modeled as r_j independent principal components in the jth subspace.

Decoding samples from the feature distribution. Using the CIFAR-10 and CelebA datatsets, we visualize images decoded from samples of learned feature subspace. For the CIFAR-10 dataset, for each class j, we first compute the top four principal components of the learned features $\mathbf{Z}_j$ (via SVD). For each class j, we then compute $|\langle \boldsymbol{z}_j^i, \boldsymbol{v}_j^l \rangle|$, the cosine similarity between the l-th principal direction $\boldsymbol{v}_j^l$ and feature sample $\boldsymbol{z}_j^i$. After finding the top five $\boldsymbol{z}_j^i$ according to $|\langle \boldsymbol{z}_j^i, \boldsymbol{v}_j^l \rangle|$ for each class j, we reconstruct images $\hat{\boldsymbol{x}}_j^i = g(\boldsymbol{z}_j^i)$. Each row of Figure 6 is for one principal component. We observe that images in the same row share the same visual attributes; images in different rows differ significantly in visual characteristics such as shape, background, and style. See Figure A7 of Appendix A.4 for more visualization of principal components learned for all 10 classes of CIFAR-10. These results clearly demonstrate that the principal components in each subspace of the Gaussian disentangles different visual attributes. In addition, we do not observe any mode dropping for any of the classes, although the dimensions of the classes were not known a priori.

(**a**) Horse (**b**) Ship

Figure 6. CIFAR-10 dataset. Visualization of top 5 reconstructed $\hat{\boldsymbol{x}} = g(\boldsymbol{z})$ based on the closest distance of $\boldsymbol{z}$ to each row (top 4) of principal components of data representations for class 7—'Horse' and class 8—'Ship'.

Disentangled visual attributes as principal components. For the CelebA dataset, we calculate the principal components of all learned features in the latent space. Figure 7a shows some decoded images along these principal directions. Again, these principal components seem to clearly *disentangle* visual attributes/factors such as wearing a hat, changing hair color, and wearing glasses. More examples can be found in Appendix A.6. The results are consistent with *the property of MCR^2 that promotes diversity of the learned features.*

(**a**) Disentangled attributes as principal components (**b**) Interpolation between distinct samples

Figure 7. CelebA dataset. (**a**): Sampling along three principal components that seem to correspond to different visual attributes; (**b**): Samples decoded by interpolating along the line between features of two distinct samples.

Linear interpolation between features of two distinct samples. Figure 7b shows interpolating features between pairs of training image samples of the CeleA dataset, where for two training images x_1 and x_2, we reconstruct based on their linearly interpolated feature representations by $\hat{x} = g(\alpha f(x_1) + (1-\alpha) f(x_2)), \alpha \in [0,1]$. The decoded images show continuous morphing from one sample to another in terms of visual characteristics, as opposed to merely a superposition of the two images. Similar interpolation results between two digits in the MNIST dataset can be found in Figure A3 of the Appendix A.2.

Encoded features for classification. Notice that not only is the learned decoder good for generative purposes, but the encoder is also good for discriminative tasks. In this experiment, we evaluate the discriminativeness of the learned CTRL model by testing how well the encoded features can help classify the images. We use features of the training images to compute the learned subspaces for all classes, then classify features of the test images based on a simple nearest subspace classifier. Many other encoding methods train a classifier (say, with an additional layer) after the learned features. Results in Table 3 show that our model gives competitive classification accuracy on MNIST compared to some of best VAE-based methods. We also tested the classification on CIFAR-10, and the accuracy is currently about 80.7%. As expected, the representation learned with the multi-class objective is very discriminative and good for classification tasks. Be aware that all generative models, GANs, VAEs, and ours, are not specifically engineered for classification tasks. Hence, one should not expect the classification accuracy to compete with supervised-trained classifiers yet. This demonstrates that the learned CTRL model is not only generative but also discriminative.

Table 3. Classification accuracy on MNIST compared to classifier-based VAE methods [42]. Most of these VAE-based methods require auxiliary classifiers to boost classification performance.

Method	VAE	Factor VAE	Guide-VAE	DC-VAE	CTRL-Binary	CTRL-Multi
MNIST	97.12%	93.65%	98.51%	98.71%	89.12%	98.30%

4. Open Theoretical Problems

So far, we have given theoretical intuition and derivation for the formulation of closed-loop transcription, as well as empirical evidence to showcase both the performance and potential of this formulation. In this section, we take a step back to explore the theoretical underpinnings of the closed-loop LDR transcription. We organize this section by discussing three primary objectives associated with learning an LDR representation:

1. *Learn a simple linear discriminative representation $f(\mathbf{X})$ of the data $\mathbf{X}$*, which we can reliably use to classify the data.

2. *Learn a reconstruction* $g \circ f(\boldsymbol{X}) \sim \boldsymbol{X}$ *of the so-learned representation* $f(\boldsymbol{X})$, to ensure consistency in the representation.
3. *Learn both representation and reconstruction in a closed-loop manner*, using feedback from the encoder f and decoder g to jointly solve the above two tasks.

These three objectives encompass the overarching principle of CTRL transcription, and indeed each of these objectives are tied to a wide array of mathematical and theoretical problems. We now outline some of the most important theoretical questions or hypotheses implicated by our results, which we leave for future work to study and to answer, likely by a broader range of research communities.

4.1. Distributions of the LDR Representation

Our primary mode of optimizing for a "simple representation" is through the LDR framework proposed in [2]. One important open theoretical problem is finding the right energy function to optimize in order to promote LDR. It was shown in [2] that an LDR can be learned for the multi-class data by maximizing the MCR^2 objective $\Delta R(\boldsymbol{Z})$ given in (5). This motivates the first two terms in our objective function (12): maximizing $\Delta R(\boldsymbol{Z}), \Delta R(\hat{\boldsymbol{Z}})$ promotes their representations to be LDRs.

Although the authors in [2] have shown the MCR^2 objective can promote the features learned to be in orthogonal subspaces and characterized the optimal second moments of the distributions, there remain open questions regarding the optimal distributions within the subspaces. A standing hypothesis is that the optimal distributions should be Gaussian. There is indeed already theoretical work on similar energy functions: the Brascamp–Lieb inequalities [67], where the authors study a functional similar to the rate-reduction objective which, in certain contexts, is maximized uniquely by Gaussians. Hence, an important future theoretical direction for the CTRL transcription is to exactly characterize distributional properties of the extremals (both minima and maxima) of the MCR^2 objective or its variants. Such results can further justify the use of Gaussian models (14) to characterize the learned features within the subspaces.

We also notice that the so-learned LDR features have additional striking properties, as shown by examples in Figure 7. Distinctive visual attributes of the imagery data seem to be clearly disentangled by different principal components of the distribution, and along each principal direction, one can linearly interpolate the features, whereas the original data are nonlinear and cannot be directly interpolated. These results go beyond the guarantees given by [2], and an open theoretical problem is that of studying just how the CTRL transcription learns to disentangle and linearize such visual attributes. This understanding is crucial to extend the CTRL transcription framework beyond the 2D vision domain.

4.2. Self-Consistency in the Learned Reconstruction

If the learned encoder $\boldsymbol{Z} = f(\boldsymbol{X})$ is an embedding of the data submanifolds to the subspaces, it should admit an inverse (decoding) mapping $\hat{\boldsymbol{X}} = g(\boldsymbol{Z})$. As distributional distance in the data space is hard to come by, the rate reduction $\Delta R(\boldsymbol{Z}, \hat{\boldsymbol{Z}})$ gives a well-defined distribution distance between $\boldsymbol{Z}$ and $\hat{\boldsymbol{Z}}$ which is used to enforce similarity between $\boldsymbol{X}$ and $\hat{\boldsymbol{X}}$ in our formulation. Notice that, unlike the KL-divergence or the JS-divergence, the rate reduction is well-defined for degenerate distributions and easily computable in closed-form between mixtures of (degenerate) Gaussians. The third term of Equation (12), $\sum_{j=1}^{k} \Delta R\big(\boldsymbol{Z}_j(\theta), \hat{\boldsymbol{Z}}_j(\theta, \eta)\big)$, is exactly this distributional distance, which is minimized only when the estimated second moments of $\boldsymbol{Z}_j$ and $\hat{\boldsymbol{Z}}_j$ are the same. While this distributional distance seems weaker than sample-wise ℓ^2-distance, we observe strong reconstruction performance nevertheless.

Notice that the current objectives (12) or (13) do not impose any constraints on the mappings of individual samples. That is, they do not explicitly specify how an individual sample $\boldsymbol{x}$ should be related to its decoded version $\hat{\boldsymbol{x}} = g(f(\boldsymbol{x}))$, or how their corresponding features $\boldsymbol{z}$ and $\hat{\boldsymbol{z}}$ are related. Hence, theoretically, nothing is known about relationships between individual samples and their features. However, somewhat surprisingly, experimental results with the multi-class objective (12) in next section suggest that they actually can be rather close, at least for the given training samples X. For example, see Figure 5. Of course, one could consider explicitly imposing certain sample-wise requirements in the objectives, such as enforcing x^i to be close to $\hat{x}^i = g(f(x^i))$. It has been observed empirically in GANs or VAEs that imposing such sample-wise similarity or dissimilarity would improve visual quality around samples of interest, such as the DC-VAE [42] and the OpenGAN [68]. However, theoretically, how such sample-wise distances or constraints may affect the difficulty or accuracy of learning the correct support and density of the distributions remains an open problem.

4.3. Properties of the Closed-Loop Minimax Game

Above are the two primary objectives for CTRL transcription: while the encoder f tries to maximize the expressiveness and discriminativeness of the learned LDR representation, the decoder g tries to minimize the reconstruction error and coding rates. The competing objectives of the encoder f and the decoder g naturally lead to a two-player game. In this paper, we have formulated this game as a zero-sum game, namely Equation (12). Likewise, we have also implemented the most straightforward algorithm for solving this zero-sum game: gradient descent–ascent (GDA) [52], where the minimizer and maximizer take alternating gradient steps. These simplifications into a GDA-optimized zero-sum game were made in order to create a concrete algorithm for our experimentation. However, simplifying to a zero-sum game and GDA is certainly not the only way to solve the more general game described above. This game-theoretic formulation puts CTRL transcription outside of the theoretical realm of [2], since we are no longer finding pure maximizers of $\Delta R(\boldsymbol{Z})$, but rather stable minimax equilibria.

As is the case with GANs, these equilibria may not necessarily be Nash equilibria [50], but rather the more general sense of Stackelberg [69]. So, the problem of studying minimax equilibria of (12) is likely, in its most general form, quite challenging. Nevertheless, our experiments suggest such equilibria tend to be well-behaved, e.g., having a large range of attraction. Our extensive empirical experiments and ablation studies indicate that, in general, the minimax objective converges rather stably to good equilibria for all the real datasets without any special optimization tricks or particular requirements on the networks. The only important factor for the stability of the optimization seems to be a large enough batch size (see Appendix A.10). These observations can be further corroborated with analysis on simpler models: our ongoing work suggests that if we restrict our attention to simplified data structures (e.g., X distributed on a linear subspace), then one can provide theoretical guarantees that the equilibria become efficiently and correctly solvable by the minimax formulation. Extending such analysis to more sophisticated data structures (multiple subspaces, nonlinear submanifolds) remains an exciting new directions for future research.

Despite many possible pathological solutions to the minimax game, empirically, as we have presented in the previous section (alongside many examples in the Appendix A), the solution found by the simple GDA algorithm generally strikes a good trade-off between expressiveness and parsimony of the learned model. The solution automatically determines the proper dimensions for different classes. Ablation studies in Appendix A.10 on the large ImageNet dataset further suggest that this formulation is insensitive to over-parameterization by increasing network width, as long as the batch size grows accordingly. However, a rigorous justification for such good model-selection properties remains widely open.

5. Conclusions and Future Work

This work provides a novel formulation for learning a *both generative and discriminative* representation for a multi-class, multi-dimensional, possibly nonlinear, distribution of real-world data. We have provided compelling empirical evidence that the distribution of most datasets can be effectively mapped to an LDR, a union of independent principal subspaces and principal components. The objective function is entirely based on an intrinsic information-theoretic measure, the rate reduction, without any other heuristics or regularizing terms. The objective can be achieved with a closed-loop minimax game between the two encoder and the decoder networks without any additional network(s).

The main purpose of this paper is to demonstrate the conceptual simplicity and practical potential of this new framework for distribution/representation learning, instead of striving for state-of-the-art performance with heavy engineering. Nevertheless, with our preliminary implementation, a more informative LDR of the data can be effectively learned with a simple closed-loop transcription for a variety of real-world, multi-class, multi-modal visual datasets, from small to large, from low-resolution to higher-resolution, from domain-specific to diverse categories. The so-learned encoder f already enjoys the benefits of AE/VAEs for their discriminative property and the decoder g with the benefits of GANs for their good generative visual quality. However, probably more importantly, the internal structures of the learned feature representation has now become transparent, hence *fully interpretable and controllable* (for generative purposes): visual differences between classes are naturally "disentangled" as independent subspaces, while diverse visual attributes within each class are "disentangled" as principal components within each subspace. From extensive ablation studies given in the Appendix A, we see that the rate-reduction-based objective can be stably optimized across a wide range of datasets and network architectures without any additional regularizations or engineering tricks. Both the *feedback closed-loop* and the *rate-reduction measure* play indispensable roles in fostering the ease and success of finding the CTRL transcription.

One may notice that there are many ways this simple formulation can be significantly improved or extended. Firstly, in this work, we have simply adopted networks that were designed for GANs, but they may not be optimal for the rate-reduction-type objectives. For example, our ablation study already suggests that some of the components of such networks such as spectral normalization are not quite essential. Characteristics from the white-box ReduNet [2] derived from optimizing rate reduction can be explored in the future. Secondly, notice that our rate-reduction objectives do not impose any requirements on how individual samples should be encoded or decoded although the results from the multi-class objective indicate a certain level of alignment on the individual samples. Recent studies such as DC-VAE [42] or OpenGAN [68] suggest that imposing additional regularization on individual samples may further improve decoded visual quality. Such regularization can certainly be incorporated into this new framework. Last but not the least, compared to GANs and VAEs, our method leads to an *explicit* structured model for the feature distribution: a mixture of incoherent subspace Gaussians. Such an explicit model has the potential of making many subsequent tasks easier and better: better control of feature sampling for decoding and synthesis [70], designing more robust generators and classifiers for noise and corruptions based on the low-dimensional structures identified, or even extending to the settings of incremental and online learning [71,72]. We leave all these new directions, together with all the open theoretical problems posed in Section 4, for future investigation.

Author Contributions: This work has been the result of a successful team effort. In particular, the first four authors have contributed almost equally to this work. X.D.: investigation, methodology, project administration, software, writing—original draft preparation; S.T.: investigation, methodology, software, visualization, writing—original draft preparation; M.L.: investigation, software, visualization, writing—original draft preparation; Z.W.: investigation, software, visualization, writing—original draft preparation; M.P.: formal analysis, writing—original draft preparation; K.H.R.C.: validation, writing—review and editing; P.Z.: formal analysis, writing—review and editing; Y.Y.: validation, writing—review and editing; X.Y.: resources, writing—review and editing; H.-Y.S.: resources, writing—review and editing; Y.M.: conceptualization, formal analysis, funding acquisition, methodology, supervision, writing—original draft preparation, writing—review and editing; All authors have read and agreed to the published version of the manuscript.

Funding: This research was funded by ONR grants N00014-20-1-2002 and N00014-22-1-2102, the joint Simons Foundation-NSF DMS grant #2031899, as well as partial support from Berkeley FHL Vive Center for Enhanced Reality and Berkeley Center for Augmented Cognition, Tsinghua-Berkeley Shenzhen Institute (TBSI) Research Fund, and Berkeley AI Research (BAIR).

Data Availability Statement: Data and results can be found in Section 3 and Appendix A.

Acknowledgments: Earliest ideas of this work were germinated during a hiking event of Ma's group on Berkeley hills during the summer of 2020. Former group members Chong You (now at Google) and Yichao Zhou (now at Apple) were part of a stimulating discussion on possible extensions or applications of a new rate-reduction framework being developed then. During the preparation of this work, we consulted several experts on some of the related topics. The authors would like to thank Jiantao Jiao of UC Berkeley for discussions about the theoretical conditions for learning distributions via GANs. We thank Benjamin Haeffele of Johns Hopkins University for sharing thoughts on how to learn subspaces correctly and on how to optimize the rate-reduction objectives efficiently. We would also like to thank Shankar Sastry and Manxi Wu of UC Berkeley and Chaobing Song of Univ. of Wisconsin-Madison for informative discussions on how to solve minimax games correctly and efficiently, as well as Chih-Yuan Chiu and Druv Pai of UC Berkeley for engaging discussions on theoretical directions for the CTRL transcription. Last but not the least, we would like to thank Stefano Soatto of UCLA for stimulating discussions and sometimes heated debates on how information can be efficiently and effectively encoded in deep networks.

Conflicts of Interest: The authors declare no conflicts of interest.

Appendix A

Appendix A.1. Experiment Settings and Implementation Details

Network backbones. For MNIST, we use the standard CNN models in Tables A1 and A2, following the DCGAN architecture [63]. We resize the MNIST image resolution from 28×28 to 32×32 to fit DCGAN architecture. All α in lReLU (lReLU is short for Leaky-ReLU) of the encoder are set to 0.2.

We adopt ResNet architectures for CIFAR-10 shown in Tables A3 and A4, and STL-10 shown in Tables A5 and A6. Each ResBlock up is same as Resnet, but add an up-sampler after the first conv layer. All batch normalization layers of ResBlock in the encoder are replaced with spectral normalization layer.

Finally, we use the same architecture for CelebA, LSUN-bedroom, and ImageNet-128 (see Tables A7 and A8) as all three datasets have the same 128×128 resolution. Again, each ResBlock up is same as Resnet, but add an up-sampler after the first conv layer. All batch-normalization layers in the encoder are replaced with spectral normalization layer. All experiments utilize this lightweight PyTorch library "mimicry" [73] that provides implementations of some popular state-of-the-art GANs and evaluation metrics.

Table A1. Decoder for MNIST.

$z \in \mathbb{R}^{1\times1\times128}$
4×4, stride = 1, pad = 0 deconv. BN 256 ReLU
4×4, stride = 2, pad = 1 deconv. BN 128 ReLU
4×4, stride = 2, pad = 1 deconv. BN 64 ReLU
4×4, stride = 2, pad = 1 deconv. 1 Tanh

Table A2. Encoder for MNIST.

Gray image $x \in \mathbb{R}^{32\times32\times1}$
4 × 4, stride = 2, pad = 1 conv 64 lReLU
4 × 4, stride = 2, pad = 1 conv. BN 128 lReLU
4 × 4, stride = 2, pad = 1 conv. BN 256 lReLU
4 × 4, stride = 1, pad = 0 conv 128

Table A3. Decoder for CIFAR-10.

$z \in \mathbb{R}^{128}$
dense → 4 × 4 × 256
ResBlock up 256
ResBlock up 256
ResBlock up 256
BN, ReLU, 3 × 3 conv, 3 Tanh

Table A4. Encoder for CIFAR-10.

RGB image $x \in \mathbb{R}^{32\times32\times3}$
ResBlock down 128
ResBlock down 128
ResBlock 128
ResBlock 128
ReLU
Global sum pooling
dense → 128

Table A5. Decoder for STL-10.

$z \in \mathbb{R}^{128}$
dense → 6 × 6 × 512
ResBlock up 256
ResBlock up 128
ResBlock up 64
BN, ReLU, 3 × 3 conv, 3 Tanh

Table A6. Encoder for STL-10.

RGB image $x \in \mathbb{R}^{48\times48\times3}$
ResBlock down 64
ResBlock down 128
ResBlock down 256
ResBlock down 512
ResBlock 1024
ReLU
Global sum pooling
dense $\to$ 128

Table A7. Decoder for CelebA-128, LSUN-bedroom-128, and ImageNet-128.

$z \in \mathbb{R}^{128}$
dense $\to 4 \times 4 \times 1024$
ResBlock up 1024
ResBlock up 512
ResBlock up 256
ResBlock up 128
ResBlock up 64
BN, ReLU, 3×3 conv, 3 Tanh

Table A8. Encoder for CelebA-128, LSUN-bedroom-128, and ImageNet-128.

RGB image $x \in \mathbb{R}^{128\times128\times3}$
ResBlock down 64
ResBlock down 128
ResBlock down 256
ResBlock down 512
ResBlock down 1024
ResBlock 1024
ReLU
Global sum pooling
dense $\to$ 128

Optimization and training details. Across all of our experiments, we use Adam [74] as our optimizer, with hyperparameters $\beta_1 = 0.5, \beta_2 = 0.999$. We adopt the simple gradient descent–ascent algorithm for alternating minimizing and maximizing the objectives. The initial value of learning rate is set to be 0.00015 and is scheduled with linear decay. We choose $\epsilon^2 = 0.5$ for both Equations (12) and (13) in all CTRL experiments. For all CTRL-Multi experiments on ImageNet, we only choose 10 classes. The details of the 10 classes are shown in Table A9. Most experiments are trained on RTX 3090 GPUs.

Table A9. ID and correspond category for 10 classes of ImageNet.

ID	Category
n02930766	cab, hack, taxi, taxicab
n04596742	wok
n02974003	car wheel
n01491361	tiger shark, Galeocerdo cuvieri
n01514859	hen
n09472597	volcano
n07749582	lemon
n09428293	seashore, coast, seacoast, sea-coast
n02504458	African elephant, Loxodonta africana
n04285008	sports car, sport car

Appendix A.2. MNIST

Settings. On MNIST dataset, we train our model using DCGAN [63] architecture with our proposed objectives CTRL-Multi (12) and CTRL-Binary (13). The learning rate is set to 10^{-4} and the batch size is set to 2048. We train our model with 15,000 iterations.

More results illustrating auto-encoding. Here, we give more reconstruction results, or $\hat{\boldsymbol{X}}$, from CTRL-Multi and CTRL-Binary objectives, compared to their corresponding original input $\boldsymbol{X}$. As shown in the Figure A1, for the CTRL-Binary objective, it can generate clean digit-like images but the decoded $\hat{\boldsymbol{X}}$ might resemble digits from similar but different classes to the input data $\boldsymbol{X}$ since the CTRL-Binary tends to only align the distribution of all digits.

In contrast, with the CTRL-Multi objective, the decoded $\hat{\boldsymbol{X}}$ not only are coherent with the correct class with the input data $\boldsymbol{X}$, but also show very clear one-to-one mapping between individual samples $\boldsymbol{x}$ and $\hat{\boldsymbol{x}}$, although the objective (12) does not enforce that. Comparing with the results from VAE-GAN [34] and BiGAN [38], our decoded images make less errors in reconstruction and preserve much better the individual characteristics of the original samples.

(**a**) Original $\boldsymbol{X}$ (**b**) VAE-GAN $\hat{\boldsymbol{X}}$ (**c**) BiGAN $\hat{\boldsymbol{X}}$ (**d**) CTRL-Binary $\hat{\boldsymbol{X}}$ (**e**) CTRL-Multi $\hat{\boldsymbol{X}}$

Figure A1. The comparison of the reconstruction results of different methods with the input data.

Images decoded from random samples on the learned multi-class LDR. Since our CTRL-Multi objective function maps input data of each class into a different (orthogonal) subspace in the feature space, we can generate images conditioned on each class by random sampling $\boldsymbol{z}$ in the subspace of each class and then decode them back to the input space as $\hat{\boldsymbol{x}}$.

To perform random sampling in the learned subspace, we first calculate the mean feature $\bar{\boldsymbol{z}}_j$ and the singular vectors $\boldsymbol{v}_j^i$ from the SVD (or principal components) of the learned features $\boldsymbol{Z}_j$ of the training data in the class j, where index i represents the ith principal components. We only use the top $r = 8$ principal components of each class on MNIST dataset. These statistics of the subspace can be used for guiding the random sampling. Then, we sample $\boldsymbol{z}$ randomly along the principal components and around the mean feature as

$$\boldsymbol{z}_{random\_j} = \bar{\boldsymbol{z}}_j + \alpha \sum_{i=1}^{r} n_i * \sigma_j^i * \boldsymbol{v}_j^i, \tag{A1}$$

where $\bar{z}_j$ is the mean feature of class j, σ_j^i and v_j^i are the i-th singular value and principal component of class j, n_i are i.i.d. Gaussian $\mathcal{N}(0,1)$ random variables. That is, the feature in each subspace/class is modeled by an r-dimensional multivariate Gaussian, with variances σ_j^i which characterize variances of the training data in the feature space. Here, α is a hyper-parameter that controls the sampling range. As for the visualization of random generated images $g(z_{random\_j})$ conditioned on the given class, we compare our method with some other conditional generation methods such as ACGAN [25] and InfoGAN [21] (for ACGAN and InfoGAN, we generate images conditioned on class labels with randomly sampled latent z according the procedures mentioned in their respective works). Our model can give realistic and correct conditional generation results with high diversity in each class, while other methods may make mistakes in the generation between some similar classes such as classes 3 and 5 for InfoGAN.

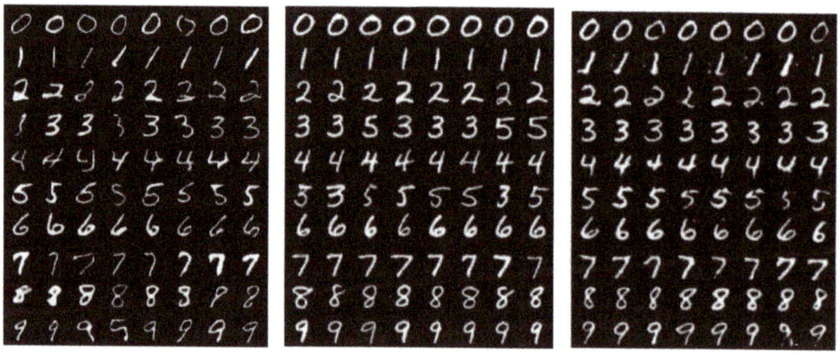

(**a**) ACGAN (**b**) InfoGAN (**c**) CTRL-Multi

Figure A2. Comparison of randomly generated images conditioned on each class.

Interpolation between samples in different classes. We randomly sample some images from each class. For each image x_1, we randomly sample another image x_2 from a different class. For such a pair of images x_1 and x_2, we reconstruct them based on their linearly interpolated feature representations by $\hat{x} = g(\alpha f(x_1) + (1-\alpha) f(x_2)), \alpha \in [0,1]$, the results of which are shown in the Figure A3. For each row in the figure from left to the right, the reconstructed images continuously morph from one digit to a different digit with a natural transition in shape rather than a simple superposition of the two images. This also confirms that space between subspaces for the digits does not represent valid digits but only shapes with digit-like strokes. Hence for generative purposes, knowing the supports of valid digits is extremely important.

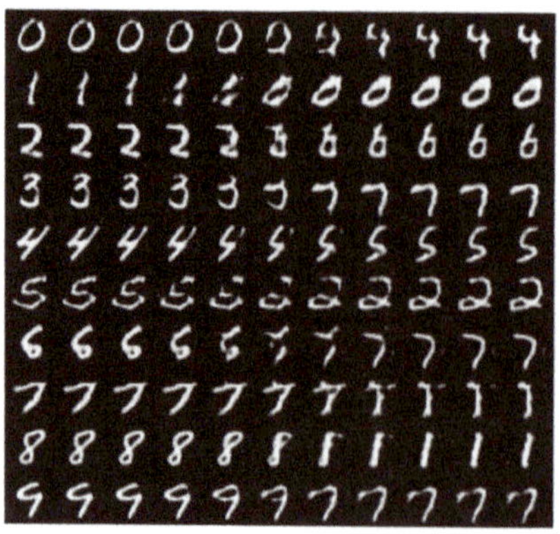

Figure A3. Images generated from the interpolation between samples in different classes.

Appendix A.3. Transformed MNIST

Settings. In this experiment, we verify that the CTRL-Multi objective can preserve diverse data modes in the learned feature embeddings. We construct a transformed MNIST dataset with five modes: normal, large (1.5×), small (0.5×), rotate 45° left, and rotate 45° right. Each image data point will be randomly transformed to one of the modes. Representative examples of such training data can be found in Figure A4a. We train the model with learning rate 1×10^{-4} and batch size 2048 for 15,000 iterations.

Auto-encoding results. Figure A4b gives the decoded results of the training data with different modes. Even though the data are now much more diverse for each class, decoder learned from the CTRL-Multi objective can still achieve high sample-wise similarity to the original images.

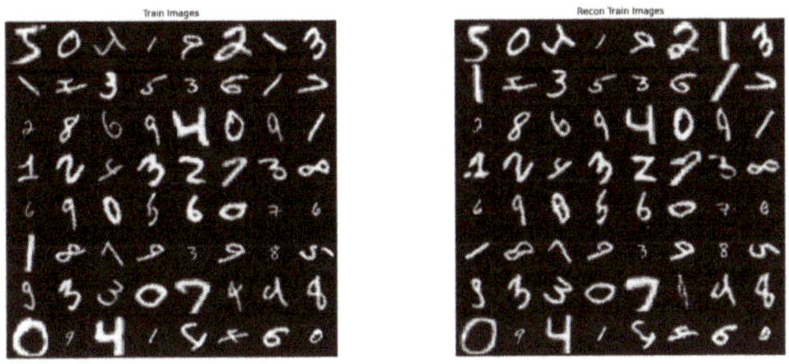

(**a**) Original X (**b**) Decoded $\hat{X}$

Figure A4. Original (training) data X and their decoded version $\hat{X}$ on the transformed MNIST.

Identifying different modes. Similar to the earlier experiments of Figure 6 for CIFAR-10 in the main paper, we find the top principal components of features of each class Z_j (via SVD) and generate new images using the learned decoder g from features of the training images aligned the best with these components.

In Figure A5, we select three classes 0, 1, 2 and visualize samples from the top $r = 8$ principal components for each class. Each row represents one principal component direction. As can be seen in the figure, the decoded images along each principal component shows a similar mode and the modes along different component directions are rather incoherent. All major modes of the original data can be identified as one of these principal component directions. This clearly shows that the CTRL-Multi objective can keep the different modes within each class of the data X_j as the principal component directions of Z_j, and these modes can also be retained in the decoded images $\hat{X}_j$.

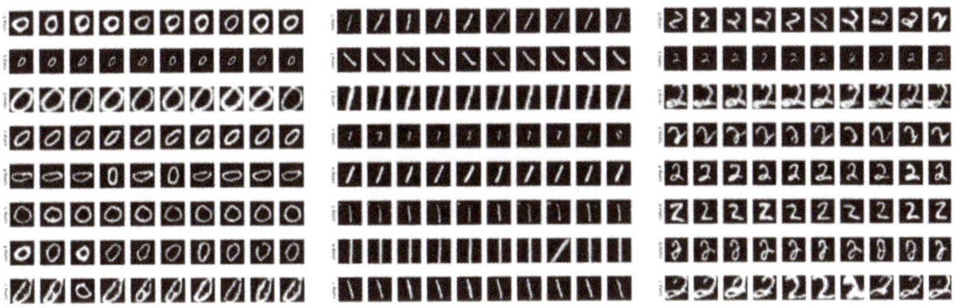

(**a**) Components of class "0" (**b**) Components of class "1" (**c**) Components of class "2"

Figure A5. The reconstructed images $\hat{X}$ from the features Z best aligned along top-8 principal components on the transformed MNIST dataset. Each row represents a different principal component.

Appendix A.4. CIFAR-10

Settings. For all experiments on CIFAR-10, we follow the common training hyper-parameters in Appendix A.1. Beyond that, for each experiment, we run 450,000 iterations with batch size 1600.

Images decoded from random samples on the CTRL-Multi. We sample z in the feature space randomly along the principal components and around the mean feature of each class Z_j as in the MNIST case, according to Equation (A1). The generated images from the sampled features are illustrated in Figure A6, one row per class. As we see, the generator learned from the CTRL-Multi objective is capable of generating diverse images for each class.

Further, for visualization of random generated images $g(z_{random\_j})$ conditioned on the given class, we compare our method with some other conditional generation method such as ACGAN [25] and InfoGAN [21]. For all three experiments, we have randomly sampled 8 images per class in CIFAR-10. For more complex datasets such as CIFAR-10, our model can give more realistic conditional generation results for different classes with high diversity within each class.

(**a**) ACGAN (**b**) InfoGAN (**c**) CTRL-Multi

Figure A6. Comparison of randomly generated images conditioned on each class.

Generating images along different PCA components for each class. For each class, we first compute the top 10 principal components (singular vectors of the SVD) of Z and then for each of the top singular vectors, we display in each row the top 10 reconstructed image $\hat{X}$ whose Z are closest to the singular vector using methods described in the main body of the paper, Section 3.3. The results are given in Figure A7. Notice that images in each row are very similar as they are sampled along the same principal component, whereas images in different rows are very different as they are orthogonal in the feature space. These results indicate that the features learned by our method can not only disentangle different classes as orthogonal subspaces but can also disentangle different visual attributes within each class as (orthogonal) principal components within each subspace.

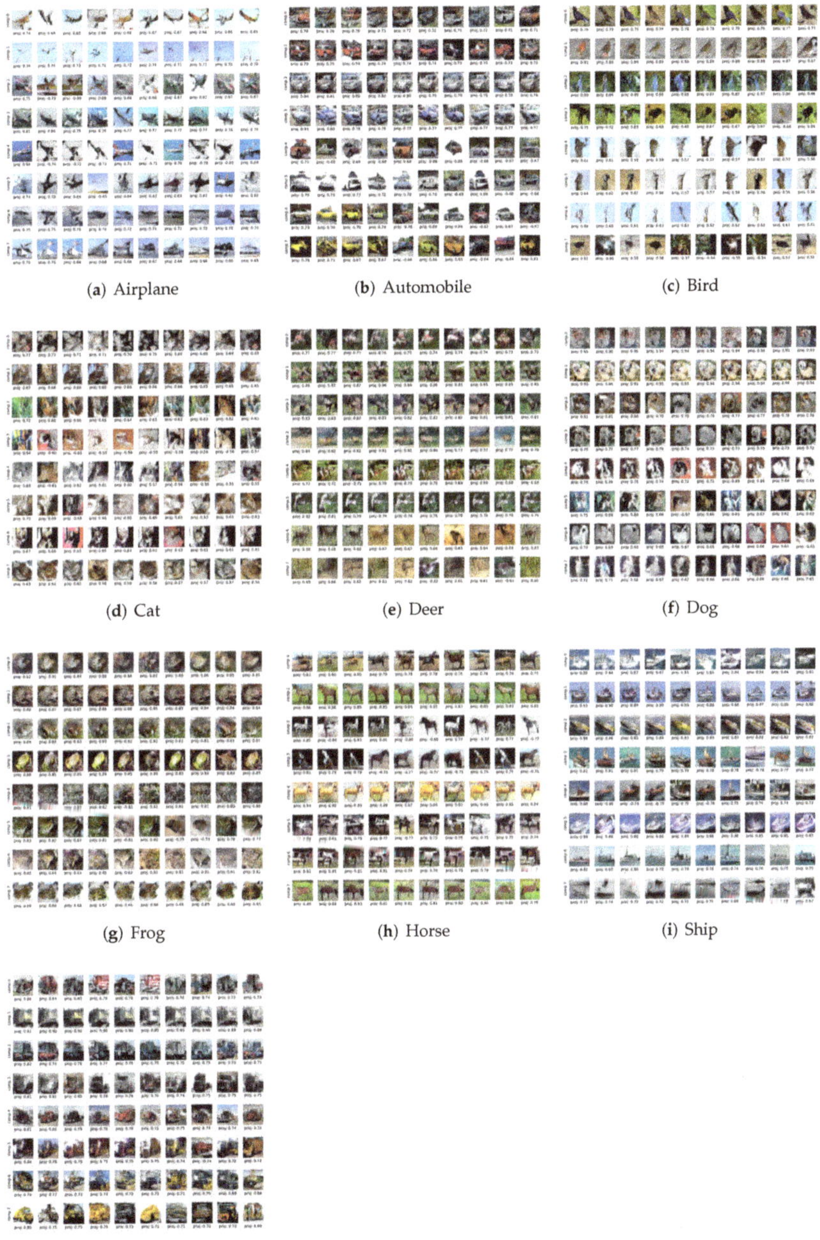

(**a**) Airplane (**b**) Automobile (**c**) Bird (**d**) Cat (**e**) Deer (**f**) Dog (**g**) Frog (**h**) Horse (**i**) Ship (**j**) Truck

Figure A7. Reconstructed images $\hat{\boldsymbol{X}}$ from features $\boldsymbol{Z}$ close to the principal components learned for the 10 classes of CIFAR-10.

Appendix A.5. STL-10

Settings. For all experiments on STL-10, we follow the common training hyper-parameters in Appendix A.1. For the CTRL-Binary setting, we train 150,000 iterations. For the CTRL-Multi setting, we initialize the weights from the 20,000-th iteration of CTRL-

Binary checkpoint and train for another 80,000 iterations (with the CTRL-Multi objective). The IS and FID scores on the STL-10 dataset are reported in Table A10, on par or even better than existing methods such as SNGAN [31] or DC-VAE [42].

Visualizing auto-encoding property for the CTRL-Binary. We visualize the original images x and their decoded $\hat{x}$ generated by the LDR model learned from the CTRL-Binary objective. The results are shown in Figure A8 for STL-10.

(**a**) Original X (**b**) Decoded $\hat{X}$

Figure A8. Visualizing the original x and corresponding decoded $\hat{x}$ results on STL-10 dataset. Note the model is trained from the CTRL-Binary objective hence sample- or class-wise correspondence is relatively poor, but the decoded image quality is very good.

Appendix A.6. Celeb-A and LSUN

To verify that our formulation works on images of higher resolution, we conduct experiments on the Celeb-A and LSUN datasets, which have a resolution of 128 × 128.

Settings. For all experiments on these datasets, we follow the common training hyperparameters in Appendix A.1. We choose a 300 batch size for Celeb-A and LSUN. Both of them are trained with the CTRL-Binary objective and for 450,000 iterations.

Generating images along different PCA components. We calculate the principal components of the learned features $\mathbf{Z}$ in the latent subspace. We manually choose three principle components which are related to hat, hair color, and glasses (see Figure A9). The three components are 9th, 19th, and 23rd respectively from the overall 128 principal components. These principal directions seem to clearly disentangle visual attributes/factors such as wearing a hat, changing hair color, and wearing glasses.

Images generated from random sampling of the feature space. We sample z randomly according to the following Gaussian model:

$$z_{random} = \bar{z} + \alpha \sum_{i=1}^{r} n_i * \sigma_i * \boldsymbol{v}_i, \tag{A2}$$

where $\bar{z}$ is the mean feature, σ_i and $\boldsymbol{v}_i$ are the ith singular value and singular vector, respectively, n_i are i.i.d. Gaussian $\mathcal{N}(0,1)$ random variables. As before α is a hyperparameter to control the sampling range. We use the top r = 100 principle components for random sampling. The random generated images are realistic and diverse (see Figure A10).

Visualizing auto-encoding property for CTRL-Binary. We visualize the original image x and their decoded $\hat{x}$ using the LDR model learned from the CTRL-Binary objective. The results are shown in Figures A11 and A12 for the Celeb-A dataset and the LSUN dataset, respectively. The CTRL-Binary objective can give very good visual quality for $\hat{x}$ but cannot ensure sample-to-sample alignment. Nevertheless, the decoded $\hat{x}$ seems to be very similar to the original x in some main visual attributes. We believe the binary objective manages to align only the dominant principal component(s) associated with the most salient visual attributes, say, pose of the face for Celeb-A or layout of the room for LSUN, between features of $\boldsymbol{X}$ and $\hat{\boldsymbol{X}}$.

(**a**) Hat (**b**) Hair Color (**c**) Glasses

Figure A9. Sampling along the 9th, 19th, and 23rd principal components of the learned features $\mathbf{Z}$ seems to manipulate the visual attributes for generated images on the CelebA dataset.

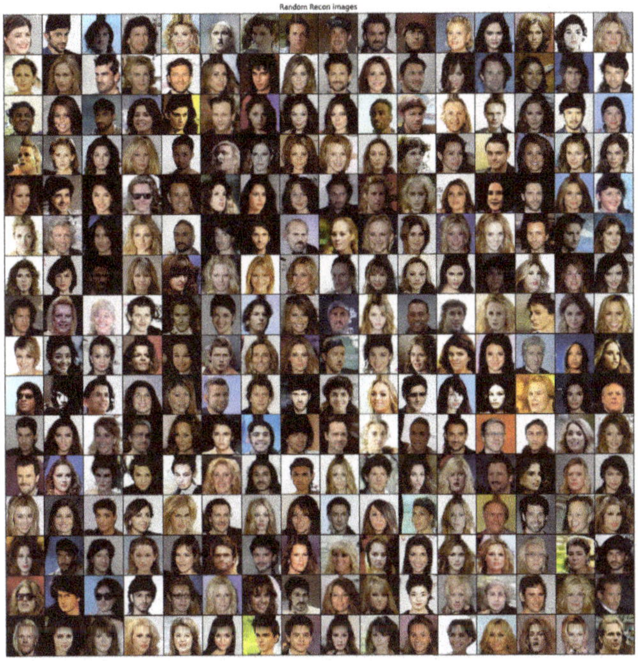

Figure A10. Images decoded from randomly sampled features, as a learned Gaussian distribution (A2), for the CelebA dataset.

(**a**) Original $\boldsymbol{X}$ (**b**) Decoded $\hat{\boldsymbol{X}}$

Figure A11. Visualizing the original $\boldsymbol{x}$ and corresponding decoded $\hat{\boldsymbol{x}}$ results on Celeb-A dataset. The LDR model is trained from the CTRL-Binary objective.

(a) Original X (b) Decoded $\hat{X}$

Figure A12. Visualizing the original x and corresponding decoded $\hat{x}$ results on LSUN-bedroom dataset. The LDR model is trained from the CTRL-Binary objective.

Appendix A.7. ImageNet

Settings. To verify that the CTRL works on large-scale datasets, we train it on the ImageNet. For all experiments on the ImageNet, we follow the common training hyper-parameters in Appendix A.1.

We first train our model with the CTRL-Binary objective with batch size of 1800 on the whole ImageNet ILSVRC 2012 dataset. The number of training iterations is 450,000.

After that, we fine-tune the pretrained model with the CTRL-Multi objective, on 10 selected classes. Information about the 10 classes can be found in Table A9. The fine-tune batch size is 1024, and we train another 35,000 iterations for it. This experiment takes 120 GPU hours on 8 A100-SXM4 GPUs. Note that our choice of batch size is substantially larger than those commonly adopted in other works while training on the ImageNet (e.g., 128 in [31]). We empirically observe that training with a larger batch size generates images of better quality and clearer class alignment. This is consistent with the proposed CTRL-Multi objective as it explicitly encourages alignment of class distributions, therefore benefiting from a larger batch that better captures overall data distributions. We leave a more rigorous study of the effect of batch size for future work.

Due to the heavy computation of such large batch size, we present the intermediate result obtained at the early iteration 35,000 whereas most existing methods run with significantly larger number of iterations. Nevertheless, the intermediate result already verify the efficacy of our framework. In addition, we present the full version of the comparison with existing generative methods in Table A10. We see the IS and FID scores for CTRL-Multi degraded a little after the finetuning. This is expected as learning a more refined separation and alignment of 10 classes is a more challenging task than 2 classes. This is consistently observed from experiments on other datasets too.

Visualizing feature similarity for CTRL-Multi. We visualize the cosine similarity among features $\mathbf{Z}$ of different classes learned from the CTRL-Multi objective in Figure A13. In addition, we provide the visualization of alignment between features $\mathbf{Z}$ and decoded features features $\hat{\mathbf{Z}}$. These results demonstrate that not only the encoder has already learnt to discriminate between classes, but also the learned $\mathbf{Z}$ and $\hat{\mathbf{Z}}$ are aligned clearly within each class.

Table A10. Comparison on CIFAR-10, STL-10, and ImageNet. ↑ means higher is better. ↓ means lower is better.

Method	CIFAR-10		STL-10		ImageNet	
	IS↑	FID↓	IS↑	FID↓	IS↑	FID↓
GAN based methods						
DCGAN [63]	6.6	-	7.8	-	-	-
SNGAN [31]	7.4	29.3	**9.1**	40.1	-	48.73
CSGAN [28]	8.1	19.6	-	-	-	-
LOGAN [29]	**8.7**	**17.7**	-	-	-	-
VAE/GAN based methods						
VAE [5]	3.8	115.8	-	-	-	-
VAE/GAN [64]	7.4	39.8	-	-	-	-
NVAE [41]	-	50.8	-	-	-	-
DC-VAE [42]	**8.2**	**17.9**	8.1	41.9	-	-
CTRL-Binary (ours)	**8.1**	**19.6**	8.4	**38.6**	7.74	**46.95**
CTRL-Multi (ours)	7.1	23.9	7.7	45.7	6.44	55.51

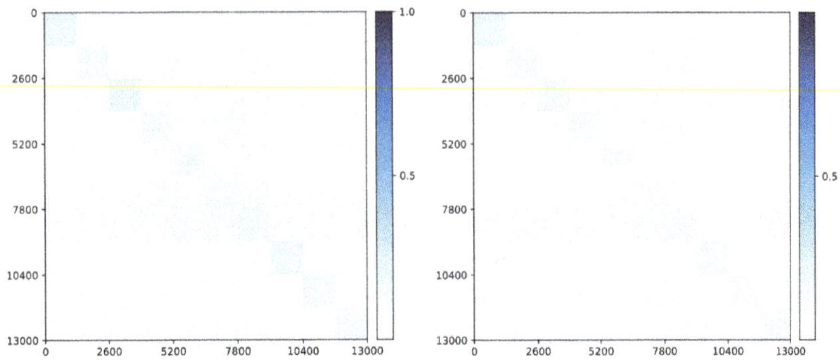

(**a**) $|\mathbf{Z}^\top \mathbf{Z}|$ (**b**) $|\mathbf{Z}^\top \hat{\mathbf{Z}}|$

Figure A13. Visualizing feature alignment: (**a**) among features $|\mathbf{Z}^\top \mathbf{Z}|$, (**b**) between features and decoded features $|\mathbf{Z}^\top \hat{\mathbf{Z}}|$. These results obtained after 200,000 iterations.

Visualizing auto-encoding property for CTRL-Multi. We visualize the original images $\boldsymbol{X}$ and their decoded $\hat{\boldsymbol{X}}$ using the LDR model fine-tuned with the CTRL-Multi objective. The results are shown in Figure A14 for the selected 10 classes in ImageNet. The CTRL-Multi objective can give good visual quality for $\hat{\boldsymbol{X}}$ as well as sample-to-sample alignment.

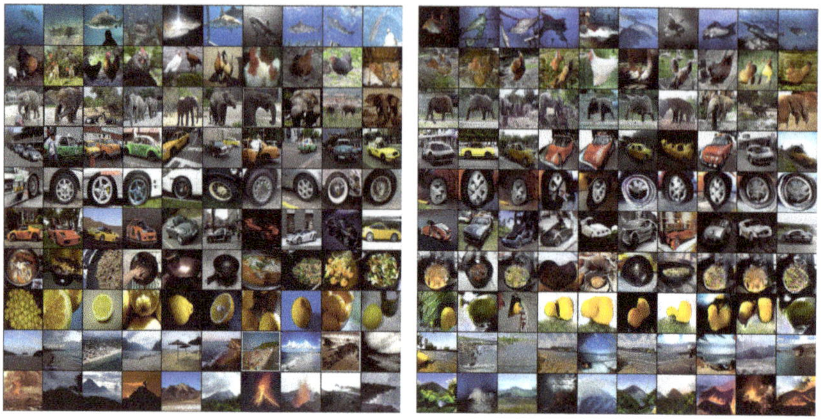

(a) Original X (b) Decoded $\hat{X}$

Figure A14. Visualizing the original X and corresponding decoded $\hat{X}$ results on ImageNet (10 classes). The LDR model is fine-tuned using the CTRL-Multi objective. These visualizations are obtained after 35,000 iterations.

Appendix A.8. Ablation Study on Closed-Loop Transcription and Objective Functions

To empirically validate the necessity and respective roles of the closed-loop transcription and the rate reduction (ΔR) objective, we conduct two sets of experiments. For the first set of experiments, we modify our closed-loop architecture by instantiating more than two networks while keeping the objective function (12) unchanged. For the second set of experiments, we keep the closed-loop architecture but replace all rate reduction (ΔR) terms in (12) with corresponding cross-entropy, or remove some of the terms. Experiments here shed insight onto how the closed-loop transcription and the rate reduction affect separately the performance, including sample-wise reconstruction, the alignment of Z and $\hat{Z}$ space, and the diversity of intra-class features.

Appendix A.8.1. The Importance of the Closed-Loop

To evaluate the importance of the closed-loop transcription, we experiment on modified versions of the closed-loop architecture (A3). Notice that many architectures have been proposed and experimented before to promote the encoder f and decoder g to be mutually inverse or cycle consistent (at least for mappings between the data and feature distributions), such as BiGAN [38], VAE-GAN [34], and CycleGAN [56]. However, the cycle consistency is typically enforced through a third discriminator network. (In the case of CycleGAN [56], one needs two additional discriminator networks, one for each domain).

Here, we experiment on whether similar ideas work with the rate-reduction objective. First, we break the closed-loop and use a separate encoder network $f^2 : \hat{X} \to \hat{Z}$ to replace the original encoder f. The revised architecture is summarized in the diagram (A4). Second, to emulate the architecture of VAE-GAN [34], we also instantiate an extra encoder network f^2 and compute the CTRL-Multi objective using $\tilde{Z}$ and $\hat{Z}$. The resulting architecture is also summarized in the diagram (A5).

$$X \xrightarrow{f(x,\theta)} Z \xrightarrow{g(z,\eta)} \hat{X} \xrightarrow{f(x,\theta)} \hat{Z}; \tag{A3}$$

$$X \xrightarrow{f^1(x,\theta^1)} Z \xrightarrow{g(z,\eta)} \hat{X} \xrightarrow{f^2(x,\theta^2)} \hat{Z}; \tag{A4}$$

$$X \xrightarrow{f^1(x,\theta^1)} Z \xrightarrow{g(z,\eta)} \hat{X}, X \xrightarrow{f^2(x,\theta^2)} \hat{Z}, \tilde{Z}. \tag{A5}$$

We run experiments on MNIST with the three different architectures, and choose the network from Table A1 for the encoder and Table A2 for the decoder, and the training hyper-parameters follow Appendix A.1. The qualitative results are shown in Figure A15. Both architectures (A4) and (A5) failed to generate meaningful images. These experiments show that directly applying rate-reduction objectives without the closed-loop or architectures that loosely enforcing cycle consistency fails to work. Instead, the closed-loop formulation allows us to use only two networks, without the need of any extra network.

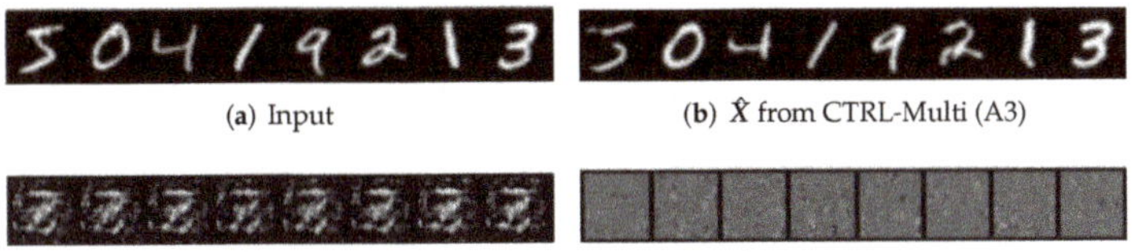

(a) Input (b) $\hat{X}$ from CTRL-Multi (A3) (c) $\hat{X}$ from architecture (A4) (d) $\hat{X}$ from architecture (A5)

Figure A15. Qualitative results for ablation study with alternative architectures to the proposed CTRL.

Appendix A.8.2. The Importance of Rate Reduction

By replacing the rate reduction (ΔR) terms in the objective function (12) with cross-entropy, we introduce a linear mapping $\mathbf{W} \in \mathbb{R}^{d \times k}$ to map $\mathbf{Z} \in \mathbb{R}^{d \times n}$ from feature space to logits $\gamma = \mathbf{Z}^\top \mathbf{W}$. We then calculate the softmax cross-entropy function on logits γ and one hot label matrix $\mathbf{Y}$. Here $\mathcal{H}(\gamma, \mathbf{Y}) = \sum_{i=1}^{n} \sum_{j=1}^{k} Y_{ij} \log \frac{e^{\gamma_{ij}}}{\sum_{j=1}^{k} e^{\gamma_{ij}}}$ is the formulation of softmax cross-entropy function and $\mathbf{Y} \in \mathbb{R}^{n \times k}$ is one hot label matrix. Then, we can replace the first two terms of (12) ($\Delta R(\mathbf{Z})$ and $\Delta R(\hat{\mathbf{Z}})$) with $\mathcal{H}(\mathbf{Z}^\top \mathbf{W}, \mathbf{Y})$ and $\mathcal{H}(\hat{\mathbf{Z}}^\top \mathbf{W}, \mathbf{Y})$. For the third term of (12), we extract j-th class one hot feature $\gamma_j = \mathbf{Z}_j^\top \mathbf{W}$, $\hat{\gamma}_j = \hat{\mathbf{Z}}_j^\top \mathbf{W}$ from $\mathbf{Z}$ and $\hat{\mathbf{Z}}$, and define the distance $\mathcal{D}(\gamma_j, \hat{\gamma}_j) = \frac{e^{\gamma_j}}{e^{\gamma_j} + e^{\hat{\gamma}_j}}$ of them. For the third term of (12), we further introduce k linear layers as discriminators $\{\mathcal{D}_j\}_{j=1}^{k}$ for each class. Then, we replace the third term with the GAN's objective function as $\sum_{j=1}^{k} \mathbb{E}[\log \mathcal{D}_j(\mathbf{Z}_j)] + \mathbb{E}[\log(1 - \mathcal{D}_j(\hat{\mathbf{Z}}_j))]$ ($\mathbb{E}[\mathbf{X}]$ denote the expectation of $\mathbf{X}$). Now, we have the cross-entropy version objective function (A6) for the closed-loop framework. We denote the closed-loop framework with cross-entropy as Closed-loop-CE.

$$\min_{\eta} \max_{\theta, \mathbf{W}, \mathcal{D}} \mathcal{T}_{\mathbf{X}}(\theta, \eta, \mathbf{W}, \mathcal{D}) \doteq \mathcal{H}(\mathbf{Z}^\top \mathbf{W}, \mathbf{Y}) + \mathcal{H}(\hat{\mathbf{Z}}^\top \mathbf{W}, \mathbf{Y}) + \sum_{j=1}^{k} \mathbb{E}[\log \mathcal{D}_j(\mathbf{Z}_j)] + \mathbb{E}[\log(1 - \mathcal{D}_j(\hat{\mathbf{Z}}_j))]. \tag{A6}$$

We run the experiments on MNIST and CIFAR10. The architectures of MNIST and CIFAR10 are given in Tables A1–A4 (In the context of this section, we use the term Decoder and Generator interchangeably; similarly for Encoder and Discriminator).

Results on MNIST. The training hyper-parameters of CTRL-Multi and Closed-loop-CE on MNIST are following Appendix A.1. Comparisons between CTRL-Multi and Closed-loop-CE are listed in Figures A16–A18.

Figure A16b,c show the reconstructed images $\hat{X}$ from Closed-loop-CE and CTRL-Multi. Both methods can give sample-wise reconstruction results due to the closed-loop transcription framework. However, comparing training images whose features are best aligned with the principal components of class '2' in Figure A17, we see that the principal components of CE features do not correspond to consistent visual attributes of the images, whereas ours do.

From the heatmaps in Figure A18a,b, we see the features learned by rate reduction possess clear orthogonal subspace structures, whereas those learned by Closed-loop-CE do not. Moreover, Figure A18c,d shows that the learned features of CTRL-Multi have higher singular values for the top principal components of each class, corresponding to a more linearized and diverse feature distribution, whereas those by Closed-loop-CE do not.

(**a**) Original X (**b**) $\hat{X}$ by Closed-loop-CE (**c**) $\hat{X}$ by CTRL-Multi

Figure A16. The comparison of sample-wise reconstruction between the Closed-loop-CE objective and the CTRL-Multi objective.

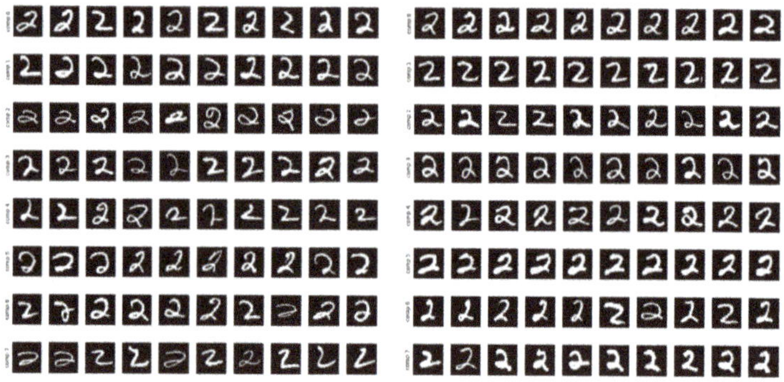

(**a**) Closed-loop-CE (**b**) CTRL-Multi

Figure A17. Training samples along different principal components of the learned features of digit '2'.

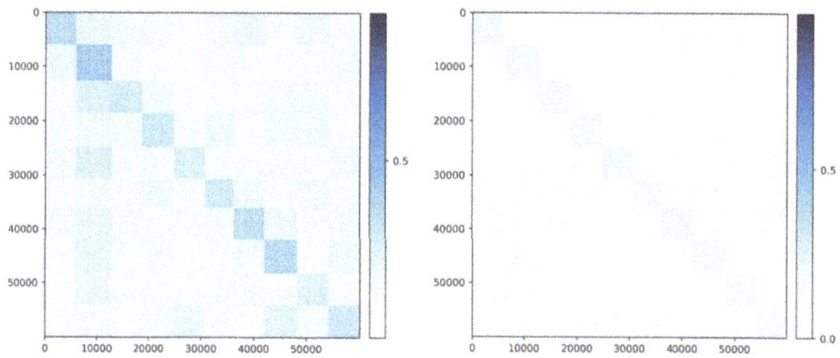

(**a**) (**b**)

Figure A18. *Cont.*

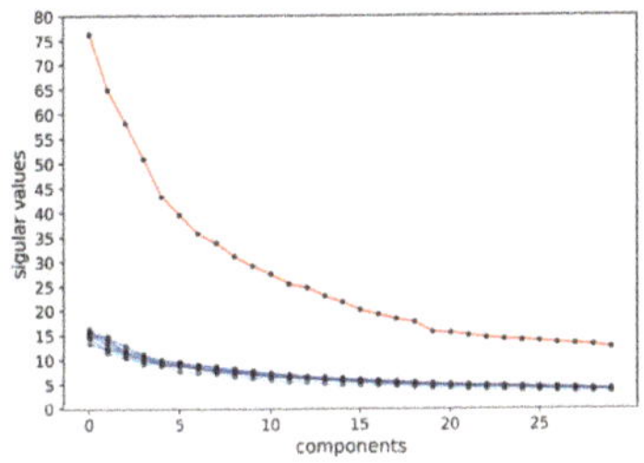

(c)

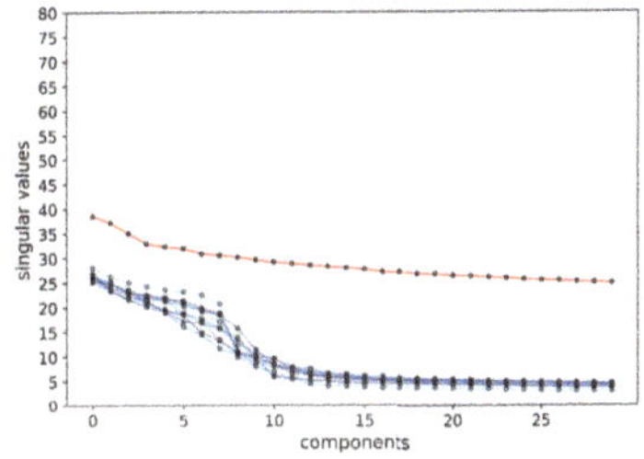

(d)

Figure A18. Comparison Closed-loop-CE and CTRL-Multi on $|\mathbf{Z}^\top \hat{\mathbf{Z}}|$ and PCA singular values. (**a**) $|\mathbf{Z}^\top \hat{\mathbf{Z}}|$ from Closed-loop-CE. (**b**) $|\mathbf{Z}^\top \hat{\mathbf{Z}}|$ from CTRL-Multi. (**c**) PCA of learned features by the Closed-loop-CE objective for each class. (**d**) PCA of learned features by the CTRL-Multi objective for each class.

Failed Attempts on CIFAR-10 with Cross Entropy. The training hyper-parameters of Closed-loop-CE on CIFAR10 follow Appendix A.1. We perform the grid search on three hyper-parameters: learning rate $\{1.5 \times 10^{-2}, 1.5 \times 10^{-3}, 1.5 \times 10^{-4}\}$, batch size (800 or 1600), and inner loop (1,2,3,4), conducting 24 experiments in total. All cases of the Closed-loop-CE fail to converge or experience model collapse on the CIFAR-10 dataset.

Appendix A.8.3. Ablation Study on the CTRL-Multi Objectives

In this section, we investigate the influence of each term of the objective function (12) and see how they affect the learned features $\mathbf{Z}, \hat{\mathbf{Z}}$ and sample-wise reconstruction. We follow the same experiment setting with CTRL-Multi on MNIST (Appendix A.1), and conduct three experiments, each with a modified version of the original objective. Objective I is the original objective with all three terms, Objective II removes the second term $\Delta R(\hat{\mathbf{Z}})$, and Objective III keeps only the third term $\Delta R(\mathbf{Z}, \hat{\mathbf{Z}})$. The results in Figure A19 show that using Objective II we can still maintain the sample-wise reconstruction property, but the image quality is lower when compared those constructed by Objective I (Figure A19b vs. Figure A19c). Objective III loses the sample-wise reconstruction property (Figure A19a vs. Figure A19d). Finally, the results from Figures A20 and A21 show that without the first two terms, the learned features $\mathbf{Z}$ and $\hat{\mathbf{Z}}$ have poor class-to-class alignment and their principal components do not show clear subspace structure with higher singular values within each class.

Table A11. Three different objective functions for CTRL.

Objective I:	$\min_\eta \max_\theta \mathcal{T}_{\mathbf{X}}(\theta,\eta) = \Delta R(\mathbf{Z}(\theta)) + \Delta R(\hat{\mathbf{Z}}(\theta,\eta)) + \sum_{j=1}^{k} \Delta R(\mathbf{Z}_j(\theta), \hat{\mathbf{Z}}_j(\theta,\eta))$.
Objective II:	$\min_\eta \max_\theta \mathcal{T}_{\mathbf{X}}(\theta,\eta) = \Delta R(\mathbf{Z}(\theta)) + \sum_{j=1}^{k} \Delta R(\mathbf{Z}_j(\theta), \hat{\mathbf{Z}}_j(\theta,\eta))$.
Objective III:	$\min_\eta \max_\theta \mathcal{T}_{\mathbf{X}}(\theta,\eta) = \sum_{j=1}^{k} \Delta R(\mathbf{Z}_j(\theta), \hat{\mathbf{Z}}_j(\theta,\eta))$.

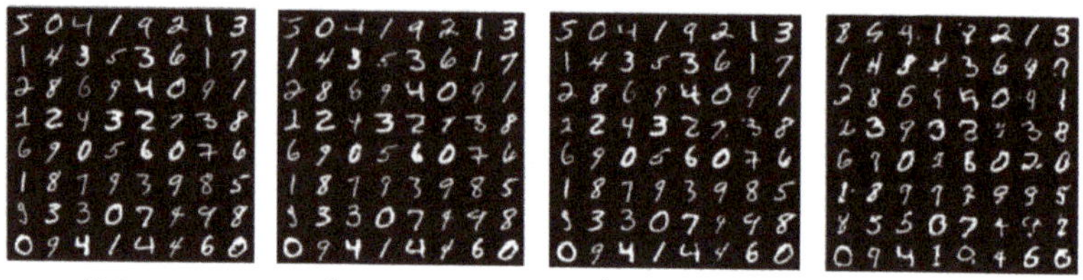

(**b**) Input (**c**) $\hat{X}$ from objective I (**d**) $\hat{X}$ from objective II (**e**) $\hat{X}$ from objective III

Figure A19. The influence of the choice of objective functions on the reconstruction: decoded images $\hat{X}$ from the objective I, II, or III.

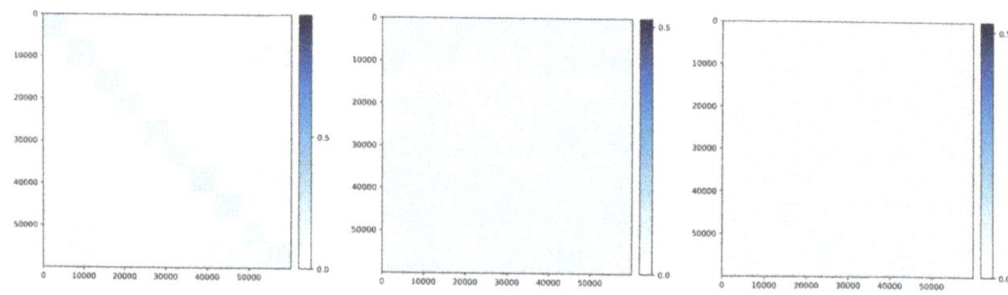

(**a**) Objective I (**b**) Objective II (**c**) Objective III

Figure A20. Correlation $|\mathbf{Z}^\top \hat{\mathbf{Z}}|$ between features $\mathbf{Z}$ and $\hat{\mathbf{Z}}$ learned with Objective I, II, or III.

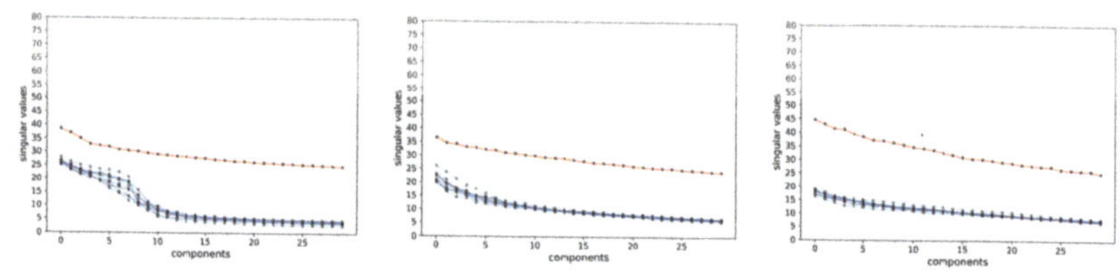

(**a**) Objective I (**b**) Objective II (**c**) Objective III

Figure A21. PCAs of the features learned with Objective I, II, or III.

Appendix A.9. Ablation Study on Sensitivity to Spectral Normalization

It is known that spectral normalization is important to improve the stability of training GANs. Here, we test our formulation with and without the spectral normalization. We follow the setting from Appendix A.1 and test on CIFAR10, using the network architecture from Tables A3 andA4. All settings of two experiments are exactly same except with or without spectral normalization. We see that our formulation is stable in both settings and generate similar images. The only difference is that the quantitative scores in terms of IS and FID is higher with the spectral normalization.

Table A12. Ablation study the influence of spectral normalization. ↑ means higher is better. ↓ means lower is better.

Backbone = SNGAN		CTRL-Binary SN = True	CTRL-Binary SN = False	CTRL-Multi SN = True	CTRL-Multi SN = False
CIFAR-10	IS ↑	8.1	6.6	7.1	5.8
	FID ↓	19.6	27.8	23.9	41.5

Appendix A.10. Ablation Study on Trade-Off between Network Width and Batch Size

Empirically, we observed that for our formulation, the larger the batch size, the better the results. To justify our use of batch size that is larger than those adopted in previous works such as [31], we conduct the following experiment which studies the training behavior of our proposed CTRL-Multi objective. Specifically, we train on the selected 10 classes of ImageNet with varying number of widest channels in our chosen architecture (specified in Appendix A.1) and batch size. We train both the encoder and decoder from scratch without fine-tuning. Other hyper-parameter settings detailed in Appendix A.7 are fixed. We present the results in Table A13. In the table, we denote training sessions that do not produce meaningful images as "failure" and those that do as "success". In the "failure" scenario, we noticed that the second term in the CTRL-Multi objective (12) would collapse to near 0 and could not be recovered, implying the decoder has essentially lost in the minimax game. In the "success" scenario, both the first terms of (12) stay close to each other and neither would collapse to near 0. The results present an interesting diagonal pattern that captures the relationship between batch size and network width. With a wider network and more channels, the network contains a greater capacity but would require a larger batch to stabilize training. This experiment justifies our use of a larger batch in our experiment in Appendix A.7 and also presents an interesting trade-off between network capacity and batch size for training.

Table A13. Ablation study on ImageNet about trade-off between batch size (BS) and network width (Channel #).

	Channel# = 1024	**Channel# = 512**	**Channel# = 256**
BS = 1800	success	success	success
BS = 1600	success	success	success
BS = 1024	failure	success	success
BS = 800	failure	failure	success
BS = 400	failure	failure	failure

Appendix A.11. Ablation Study on Feature Dimension

In this paper so far, for simplicity and uniformity, we have chosen the feature dimension $d = nz$ to be 128 for all experiments. In practice, however, the choice of feature dimension may affect the performance of the learned features: common practices suggest the larger the model, the better the performance could be. Hence, in this last section, we conduct experiments to show how the feature dimension affects the performance. It is not our intention to find the best feature dimension (nor the best network) with this work. We only want to show that there is room to improve the results presented in this paper.

The baseline experiment is conducted on CIFAR-10 with architectures from Table A2 and Table A1, training hyper-parameters are following the setting in Appendix A.1. Here, we change the feature dimension nz, batch size, and learning rate to 512, 8196, and 0.5×10^{-4} respectively. Figure A22 shows the comparison of (randomly selected, not cherry-picked) reconstructed images with the original ones. We observe a significant improvement in visual quality over the results with a lower feature dimension. The IS and FID scores reported in Table A14 also confirm the improvement.

Table A14. IS and FID scores of images reconstructed by LDR models learned with different feature dimensions. ↑ means higher is better. ↓ means lower is better.

		dim = 128		**dim = 512**	
		CTRL-Binary	**CTRL-Multi**	**CTRL-Binary**	**CTRL-Multi**
CIFAR-10	IS ↑	8.1	7.1	8.4	8.2
	FID ↓	19.6	23.6	18.7	20.5

Figure A22. Reconstruction results by LDR models learned with different feature dimensions.

References

1. Lee, J.M. *Introduction to Smooth Manifolds*; Springer: Berlin/Heidelberg, Germany, 2002.
2. Chan, K.H.R.; Yu, Y.; You, C.; Qi, H.; Wright, J.; Ma, Y. ReduNet: A White-box Deep Network from the Principle of Maximizing Rate Reduction. *arXiv* **2021**, arXiv:2105.10446.
3. Kramer, M.A. Nonlinear principal component analysis using autoassociative neural networks. *AICHE J.* **1991**, *37*, 233–243. [CrossRef]
4. Hinton, G.E.; Zemel, R.S. Autoencoders, Minimum Description Length and Helmholtz Free Energy. In Proceedings of the 6th International Conference on Neural Information Processing Systems (NIPS'93), Siem Reap, Cambodia, 13–16 December 1993; Morgan Kaufmann Publishers Inc.: San Francisco, CA, USA, 1993; pp. 3–10.
5. Kingma, D.P.; Welling, M. Auto-encoding variational Bayes. *arXiv* **2013**, arXiv:1312.6114.
6. Zhao, S.; Song, J.; Ermon, S. InfoVAE: Information maximizing variational autoencoders. *arXiv* **2017**, arXiv:1706.02262.
7. Wang, Z.; Bovik, A.; Sheikh, H.; Simoncelli, E. Image quality assessment: From error visibility to structural similarity. *IEEE Trans. Image Process.* **2004**, *13*, 600–612. [CrossRef] [PubMed]
8. Tu, Z. Learning Generative Models via Discriminative Approaches. In Proceedings of the Conference on Computer Vision and Pattern Recognition, Minneapolis, MN, USA, 18–23 June 2007; pp. 1–8. doi: 10.1109/CVPR.2007.383035. [CrossRef]
9. Goodfellow, I.; Pouget-Abadie, J.; Mirza, M.; Xu, B.; Warde-Farley, D.; Ozair, S.; Courville, A.; Bengio, Y. Generative adversarial nets. In *Advances in Neural Information Processing Systems*; MIT Press: Cambridge, MA, USA, 2014; pp. 2672–2680.
10. Arjovsky, M.; Chintala, S.; Bottou, L. Wasserstein generative adversarial networks. In Proceedings of the International Conference on Machine Learning, Sydney, Australia, 6–11 August 2017; pp. 214–223.
11. Salmona, A.; Delon, J.; Desolneux, A. Gromov-Wasserstein Distances between Gaussian Distributions. *arXiv* **2021**, arXiv:2104.07970.
12. Wright, J.; Ma, Y. *High-Dimensional Data Analysis with Low-Dimensional Models: Principles, Computation, and Applications*; Cambridge University Press: Cambridge, UK, 2021.
13. Yu, Y.; Chan, K.H.R.; You, C.; Song, C.; Ma, Y. Learning Diverse and Discriminative Representations via the Principle of Maximal Coding Rate Reduction. In *Advances in Neural Information Processing Systems*; MIT Press: Cambridge, MA, USA, 2020.
14. Bengio, Y.; Courville, A.; Vincent, P. Representation learning: A review and new perspectives. *IEEE Trans. Pattern Anal. Mach. Intell.* **2013**, *35*, 1798–1828. [CrossRef]
15. Srivastava, A.; Valkoz, L.; Russell, C.; Gutmann, M.U.; Sutton, C. VeeGAN: Reducing mode collapse in GANs using implicit variational learning. In *Advances in Neural Information Processing Systems*; MIT Press: Cambridge, MA, USA, 2017; pp. 3310–3320.
16. Mirza, M.; Osindero, S. Conditional generative adversarial nets. *arXiv* **2014**, arXiv:1411.1784.
17. Sohn, K.; Lee, H.; Yan, X. Learning structured output representation using deep conditional generative models. In *Advances in Neural Information Processing Systems*; MIT Press: Cambridge, MA, USA, 2015; pp. 3483–3491.
18. Mathieu, M.F.; Zhao, J.J.; Zhao, J.; Ramesh, A.; Sprechmann, P.; LeCun, Y. Disentangling factors of variation in deep representation using adversarial training. In *Advances in Neural Information Processing Systems*; MIT Press: Cambridge, MA, USA, 2016; pp. 5040–5048.
19. Van den Oord, A.; Kalchbrenner, N.; Espeholt, L.; Vinyals, O.; Graves, A.; Kavukcuoglu, K. Conditional image generation with PixelCNN decoders. In *Advances in Neural Information Processing Systems*; MIT Press: Cambridge, MA, USA, 2016; pp. 4790–4798.
20. Wang, T.C.; Liu, M.Y.; Zhu, J.Y.; Tao, A.; Kautz, J.; Catanzaro, B. High-resolution image synthesis and semantic manipulation with conditional GANs. In Proceedings of the IEEE Conference on Computer Vision and Pattern Recognition, Salt Lake City, UT, USA, 18–22 June 2018; pp. 8798–8807.
21. Chen, X.; Duan, Y.; Houthooft, R.; Schulman, J.; Sutskever, I.; Abbeel, P. InfoGAN: Interpretable Representation Learning by Information Maximizing Generative Adversarial Nets. In *Advances in Neural Information Processing Systems*; MIT Press: Cambridge, MA, USA, 2016; pp. 2172–2180.
22. Tang, S.; Zhou, X.; He, X.; Ma, Y. Disentangled Representation Learning for Controllable Image Synthesis: An Information-Theoretic Perspective. In Proceedings of the 25th International Conference on Pattern Recognition (ICPR), Milan, Italy, 10–15 January 2021; pp. 10042–10049. [CrossRef]
23. Li, K.; Malik, J. Implicit Maximum Likelihood Estimation. *arXiv* **2018**, arXiv:1809.09087.

24. Li, K.; Peng, S.; Zhang, T.; Malik, J. Multimodal Image Synthesis with Conditional Implicit Maximum Likelihood Estimation. *Int. J. Comput. Vis.* **2020**, *128*, 2607–2628. [CrossRef]
25. Odena, A.; Olah, C.; Shlens, J. Conditional image synthesis with auxiliary classifier GANs. In Proceedings of the International Conference on Machine Learning, Sydney, Australia, 6–11 August 2017; pp. 2642–2651.
26. Dumoulin, V.; Shlens, J.; Kudlur, M. A learned representation for artistic style. *arXiv* **2016**, arXiv:1610.07629.
27. Brock, A.; Donahue, J.; Simonyan, K. Large scale GAN training for high fidelity natural image synthesis. *arXiv* **2018**, arXiv:1809.11096.
28. Wu, Y.; Rosca, M.; Lillicrap, T. Deep compressed sensing. In Proceedings of the International Conference on Machine Learning, PMLR, Long Beach, CA, USA, 9–15 June 2019; pp. 6850–6860.
29. Wu, Y.; Donahue, J.; Balduzzi, D.; Simonyan, K.; Lillicrap, T. Logan: Latent optimisation for generative adversarial networks. *arXiv* **2019**, arXiv:1912.00953.
30. Papyan, V.; Han, X.; Donoho, D.L. Prevalence of Neural Collapse during the terminal phase of deep learning training. *arXiv* **2020**, arXiv:2008.08186.
31. Miyato, T.; Kataoka, T.; Koyama, M.; Yoshida, Y. Spectral Normalization for Generative Adversarial Networks. In Proceedings of the International Conference on Learning Representations, Vancouver, BC, Canada, 30 April–3 May 2018.
32. Lin, Z.; Khetan, A.; Fanti, G.; Oh, S. Pacgan: The power of two samples in generative adversarial networks. In *Advances in Neural Information Processing Systems*; MIT Press: Cambridge, MA, USA, 2018; pp. 1498–1507.
33. Feizi, S.; Farnia, F.; Ginart, T.; Tse, D. Understanding GANs in the LQG Setting: Formulation, Generalization and Stability. *IEEE J. Sel. Areas Inf. Theory* **2020**, *1*, 304–311. [CrossRef]
34. Larsen, A.B.L.; Sønderby, S.K.; Larochelle, H.; Winther, O. Autoencoding beyond pixels using a learned similarity metric. *arXiv* **2015**, arXiv:1512.09300.
35. Rosca, M.; Lakshminarayanan, B.; Warde-Farley, D.; Mohamed, S. Variational Approaches for Auto-Encoding Generative Adversarial Networks. *arXiv* **2017**, arXiv:1706.04987.
36. Bao, J.; Chen, D.; Wen, F.; Li, H.; Hua, G. CVAE-GAN: Fine-grained image generation through asymmetric training. In Proceedings of the IEEE International Conference on Computer Vision, Venice, Italy, 22–29 October 2017; pp. 2745–2754.
37. Huang, H.; He, R.; Sun, Z.; Tan, T.; Li, Z. IntroVAE: Introspective Variational Autoencoders for Photographic Image Synthesis. In *Advances in Neural Information Processing Systems*; MIT Press: Cambridge, MA, USA, 2018; Volume 31.
38. Donahue, J.; Krähenbühl, P.; Darrell, T. Adversarial feature learning. *arXiv* **2016**, arXiv:1605.09782.
39. Dumoulin, V.; Belghazi, I.; Poole, B.; Mastropietro, O.; Lamb, A.; Arjovsky, M.; Courville, A. Adversarially learned inference. *arXiv* **2016**, arXiv:1606.00704.
40. Ulyanov, D.; Vedaldi, A.; Lempitsky, V. It takes (only) two: Adversarial generator-encoder networks. In Proceedings of the Thirty-Second AAAI Conference on Artificial Intelligence, New Orleans, LA, USA, 2–7 February 2018.
41. Vahdat, A.; Kautz, J. Nvae: A deep hierarchical variational autoencoder. *arXiv* **2020**, arXiv:2007.03898.
42. Parmar, G.; Li, D.; Lee, K.; Tu, Z. Dual contradistinctive generative autoencoder. In Proceedings of the IEEE/CVF Conference on Computer Vision and Pattern Recognition, Nashville, TN, USA, 21–24 June 2021; pp. 823–832.
43. Bacharoglou, A. Approximation of probability distributions by convex mixtures of Gaussian measures. *Proc. Am. Math. Soc.* **2010**, *138*, 2619–2619. [CrossRef]
44. Hastie, T. *Principal Curves and Surfaces*; Technical Report; Stanford University: Stanford, CA, USA, 1984.
45. Hastie, T.; Stuetzle, W. Principal Curves. *J. Am. Stat. Assoc.* **1987**, *84*, 502–516. [CrossRef]
46. Vidal, R.; Ma, Y.; Sastry, S. *Generalized Principal Component Analysis*; Springer: Berlin/Heidelberg, Germany, 2016.
47. Ma, Y.; Derksen, H.; Hong, W.; Wright, J. Segmentation of multivariate mixed data via lossy data coding and compression. *PAMI* **2007**, *29*, 9. [CrossRef]
48. Jolliffe, I. *Principal Component Analysis*; Springer: New York, NY, USA, 1986.
49. Hong, D.; Sheng, Y.; Dobriban, E. Selecting the number of components in PCA via random signflips. *arXiv* **2020**, arXiv:2012.02985.
50. Farnia, F.; Ozdaglar, A.E. GANs May Have No Nash Equilibria. *arXiv* **2020**, arXiv:2002.09124.
51. Dai, Y.H.; Zhang, L. Optimality Conditions for Constrained Minimax Optimization. *arXiv* **2020**, arXiv:2004.09730.
52. Korpelevich, G.M. The extragradient method for finding saddle points and other problems. *Matecon* **1976**, *12*, 747–756.
53. Fiez, T.; Ratliff, L.J. Gradient Descent-Ascent Provably Converges to Strict Local Minmax Equilibria with a Finite Timescale Separation. *arXiv* **2020**, arXiv:2009.14820.
54. Bai, S.; Kolter, J.Z.; Koltun, V. Deep Equilibrium Models. *arXiv* **2019**, arXiv:1909.01377.
55. Ghaoui, L.E.; Gu, F.; Travacca, B.; Askari, A. Implicit Deep Learning. *arXiv* **2019**, arXiv:1908.06315.
56. Zhu, J.Y.; Park, T.; Isola, P.; Efros, A.A. Unpaired image-to-image translation using cycle-consistent adversarial networks. In Proceedings of the IEEE International Conference on Computer Vision, Venice, Italy, 22–29 October 2017; pp. 2223–2232.
57. LeCun, Y.; Bottou, L.; Bengio, Y.; Haffner, P. Gradient-based learning applied to document recognition. *Proc. IEEE* **1998**, *86*, 2278–2324. [CrossRef]
58. Krizhevsky, A.; Hinton, G. Learning Multiple Layers of Features from Tiny Images. 2009. Available online: https://www.cs.toronto.edu/~kriz/learning-features-2009-TR.pdf (accessed on 9 February 2022).
59. Coates, A.; Ng, A.; Lee, H. An analysis of single-layer networks in unsupervised feature learning. In Proceedings of the Fourteenth International Conference on Artificial Intelligence and Statistics, Ft. Lauderdale, FL, USA, 11–13 April 2011; pp. 215–223.

60. Liu, Z.; Luo, P.; Wang, X.; Tang, X. Deep Learning Face Attributes in the Wild. In Proceedings of the International Conference on Computer Vision (ICCV), Santiago, Chile, 7–13 December 2015.
61. Yu, F.; Seff, A.; Zhang, Y.; Song, S.; Funkhouser, T.; Xiao, J. Lsun: Construction of a large-scale image dataset using deep learning with humans in the loop. *arXiv* **2015**, arXiv:1506.03365.
62. Russakovsky, O.; Deng, J.; Su, H.; Krause, J.; Satheesh, S.; Ma, S.; Huang, Z.; Karpathy, A.; Khosla, A.; Bernstein, M.; et al. Imagenet large scale visual recognition challenge. *Int. J. Comput. Vis.* **2015**, *115*, 211–252. [CrossRef]
63. Radford, A.; Metz, L.; Chintala, S. Unsupervised representation learning with deep convolutional generative adversarial networks. *arXiv* **2015**, arXiv:1511.06434.
64. Larsen, A.B.L.; Sønderby, S.K.; Larochelle, H.; Winther, O. Autoencoding beyond pixels using a learned similarity metric. In Proceedings of the International Conference on Machine Learning, PMLR, New York City, NY, USA, 19–24 June 2016; pp. 1558–1566.
65. Salimans, T.; Goodfellow, I.; Zaremba, W.; Cheung, V.; Radford, A.; Chen, X. Improved techniques for training GANs. In *Advances in Neural Information Processing Systems*, MIT Press: Cambridge, MA, USA, 2016; pp. 2234–2242.
66. Heusel, M.; Ramsauer, H.; Unterthiner, T.; Nessler, B.; Hochreiter, S. GANs trained by a two time-scale update rule converge to a local nash equilibrium. In *Advances in Neural Information Processing Systems*; MIT Press: Cambridge, MA, USA, 2017; pp. 6626–6637.
67. Jonathan Bennett, J.; Carbery, A.; Christ, M.; Tao, T. The Brascamp-Lieb Inequalities: Finiteness, Structure and Extremals. *Geom. Funct. Anal.* **2007**, *17*, 1343–1415. [CrossRef]
68. Ditria, L.; Meyer, B.J.; Drummond, T. OpenGAN: Open Set Generative Adversarial Networks. *arXiv* **2020**, arXiv:2003.08074.
69. Fiez, T.; Ratliff, L.J. Local Convergence Analysis of Gradient Descent Ascent with Finite Timescale Separation. In Proceedings of the International Conference on Learning Representations, Virtual, 3–7 May 2021.
70. Härkönen, E.; Hertzmann, A.; Lehtinen, J.; Paris, S. Ganspace: Discovering interpretable GAN controls. *arXiv* **2020**, arXiv:2004.02546.
71. Wu, Z.; Baek, C.; You, C.; Ma, Y. Incremental Learning via Rate Reduction. In Proceedings of the IEEE Computer Society Conference on Computer Vision and Pattern Recognition, Nashville, TN, USA, 20–25 June 2021.
72. Tong, S.; Dai, X.; Wu, Z.; Li, M.; Yi, B.; Ma, Y. Incremental Learning of Structured Memory via Closed-Loop Transcription. *arXiv* **2022**, arXiv:2202.05411.
73. Lee, K.S.; Town, C. Mimicry: Towards the Reproducibility of GAN Research. *arXiv* **2020**, arXiv:2005.02494.
74. Kingma, D.P.; Ba, J. Adam: A method for stochastic optimization. *arXiv* **2014**, arXiv:1412.6980.

entropy

Article

An Information Theoretic Interpretation to Deep Neural Networks †

Xiangxiang Xu [1], Shao-Lun Huang [1,*], Lizhong Zheng [2] and Gregory W. Wornell [2]

[1] Data Science and Information Technology Research Center, Tsinghua–Berkeley Shenzhen Institute, Shenzhen 518055, China; xuxx@mit.edu
[2] Department of Electrical Engineering and Computer Science, Massachusetts Institute of Technology, Cambridge, MA 02139, USA; lizhong@mit.edu (L.Z.); gww@mit.edu (G.W.W.)
* Correspondence: shaolun.huang@sz.tsinghua.edu.cn

† This work was presented in part at the 2019 IEEE International Symposium on Information Theory (ISIT), Paris, France, 7–12 July 2019.

Abstract: With the unprecedented performance achieved by deep learning, it is commonly believed that deep neural networks (DNNs) attempt to extract informative features for learning tasks. To formalize this intuition, we apply the local information geometric analysis and establish an information-theoretic framework for feature selection, which demonstrates the information-theoretic optimality of DNN features. Moreover, we conduct a quantitative analysis to characterize the impact of network structure on the feature extraction process of DNNs. Our investigation naturally leads to a performance metric for evaluating the effectiveness of extracted features, called the H-score, which illustrates the connection between the practical training process of DNNs and the information-theoretic framework. Finally, we validate our theoretical results by experimental designs on synthesized data and the ImageNet dataset.

Keywords: deep neural network; information theory; local information geometry; feature extraction

Citation: Xu, X.; Huang, S.-L.; Zheng, L.; Wornell, G.W. An Information Theoretic Interpretation to Deep Neural Networks. *Entropy* **2022**, *24*, 135. https://doi.org/10.3390/e24010135

Academic Editor: Raúl Alcaraz

Received: 7 December 2021
Accepted: 12 January 2022
Published: 17 January 2022

1. Introduction

Due to the striking performance of deep learning in various application fields, deep neural networks (DNNs) have gained great attention in modern computer science. While it is a common understanding that the features extracted from the hidden layers of DNN are "informative" for learning tasks, the mathematical meaning of informative features in DNN is generally not clear. From the practical perspective, DNN models have obtained unprecedented performance in varying tasks, such as image recognition [1], language processing [2,3], and games [4,5]. However, the understanding of the feature extraction behind these models is relatively lacking, which poses challenges for their application in security-sensitive tasks, such as the autonomous vehicle.

To address this problem, there have been numerous research efforts, including both experimental and theoretical studies [6]. The experimental studies usually focus on some empirical properties of the feature extracted by DNNs, by visualizing the feature [7] or testing its performance on specific training settings [8] or learning tasks [9]. Though such empirical methods have provided some intuitive interpretations, the performance can highly depend on the data and network architecture used. For example, while the feature visualization works well on convolutional neural networks, its application to other networks is typically less effective [10].

In contrast, theoretical studies focus on the analytical properties of the extracted feature or the learning process in DNNs. Due to the complicated structure of DNNs, existing studies were often restricted to the networks of specific structures, e.g., network with infinite width [11] or two-layer network [12,13], to characterize the theoretical behaviors. However, the interpretation of the optimal feature remains unclear, which limits their further applications. To obtain better interpretability, tools and measures from information theory [14] have recently been applied to connect DNNs with general information processing problems [15]. For instance, the information bottleneck [16,17] employs the mutual information as the metric to quantify the informativeness of features in DNN, and other information metrics, such as the Kullback–Leibler (KL) divergence [18] and Weissenstein distance [19], are also used in different problems. However, there is still a disconnection between these information metrics and the performance objectives of the inference tasks that DNNs want to solve [20]. Therefore, it is, in general, difficult to match the DNN learning with the optimization of a particular information metric.

This paper aims to provide an information-theoretic interpretation to the feature extraction process in DNNs, to bridge the gap between the practical deep learning implementations and information-theoretic characterizations. To this end, we first propose an information-theoretic feature selection framework, which establishes an information metric to measure the performance of each given feature in inference tasks. In addition, we demonstrate that the optimal features extracted by DNNs coincide with the solutions of the information-theoretic feature selection problem, which share the same performance metric. Therefore, our results give an explicit interpretation of the learning goal of the back-propagation (BackProp) and stochastic gradient descent (SGD) operations in deep learning [21], which also lead to a performance metric for evaluating the effectiveness of the extracted features. Finally, we validate our theoretic characterizations using numerical experiments on both synthesized data and the ImageNet [22] dataset for image classification.

2. Preliminaries and Methods

2.1. Methodological Background

The main method used in our development is local information geometry [23,24], which characterizes the local geometric properties of the probability distribution space. The local information geometric method is closely related to the conventional Hirschfeld–Gebelein–Rényi (HGR) maximal correlation [25–27] problem, which has attracted increasing interest in the information theory community [28–33], and has also been applied in data analysis [34] and privacy studies [35].

Specifically, we use the local information geometric method to construct and investigate an information-theoretic feature selection problem in Section 3.1, which leads to an information metric of features and also demonstrates an SVD (singular value decomposition) structure of the feature selection process. Following the same analysis framework, we characterize the optimal feature extracted by DNNs in Section 3.2, and demonstrate that the same SVD structure is shared by DNNs. Based on the established connection, we then propose an effectiveness measure for DNNs, with details presented in Section 3.3.

2.2. Notations

Throughout this paper, we use X, $\mathcal{X}$, P_X, and x to represent a discrete random variable, the range, the probability distribution, and the value of X. In addition, for any function $s(X) \in \mathbb{R}^k$ of X, we use μ_s to denote the mean of $s(X)$, and "$\tilde{~}$" to denote the centered variable with mean subtracted, e.g., $\tilde{s}(X) \triangleq s(X) - \mu_s$. Moreover, we use $\|\cdot\|$ and $\|\cdot\|_F$ to denote the ℓ_2-norm and the Frobenius norm, respectively. All logarithms in our analyses are base e, i.e., natural.

2.3. Local Information Geometry

The following concepts from local information geometry would be useful in our development.

Definition 1 (ϵ-Neighborhood)**.** *Let $\mathcal{P}^{\mathcal{X}}$ denote the space of distributions on some finite alphabet $\mathcal{X}$, and let* $\mathrm{relint}(\mathcal{P}^{\mathcal{X}})$ *denote the subset of strictly positive distributions. For a given $\epsilon > 0$, the ϵ-neighborhood of a distribution $P_X \in \mathrm{relint}(\mathcal{P}^{\mathcal{X}})$ is defined by the χ^2-divergence as*

$$\mathcal{N}_{\epsilon}^{\mathcal{X}}(P_X) \triangleq \left\{ P \in \mathcal{P}^{\mathcal{X}} : \sum_{x \in \mathcal{X}} \frac{(P(x) - P_X(x))^2}{P_X(x)} \leq \epsilon^2 \right\}.$$

Definition 2 (ϵ-Dependence)**.** *The random variables X, Y are called ϵ-dependent if $P_{XY} \in \mathcal{N}_{\epsilon}^{\mathcal{X}\times\mathcal{Y}}(P_X P_Y)$.*

Definition 3 (ϵ-Attribute)**.** *A random variable U is called an ϵ-attribute of X if $P_{X|U}(\cdot|u) \in \mathcal{N}_{\epsilon}^{\mathcal{X}}(P_X)$, for all $u \in \mathcal{U}$.*

We will focus on the small ϵ regime, which we refer to as the *local analysis regime*. In addition, for any $P \in \mathcal{P}^{\mathcal{X}}$, we define the *information vector* ϕ and *feature function* $L(x)$ corresponding to P, with respect to a reference distribution $P_X \in \mathrm{relint}(\mathcal{P}^{\mathcal{X}})$, as

$$\phi(x) \triangleq \frac{P(x) - P_X(x)}{\sqrt{P_X(x)}}, \quad L(x) \triangleq \frac{\phi(x)}{\sqrt{P_X(x)}}. \tag{1}$$

This gives a three way correspondence $P \leftrightarrow \phi \leftrightarrow L$ for all distributions in $\mathcal{N}_{\epsilon}^{\mathcal{X}}(P_X)$, which will be useful in our derivations.

2.4. Modal Decomposition

Given a pair of discrete random variables X, Y with the joint distribution $P_{XY}(x, y)$, the $|\mathcal{Y}| \times |\mathcal{X}|$ matrix $\tilde{\mathbf{B}}$ is defined as

$$\tilde{\mathbf{B}}(y, x) \triangleq \frac{P_{XY}(x, y) - P_X(x)P_Y(y)}{\sqrt{P_X(x)P_Y(y)}}, \tag{2}$$

where $\tilde{\mathbf{B}}(y, x)$ is the (y, x)th entry of $\tilde{\mathbf{B}}$. The matrix $\tilde{\mathbf{B}}$ is referred to as the canonical dependence matrix (CDM) [24]. The SVD of $\tilde{\mathbf{B}}$ is referred to as the *modal decomposition* [24] of the joint distribution P_{XY}, which has the following property [18].

Lemma 1. *The SVD of $\tilde{\mathbf{B}}$ can be written as $\tilde{\mathbf{B}} = \sum_{i=1}^{K} \sigma_i \boldsymbol{\psi}_i^Y (\boldsymbol{\psi}_i^X)^{\mathrm{T}}$, where $K \triangleq \min\{|\mathcal{X}|, |\mathcal{Y}|\}$, and σ_i denotes the ith singular value with the ordering $1 \geq \sigma_1 \geq \cdots \geq \sigma_K = 0$, and $\boldsymbol{\psi}_i^Y$ and $\boldsymbol{\psi}_i^X$ are the corresponding left and right singular vectors with $\psi_K^X(x) = \sqrt{P_X(x)}$ and $\psi_K^Y(y) = \sqrt{P_Y(y)}$.*

This SVD decomposes the feature spaces of X, Y into maximally correlated features. To see that, consider the generalized canonical correlation analysis (CCA) problem:

$$\max_{\substack{\mathbb{E}[f_i(X)] = \mathbb{E}[g_i(Y)] = 0 \\ \mathbb{E}[f_i(X) f_j(X)] = \mathbb{E}[g_i(Y) g_j(Y)] = \delta_{ij}}} \sum_{i=1}^{k} \mathbb{E}[f_i(X) g_i(Y)], \tag{3}$$

where δ_{ij} denotes the Kronecker delta function. It can be shown that for any $1 \leq k \leq K - 1$, the optimal features are $f_i(x) = \psi_i^X(x)/\sqrt{P_X(x)}$, and $g_i(y) = \psi_i^Y(y)/\sqrt{P_Y(y)}$, for $i = 0, \ldots, K-1$, where $\psi_i^X(x)$ and $\psi_i^Y(y)$ are the xth and yth entries of $\boldsymbol{\psi}_i^X$ and $\boldsymbol{\psi}_i^Y$, respectively [18]. The special case $k = 1$ corresponds to the HGR maximal correlation [25–27], and the optimal features can be computed from the ACE (Alternating Conditional Expectation) algorithm [36].

2.5. Deep Neural Networks

The architecture of deep neural networks (under log-loss) can be depicted as Figure 1, where X is the input data, e.g., images, audios, or natural languages. Moreover, Y is the objective to predict, which can represent a discrete label in classification tasks, or represent target natural languages in machine translations [37]. Specifically, for given data X, the network produces a (trainable) feature mapping to generate k-dimensional feature $s(x) = (s_1, \ldots, s_k)^{\mathrm{T}}$. In practice, the feature mapping block (depicted as the gray block in Figure 1) is typically composed of hundreds and thousands of functional components (e.g., residual block [1]) with different types of layers, and may contain recurrent structure, e.g., LSTM (Long Short-Term Memory) [38]. In general, the internal structure of the feature mapping can have various different types of designs, depending on the learning tasks.

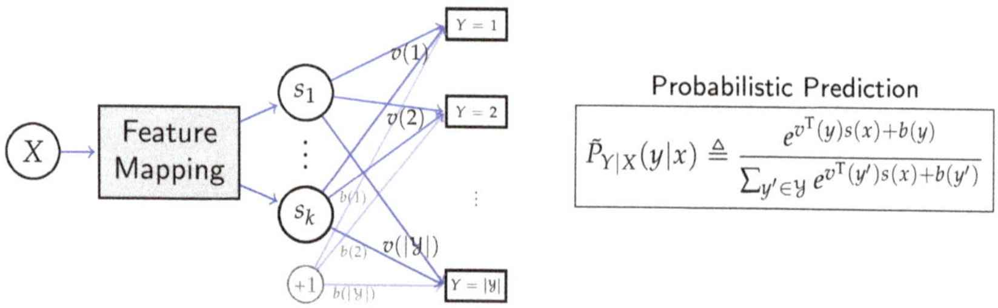

Figure 1. A deep neural network that uses data X to predict Y. All hidden layers together map the input data X to k-dimensional feature $s(x) = (s_1, \ldots, s_k)^{\mathrm{T}}$. Then, the probabilistic prediction $\tilde{P}_{Y|X}$ of Y is computed from $s(x), v(y)$, and $b(y)$, where v and bias b are the weights and bias in the last layer.

After obtaining the feature $s(X)$, the Y is then predicted by the probability distribution $\tilde{P}^{(s,v,b)}_{Y|X}$ of the form

$$\tilde{P}^{(s,v,b)}_{Y|X}(y|x) \triangleq \frac{e^{v^{\mathrm{T}}(y)s(x)+b(y)}}{\sum_{y' \in \mathcal{Y}} e^{v^{\mathrm{T}}(y')s(x)+b(y')}}, \tag{4}$$

which is obtained by applying the softmax function [39] on $v^{\mathrm{T}}(y)s(x) + b(y)$, where $v(\cdot)$ and $b(\cdot)$ are the weights and biases in the last layer, respectively (this is equivalent to the common practice that denotes weight and biases by the matrix $[v(1), \ldots, v(|\mathcal{Y}|)]^{\mathrm{T}}$ and the vector $[b(1), \ldots, b(|\mathcal{Y}|)]^{\mathrm{T}}$, respectively. However, as we will show later, expressing weights v and biases b as mappings of y can better illustrate their roles in feature selection). We will use $\tilde{P}_{Y|X}$ to refer to $\tilde{P}^{(s,v,b)}_{Y|X}$ when there is no ambiguity.

Then, for a given training set of labeled samples (x_i, y_i), for $i = 1, \ldots, N$, all the parameters in the network, including v, b, as well as those in the feature mapping block, are chosen to maximize the log-likelihood function (or, equivalently, minimize the log-loss)

$$\frac{1}{N}\sum_{i=1}^{N} \log \tilde{P}_{Y|X}(y_i|x_i). \tag{5}$$

The procedure of choosing such parameters is called the training of network, which can be performed by stochastic gradient descent (SGD) or its variants [21]. With a trained network, the label $\hat{y}$ for a new data sample x can be predicted by the maximum a posteriori (MAP) estimation, i.e., $\hat{y} = \arg\max_{y \in \mathcal{Y}} \tilde{P}_{Y|X}(y|x)$. Specifically, when we make predictions for samples in a test dataset, the proportion of samples with correct prediction (i.e., $\hat{y} = y$) over all samples is called the test accuracy.

3. Results

3.1. Information-Theoretic Feature Selection

Suppose that, given random variables X, Y with joint distribution P_{XY}, we want to infer about an attribute V of Y from observed i.i.d. samples $x_1, \ldots, x_n$ of X. When the statistical model $P_{X|V}$ is known, the optimal decision rule is the log-likelihood ratio test, where the log-likelihood function can be viewed as the optimal feature for inference. However, in many practical situations [18], it is hard to identify the model of the targeted attribute, and it is necessary to select low-dimensional informative features of X for inference tasks before knowing the model. An information-theoretic formulation of such feature selection problem is the universal feature selection problem [24], which we formalize as follows.

To begin, for an attribute V, we refer to $\mathcal{C}_\mathcal{Y} = \big\{ \mathcal{V}, \{P_V(v), v \in \mathcal{V}\}, \{\boldsymbol{\phi}_v^{Y|V}, v \in \mathcal{V}\}\big\}$, as the *configuration* of V, where $\phi_v^{Y|V} \leftrightarrow P_{Y|V}(\cdot|v)$ is the information vector specifying the corresponding conditional distribution $P_{Y|V}(\cdot|v)$. The configuration of V models the statistical correlation between V and Y. In the sequel, we focus on the local analysis regime, for which we assume that all the attributes V of our interests to detect are ϵ-attributes of Y. As a result, the corresponding configuration satisfies $\|\boldsymbol{\phi}_v^{Y|V}\| \leq \epsilon$, for all $v \in \mathcal{V}$. We refer to such configurations as *ϵ-configurations*. The configuration of V is unknown in advance but assumed to be generated from a *rotational invariant ensemble (RIE)*.

Definition 4 (RIE). *Two configurations $\mathcal{C}_\mathcal{Y}$ and $\tilde{\mathcal{C}}_\mathcal{Y}$ defined as*

$$\mathcal{C}_\mathcal{Y} \triangleq \big\{ \mathcal{V}, \{P_V(v), v \in \mathcal{V}\}, \{\boldsymbol{\phi}_v^{Y|V}, v \in \mathcal{V}\}\big\},$$

$$\tilde{\mathcal{C}}_\mathcal{Y} \triangleq \big\{ \mathcal{V}, \{P_V(v), v \in \mathcal{V}\}, \{\tilde{\boldsymbol{\phi}}_v^{Y|V}, v \in \mathcal{V}\}\big\}$$

are called rotationally equivalent, if there exists a unitary matrix $\mathbf{Q}$ such that $\tilde{\boldsymbol{\phi}}_v^{Y|V} = \mathbf{Q}\,\boldsymbol{\phi}_v^{Y|V}$, for all $v \in \mathcal{V}$. Moreover, a probability measure defined on a set of configurations is called an RIE, if all rotationally equivalent configurations have the same measure.

The RIE can be interpreted as assigning a uniform measure to the attributes with the same level of distinguishability. To infer about the attribute V, we construct a k-dimensional feature vector $h^k = (h_1, \ldots, h_k)$, for some $1 \leq k \leq K-1$, of the form

$$h_i = \frac{1}{n}\sum_{l=1}^{n} f_i(x_l), \quad i = 1, \ldots, k, \tag{6}$$

for some choices of feature functions f_i. Our goal is to determine the f_i such that the optimal decision rule based on h^k achieves the smallest possible error probability, where the performance is averaged over the possible $\mathcal{C}_\mathcal{Y}$ generated from an RIE. In turn, we denote $\xi_i^X \leftrightarrow f_i$ as the corresponding information vector, and define the matrix $\Xi^X \triangleq [\boldsymbol{\xi}_1^X \ \cdots \ \boldsymbol{\xi}_k^X]$.

Theorem 1 (Universal Feature Selection). *For $v, v' \in \mathcal{V}$, let $E_{h^k}(v, v')$ be the error exponent associated with the pairwise error probability distinguishing v and v' based on h^k, then the expected error exponent over a given RIE defined on the set of ϵ-configurations is given by*

$$\mathbb{E}\big[E_{h^k}(v, v')\big] = \frac{C_0}{2} \cdot \left\| \tilde{\mathbf{B}}\Xi^X \big((\Xi^X)^\mathrm{T}\Xi^X\big)^{-\frac{1}{2}} \right\|_\mathrm{F}^2 + o(\epsilon^2), \tag{7}$$

where $C_0 \triangleq \frac{1}{4|\mathcal{Y}|} \cdot \mathbb{E}\Big[\|\boldsymbol{\phi}_v^{Y|V} - \boldsymbol{\phi}_{v'}^{Y|V}\|^2\Big]$ is independent of the choices of f_i's, and the expectations $\mathbb{E}[\cdot]$ are taken over this RIE.

Proof. See Appendix A. □

As a result of (7), designing the ξ_i^X as the singular vectors ψ_i^X of $\tilde{\mathbf{B}}$, for $i = 1, \ldots, k$, optimizes (7) for all RIEs, pairs of (v, v'), and ϵ-configurations. Thus, the feature functions corresponding to ψ_i^X are *universally optimal* for inferring the unknown attribute V. Moreover, (7) naturally leads to an information metric $\left\| \tilde{\mathbf{B}} \Xi^X \left((\Xi^X)^{\mathrm{T}} \Xi^X \right)^{-\frac{1}{2}} \right\|_{\mathrm{F}}^2$ for any feature Ξ^X of X, measured by projecting the normalized Ξ^X through a linear projection $\tilde{\mathbf{B}}$. This information metric quantifies how informative a feature of X is when solving inference problems with respect to Y and is optimized when designing features by singular vectors of $\tilde{\mathbf{B}}$. Thus, we can interpret the universal feature selection as solving the most informative features for data inferences via the SVD of $\tilde{\mathbf{B}}$, which also coincides with the maximally correlated features in (3). Later, we will show that the feature selection in DNNs shares the same information metric as universal feature selection in the local analysis regime.

3.2. Feature Extraction in Deep Neural Networks

3.2.1. Network with Ideal Expressive Power

For convenience of analysis, we first consider the ideal case where the neural network can express any feature mapping $s(\cdot)$ as desired. While this assumption can be rather strong, the existence of such ideal networks is guaranteed by the universal approximation theorem [40]. In addition, one goal of practical network designs is to approximate the ideal networks and obtain sufficient expressive power. For such networks, we will show that when X, Y are ϵ-dependent, the extracted feature $s(x)$ and weights $v(y)$ coincide with the solutions of the universal feature selection.

To begin, we use P_{XY} to denote the joint empirical distribution of the labeled samples $(x_i, y_i), i = 1, \ldots, N$, and P_X, P_Y to denote the corresponding marginal distributions. Then, the objective function of (5) is the empirical average of the log-likelihood function

$$\frac{1}{N} \sum_{i=1}^{N} \log \tilde{P}_{Y|X}(y_i|x_i) = \mathbb{E}_{P_{XY}} \left[\log \tilde{P}_{Y|X}(Y|X) \right].$$

Therefore, maximizing this empirical average is equivalent as minimizing the KL divergence:

$$(s^*, v^*, b^*) = \underset{(s,v,b)}{\arg\min}\ D(P_{XY} \| P_X \, \tilde{P}_{Y|X}^{(s,v,b)}). \tag{8}$$

This can be interpreted as finding the best fitting to empirical joint distribution P_{XY} by distributions of the form $P_X \, \tilde{P}_{Y|X}^{(s,v,b)}$. In our development, it is more convenient to denote the bias by $d(y) = b(y) - \log P_Y(y)$, for $y \in \mathcal{Y}$. Then, the following lemma illustrates the explicit constraint on the problem (8) in the local analysis regime.

Lemma 2. *If X, Y are ϵ-dependent, then the optimal v, d for (8) satisfy*

$$|\tilde{v}^{\mathrm{T}}(y) s(x) + \tilde{d}(y)| = O(\epsilon), \quad \textit{for all } x \in \mathcal{X},\ y \in \mathcal{Y}. \tag{9}$$

Proof. See Appendix B. □

In turn, we take (9) as the constraint for solving the problem (8) in the local analysis regime. Moreover, we define the information vectors for zero-mean vectors $\tilde{s}, \tilde{v}$ as $\xi^X(x) = \sqrt{P_X(x)}\, \tilde{s}(x)$, $\xi^Y(y) = \sqrt{P_Y(y)}\, \tilde{v}(y)$, and define matrices

$$\Xi^Y \triangleq \left[\xi^Y(1) \quad \cdots \quad \xi^Y(|\mathcal{Y}|) \right]^{\mathrm{T}}, \quad \Xi^X \triangleq \left[\xi^X(1) \quad \cdots \quad \xi^X(|\mathcal{X}|) \right]^{\mathrm{T}}.$$

Lemma 3. *The KL divergence (8) in the local analysis regime (9) can be expressed as*

$$D(P_{XY} \| P_X \, \tilde{P}_{Y|X}^{(s,v,b)}) = \frac{1}{2} \| \tilde{\mathbf{B}} - \Xi^Y (\Xi^X)^{\mathrm{T}} \|_{\mathrm{F}}^2 + \frac{1}{2} \eta^{(v,b)}(s) + o(\epsilon^2), \tag{10}$$

where $\eta^{(v,b)}(s) \triangleq \mathbb{E}_{P_Y}\left[(\mu_s^{\mathrm{T}}\tilde{v}(Y) + \tilde{d}(Y))^2\right]$.

Proof. See Appendix C. □

Lemma 3 reveals key insights for feature selection in neural networks. To see this, we consider the following two learning problems: learning the optimal weight v for given s and learning the optimal feature s for given v.

For the case that s is fixed, we can optimize (10) with Ξ^X fixed and obtain the following optimal weights:

Theorem 2. *For fixed Ξ^X and μ_s, the optimal Ξ^{Y*} to minimize (10) is given by*

$$\Xi^{Y*} = \tilde{\mathbf{B}}\,\Xi^X\left((\Xi^X)^{\mathrm{T}}\Xi^X\right)^{-1}, \tag{11}$$

and the optimal weights $\tilde{v}^*$ and bias $\tilde{d}^*$ are

$$\tilde{v}^*(y) = \mathbb{E}_{P_{X|Y}}\left[\mathbf{\Lambda}_{\tilde{s}(X)}^{-1}\,\tilde{s}(X)\,\middle|\,Y = y\right], \quad \tilde{d}^*(y) = -\mu_s^{\mathrm{T}}\,\tilde{v}(Y). \tag{12}$$

where $\mathbf{\Lambda}_{\tilde{s}(X)}$ denotes the covariance matrix of $\tilde{s}(X)$.

Proof. See Appendix D. □

Specifically, when $s(x) = x$, Theorem 2 gives the optimal weights for softmax regression. Note that Equation (11) can be viewed as a projection of the input feature $\tilde{s}(x)$, to a feature $v(y)$ computable from the value of y, which is the most correlated feature to $\tilde{s}(x)$. The solution is given by the operation that left multiplies $\tilde{\mathbf{B}}$ matrix, which we refer to as *forward feature projection*.

Remark 1. *While we assume the continuous input $s(x)$ is a function of a discrete variable X, we only need the labeled samples between s and Y to compute the weights and bias from the conditional expectation (12), and the correlation between X and s is irrelevant. Thus, our analysis for weights and bias can be applied to continuous input networks by just ignoring X and taking s as the real input to network.*

We then consider the "backward feature projection" problem, which attempts to find informative feature $s^*(X)$ to minimize the loss (10) with given weights and bias. In particular, we can show that the solution of this backward feature projection is precisely symmetric to the forward one.

Theorem 3. *For fixed Ξ^Y and $\tilde{d}$, the optimal Ξ^{X*} to minimize (10) is given by*

$$\Xi^{X*} = \tilde{\mathbf{B}}^{\mathrm{T}}\,\Xi^Y\left((\Xi^Y)^{\mathrm{T}}\Xi^Y\right)^{-1}, \tag{13}$$

and the optimal feature function s^*, which are decomposed to $\tilde{s}^*$ and μ_s^*, is given by

$$\begin{aligned} \tilde{s}^*(x) &= \mathbb{E}_{P_{Y|X}}\left[\mathbf{\Lambda}_{\tilde{v}(Y)}^{-1}\,\tilde{v}(Y)\middle|X = x\right], \\ \mu_s^* &= -\mathbf{\Lambda}_{\tilde{v}(Y)}^{-1}\,\mathbb{E}_{P_Y}\left[\tilde{v}(Y)\,\tilde{d}(Y)\right], \end{aligned} \tag{14}$$

where $\mathbf{\Lambda}_{\tilde{v}(Y)}$ denotes the covariance matrix of $\tilde{v}(Y)$.

Proof. See Appendix D. □

Finally, when both s and (v, b) (and hence Ξ^X, Ξ^Y, d) can be designed, the optimal (Ξ^Y, Ξ^X) corresponds to the low rank factorization of $\tilde{\mathbf{B}}$, and the solutions coincide with the universal feature selection.

Theorem 4. *The optimal solutions for weights and bias to minimize (10) are given by* $\tilde{d}(y) = -\mu_s^{\mathrm{T}}\tilde{v}(y)$, *and* $(\Xi^Y, \Xi^X)^*$ *chosen as the largest k left and right singular vectors of* $\tilde{\mathbf{B}}$.

Proof. See Appendix E. □

Therefore, we conclude that the learning of neural networks, when both s and (v, b) are designable, is to extract the most correlated aspects of the input data X and the label Y that are informative features for data inferences from universal feature selection.

In the practical learning process of DNN, the BackProp updates the weights of the softmax layer and those on the previous layer(s) in an iterative manner. As we have illustrated in Lemma 3, such iterative updates will converge to the same solution as the alternating between the forward feature projection (11) and the backward feature projection (13), which is indeed the power method to solve the SVD for $\tilde{\mathbf{B}}$ [41], also known as the Alternating Conditional Expectation (ACE) algorithm [36].

Remark 2. *From Theorem 4, for a neural network with sufficient expressive power, the trained feature depends only on the distribution of input data rather than the training process. It is worth mentioning that this result does not contradict the practice that trained weights in hidden layers can be different during each training run. In fact, due to the over-parameterized nature of practical network designs, there exist multiple choices of weights in hidden layers to express the same optimal feature* $s(x)$.

3.2.2. Network with Restricted Expressive Power

The analysis of the previous section has considered neural networks with ideal expressive power, where the feature $s(X)$ can be selected as any desired function. In general, however, the form of feature functions that can be generalized is often limited by the network structure. In the following, we consider networks with restricted expressive power to characterize the impacts of network structure on the extracted feature.

For illustration, we consider the neural network with a hidden layer of k nodes, and a zero-mean continuous input $t = [t_1 \ \cdots \ t_m]^{\mathrm{T}} \in \mathbb{R}^m$ to this hidden layer, where t is assumed to be a function $t(x)$ of some discrete variable X. Our goal is to analyze the weights and bias in this layer with labeled samples $(t(x_i), y_i)$. Assume the activation function of the hidden layer is a generally smooth function $\sigma(\cdot)$, then the output $s_z(X)$ of the z-th hidden node is

$$s_z(x) = \sigma\Big(w^{\mathrm{T}}(z)t(x) + c(z)\Big), \quad \text{for } z = 1, \ldots, k, \ x \in \mathcal{X}, \tag{15}$$

where $w(z) \in \mathbb{R}^m$ and $c(z) \in \mathbb{R}$ are the weights and bias from input layer to hidden layer as shown in Figure 2. We denote $s = [s_1 \ \cdots \ s_k]^{\mathrm{T}}$ as the input vector to the output classification layer.

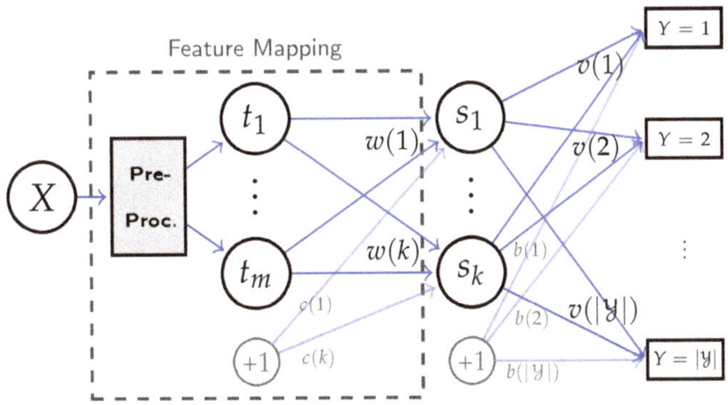

Figure 2. A multi-layer neural network, where the expressive power of the feature mapping $s(\cdot)$ is restricted by the hidden representation t. All hidden layers previous to t are fixed, represented by the "pre-processing" module.

To interpret the feature selection in hidden layers, we fix $(v(y), b(y))$ at the output layer and consider the problem of designing $(w(z), c(z))$ to minimize the loss function (8) at the output layer. Ideally, we should have picked $w(z)$ and $c(z)$ to generate $s(x)$ to match $s^*(x)$ from (14), which minimizes the loss. However, here we have the constraint that $s(x)$ must take the form of (15) and, intuitively, the network should select $w(z), c(z)$ so that $s(x)$ is close to $s^*(x)$. Our goal is to quantify the notion of such closeness.

To develop insights on feature selection in hidden layers, we again focus on the local analysis regime, where the weights and bias are assumed to satisfy the local constraint

$$\left|\tilde{v}^{\mathrm{T}}(y)s(x) + \tilde{d}(y)\right| = O(\epsilon), \quad \left|w^{\mathrm{T}}(z)\tilde{t}(x)\right| = O(\epsilon), \ \forall x, y, z. \tag{16}$$

Then, since t is zero-mean, we can express (15) as

$$s_z(x) = \sigma\left(w^{\mathrm{T}}(z)t(x) + c(z)\right) = w^{\mathrm{T}}(z)\tilde{t}(x) \cdot \sigma'(c(z)) + \sigma(c(z)) + o(\epsilon), \tag{17}$$

Moreover, we define a matrix $\tilde{\mathbf{B}}_1$ with the (z, x)th entry $\tilde{\mathbf{B}}_1(z, x) = \frac{\sqrt{P_X(x)}}{\sigma'(c(z))}\tilde{s}^*_z(x)$, which can be interpreted as a generalized CDM for the hidden layer. Furthermore, we denote $\xi_1^X(x) = \sqrt{P_X(x)}\,\tilde{t}(x)$ as the information vector of $\tilde{t}(x)$ with the matrix Ξ_1^X defined as $\Xi_1^X \triangleq \left[\xi_1^X(1) \quad \cdots \quad \xi_1^X(|\mathcal{X}|)\right]^{\mathrm{T}}$, and we also define

$$\mathbf{W} \triangleq \left[w(1) \quad \cdots \quad w(k)\right]^{\mathrm{T}}, \tag{18}$$

$$\mathbf{J} \triangleq \mathrm{diag}\{\sigma'(c(1)), \sigma'(c(2)), \cdots, \sigma'(c(k))\}. \tag{19}$$

The following theorem characterizes the loss (8).

Theorem 5. *Given the weights and bias (v, b) at the output layer, and for any input feature s, we denote $\mathcal{L}(s)$ as the loss (8) evaluated with respect to (v, b) and s. Then, with the constraints (16)*

$$\mathcal{L}(s) - \mathcal{L}(s^*) = \frac{1}{2}\left\|\Theta\tilde{\mathbf{B}}_1 - \Theta\mathbf{W}(\Xi_1^X)^{\mathrm{T}}\right\|_{\mathrm{F}}^2 + \frac{1}{2}\kappa^{(v,b)}(s, s^*) + o(\epsilon^2), \tag{20}$$

where $\Theta \triangleq ((\Xi^Y)^{\mathrm{T}}\Xi^Y)^{1/2}\mathbf{J}$, and the term $\kappa^{(v,b)}(s, s^*) = (\mu_s - \mu_{s^*})^{\mathrm{T}}\mathbf{\Lambda}_{\tilde{v}(Y)}(\mu_s - \mu_{s^*})$.

Proof. See Appendix F. □

Equation (20) quantifies the closeness between s and s^* in terms of the loss (8). Then, our goal is to minimize (20), which can be separated to two optimization problems:

$$\mathbf{W}^* = \arg\min_{\mathbf{W}} \left\| \Theta\tilde{\mathbf{B}}_1 - \Theta\mathbf{W}\left(\Xi_1^X\right)^{\mathrm{T}} \right\|_{\mathrm{F}}^2, \tag{21}$$

$$\mu_s^* = \arg\min_{\mu_s} \kappa^{(v,b)}(s, s^*). \tag{22}$$

Note that the optimization problem (21) is similar to the one that appeared in Lemma 3, and the optimal solution is given by $\mathbf{W}^* = \tilde{\mathbf{B}}_1\Xi_1^X\left(\left(\Xi_1^X\right)^{\mathrm{T}}\Xi_1^X\right)^{-1}$. Therefore, solving the optimal weights in the hidden layer can be interpreted as projecting $\tilde{s}^*(x)$ to the subspace of feature functions spanned by $t(x)$ to find the closest expressible function. In addition, the problem (22) is to choose μ_s (and hence the bias $c(z)$) to minimize the quadratic term similar to $\eta^{(v,b)}(s)$ in (10). Similar to the analyses of parameters in the last layer, we can obtain analytical solutions for hidden layer parameters, e.g., μ_s^* and w^*, with detailed discussions provided in Appendix G.

Overall, we observe the correspondence between (11), (14), and (21), (22), and interpret both operations as feature projections. Our argument can be generalized to any intermediate layer in a multi-layer network, with all the previous layers viewed as the fixed pre-processing that specifies $t(x)$, and all the layers after determining s^*. Then, the iterative procedure in back-propagation can be viewed as alternating projection finding the fixed-point solution over the entire network. This final fixed-point solution, even under the local assumption, might not be the SVD solution as in Theorem 4. This is because the limited expressive power of the network often makes it impossible to generate the desired feature function. In such cases, the concept of feature projection can be used to quantify this gap, and thus to measure the quality of the selected features.

3.3. Scoring Neural Networks

Given a learning problem, it is useful to tell whether or not some extracted features are informative [42]. Our previous development naturally gives rise to a performance metric.

Definition 5. *Given a feature $s(x) \in \mathbb{R}^k$ and weight $v(y) \in \mathbb{R}^k$ with the corresponding information matrices Ξ^X and Ξ^Y, the H-score $H(s,v)$ is defined as*

$$H(s,v) \triangleq \frac{1}{2}\|\tilde{\mathbf{B}}\|_{\mathrm{F}}^2 - \frac{1}{2}\|\tilde{\mathbf{B}} - \Xi^Y(\Xi^X)^{\mathrm{T}}\|_{\mathrm{F}}^2 = \mathbb{E}_{P_{XY}}\left[\tilde{s}^{\mathrm{T}}(X)\,\tilde{v}(Y)\right] - \frac{1}{2}\operatorname{tr}\left(\mathbf{\Lambda}_{\tilde{s}(X)}\mathbf{\Lambda}_{\tilde{v}(Y)}\right). \tag{23}$$

In addition, for given $s(x)$, we define the single-sided H-score $H(s)$ as

$$\begin{aligned} H(s) &\triangleq \max_v H(s,v) && (24)\\ &= \frac{1}{2}\|\tilde{\mathbf{B}}\|_{\mathrm{F}}^2 - \frac{1}{2}\|\tilde{\mathbf{B}} - \tilde{\mathbf{B}}\,\Xi^X\left((\Xi^X)^{\mathrm{T}}\Xi^X\right)^{-1}(\Xi^X)^{\mathrm{T}}\|_{\mathrm{F}}^2 && (25)\\ &= \frac{1}{2}\|\tilde{\mathbf{B}}\Xi^X\left((\Xi^X)^{\mathrm{T}}\Xi^X\right)^{-\frac{1}{2}}\|_{\mathrm{F}}^2 = \frac{1}{2}\mathbb{E}_{P_Y}\left[\left\|\mathbb{E}_{P_{X|Y}}\left[\mathbf{\Lambda}_{\tilde{s}(X)}^{-1/2}\,\tilde{s}(X)\,\middle|\,Y\right]\right\|^2\right]. && (26)\end{aligned}$$

H-score can be used to measure the quality of features generated at any intermediate layer of the network. It is related to (20) when choosing the optimal bias and Θ as the identity matrix. This can be understood as taking the output of this layer $s(x)$ and directly feeding it to a softmax output layer with $v(y)$ used as the weights, and $H(s,v)$ measures the resulting performance. Note that $v(y)$ here can be an arbitrary function of Y, not necessarily the weights on the next layer computed by the network. When the optimal $v^*(y)$ as defined in (12) is used, the resulting performance becomes the one-sided H-score $H(s)$, which measures the quality of $s(x)$. In addition, by comparing (26) with (7), the performance measure $H(s)$ also coincides with the information metric (7), up to a scale factor.

Specifically, for a given dataset and a feature extractor that generate $s(\cdot)$, the H-score $H(s)$ can be efficiently computed from the second equation of (26). In addition, when we use H-score to compare the performance of different feature extractors (models), the model complexity has to be taken into account to reduce overfitting. To this end, we adopt Akaike information criterion (AIC) and define *AIC-corrected H-score*

$$H_{\mathrm{AIC}}(s) \triangleq H(s) - \frac{n_{\mathrm{p}}}{n_{\mathrm{s}}} \tag{27}$$

for comparing different models, where n_{p} and n_{s} represent the number of parameters in the model and the training sample size, respectively.

In current practice, the cross-entropy $\mathbb{E}_{P_{XY}}\left[\log \tilde{P}^{(v,b)}_{Y|X}\right]$ is often used as the performance metric. One can, in principle, also use log-loss to measure the effectiveness of the selected feature at the output of an intermediate layer [42]. However, one problem of this metric is that, for a given problem, it is not clear what value of log-loss one should expect, as the log-loss is generally unbounded. In contrast, the H-score can be directly computed from the data samples and has a clear upper bound. Indeed, it follows from Lemma 1 that, for k-dimensional feature s and weights v, we have the sequence of inequalities

$$H(s,v) \leq H(s) \leq \frac{1}{2}\sum_{i=1}^{k}\sigma_i^2 \leq \frac{k}{2}, \tag{28}$$

where σ_i indicates the ith singular value of $\tilde{\mathbf{B}}$.

In particular, the first "$\leq$" follows from the definition (24), and the gap between $H(s,v)$ and $H(v)$ measures the optimality of the weights v; the second "$\leq$" follows from the first equality of (26), and the gap between two sides characterizes the difference between the chosen feature and the optimal solution, which is a useful measure of how restrictive (lack of expressive power) the network structure is; the last "$\leq$" follows from the fact that $\sigma_i \leq 1$ (cf. Lemma 1), which measures the dependency between data variable and label for the given dataset. In Section 3.4.3, we validate this metric on real data.

3.4. Experiments

This section presents experiments for validating our theoretical characterizations, with corresponding code available at https://github.com/XiangxiangXu/dnn (accessed on 7 December 2021). Specifically, all DNN models used in Section 3.4.3 are available at https://keras.io/applications/ (accessed on 7 December 2021).

3.4.1. Experimental Validation of Theorem 4

We first validate Theorem 4, the optimal feature extracted by network with ideal expressive power. Here, we consider the discrete data with alphabet sizes, $|\mathcal{X}| = 8$ and $|\mathcal{Y}| = 6$, and construct the network as shown in Figure 3. Specifically, the network input is the one-hot encoding of X, i.e., $[\mathbb{1}_X(1), \ldots, \mathbb{1}_X(|\mathcal{X}|)]^{\mathrm{T}}$, where $\mathbb{1}_X(x)$ takes one if and only if $X = x$, and takes zero otherwise. Then, the feature $s(X)$ is generated by a linear layer, with sigmoid function used as the activation function. For ease of comparison and presentation, we set feature dimension to $k = 1$, since otherwise the optimal feature (cf. Theorem 4) lies in a subspace and is non-unique. It can be verified that this network has ideal expressive power, i.e., with proper weights in the first layer, $s(X)$ can express any desired function up to scaling and shifting.

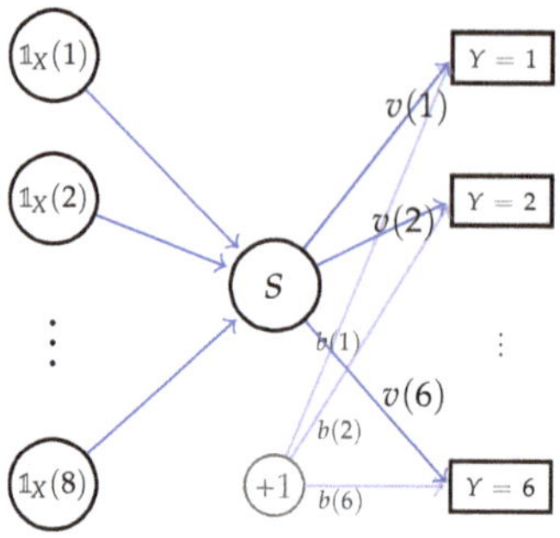

Figure 3. A simple neural network with ideal expressive power, which can generate any $k = 1$ dimensional feature s of X by tuning the weights in the first layer.

To compare the result trained by the neural network and that in Theorem 4, we first randomly generate a distribution P_{XY}, and then draw independently $n = 100{,}000$ pairs of (X, Y) samples. We then train the network using batch gradient descent, where we have applied Nesterov momentum [43] with the momentum hyperparameter being 0.9. In addition, we set the learning rate to 4 with a decay factor of 0.01 and clip gradients with norm exceeding 0.5. After training, the learned values of $s(x), v(y)$ and $b(y)$ are shown in Figure 4 and compared with theoretical results. From the figure, we can observe that the training results match our theoretical analyses.

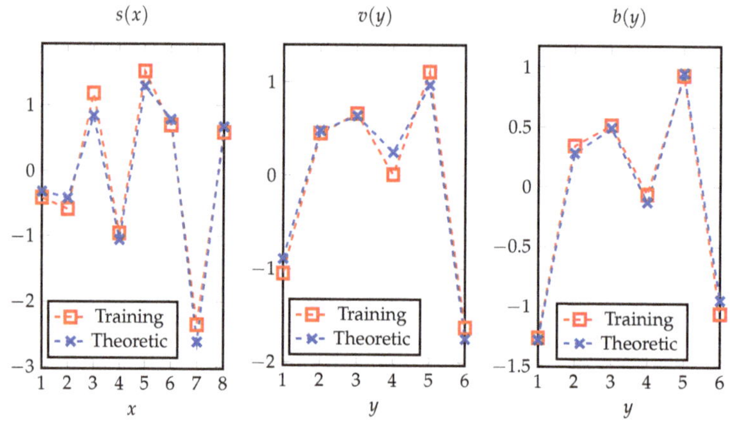

Figure 4. The trained feature s, weights v, and bias b of the network in Figure 3, which are compared with the corresponding theoretical results to show their coincidences.

3.4.2. Experimental Validation of Theorem 5

In addition, we validate Theorem 5 by the neural network depicted in Figure 5, with the same settings of X, Y. Specifically, the number of neurons in hidden layers are set to $m = 4$ and $k = 3$, where $t(X)$ is randomly generated from X, and we have chosen sigmoid function as the activation function $\sigma(\cdot)$ to generate $s(x)$. We then fix the weights and bias at the output layer and train the weights $w(1), w(2), w(3)$ and bias c in the hidden layer to optimize the log-loss. Specifically, we use the batch gradient descent with the Nesterov momentum hyperparameter being 0.9. In addition, we set the learning rate to 4 with a decay factor of 10^{-6} and clip gradients with norm exceeding 0.1. After training, Figure 6 shows the matching between the learned results and the corresponding theoretical values.

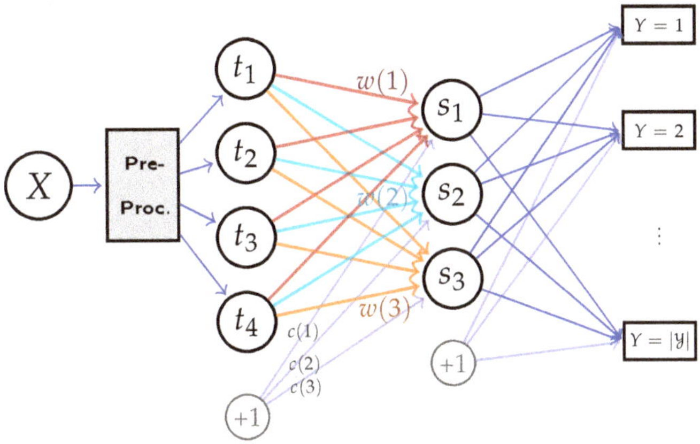

Figure 5. The designed network for validating the impact of network structure on feature extraction, with $m = 4$ and $k = 3$ neurons in two hidden layers. Our goal is to compare the learned weights $w(1), w(2), w(3)$ and bias c in the hidden layer with our theoretic characterizations in Section 3.2.2.

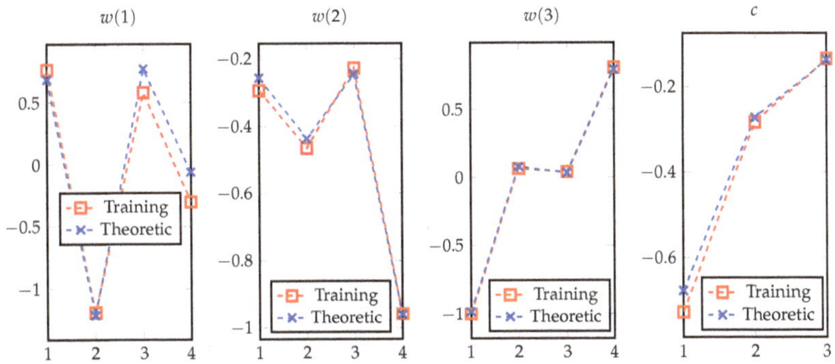

Figure 6. The trained weights w and bias c of the network in Figure 5, which are compared with the corresponding theoretical results to show their coincidences.

3.4.3. Experimental Validation of H-Score

To validate H-score as a performance measure for extracted features, we compare the H-score and classification accuracy of DNNs on image classification tasks. Specifically, we use the ImageNet Large Scale Visual Recognition Challenge 2012 (ILSVRC2012) [22] dataset as the dataset and extract features using several deep neural networks with representative architectures designs [44–49]. After training the feature extractors on the ILSVRC2012 training set, we then compute the H-score of the feature in the last hidden layer, as well as the classification accuracies on ILSVRC2012 validation set (here, we use ILSVRC2012 validation set for testing, as the labels in ILSVRC2012 testing set have not been publicly released). The results are summarized in Table 1, where $H_{\text{AIC}}(s)$ is the AIC-corrected H-score as defined in (27), with n_{p} being the number of model parameters, and $n_{\text{s}} = 1{,}300{,}000$ corresponding to the number of training samples in ImageNet. The AIC-corrected H-score is consistent with the classification accuracy, which validates the effectiveness of H-score as a measurement of neural networks.

Table 1. Classification accuracy and H-score for different DNN models on ImageNet dataset, where "Paras" indicates the number of parameters (in millions) in the model and H_{AIC} represents the AIC-corrected H-score.

DNN Model	Paras [$\times 10^6$]	$H(s)$	$H_{\text{AIC}}(s)$	Accuracy [%]
VGG16 [44]	138.4	148.3	41.9	64.2
VGG19 [44]	143.7	152.7	42.2	64.7
MobileNet [45]	4.3	45.9	42.6	68.4
DenseNet121 [46]	8.1	59.5	53.3	71.4
DenseNet169 [46]	14.3	81.2	70.2	73.6
DenseNet201 [46]	20.2	89.1	73.5	74.4
Xception [47]	22.9	179.8	162.2	77.5
InceptionV3 [48]	23.9	181.2	162.9	76.3
InceptionResNetV2 [49]	55.9	241.1	198.1	79.1

4. Discussion

Our characterization gives an information-theoretic interpretation of the feature extraction process in DNNs, which also provides a practical performance measure for scoring neural networks. Different from empirical studies focusing on specific datasets [7], our development is based on the probability distribution space, which is more general and can also provide theoretic insights. Moreover, the information-theoretic framework allows us to obtain direct operational meaning and better interpretations for the solutions, compared with optimization-based theoretical characterizations, e.g., [11,13].

As a first step in establishing a rigorous framework for DNN analysis, the present work can be extended in both theoretical and practical aspects. From the theoretical perspective, one extension is to investigate the analytical properties for general DNNs, using the theoretic insights obtained from local analysis regime. For example, it was shown in [50] that the symmetry between feature and weights in DNNs established in the local analysis regime (cf. Section 3.2.1) also holds for general probability distributions. Another extension is to apply the framework to investigate the optimal feature for structured data or network, e.g., data with sparsity structure [51].

From the practical perspective, in addition to the demonstrated example of evaluating existing DNN models (cf. Section 3.4.3), the H-score can also be used as an objective function in designing learning algorithms. In particular, such usages have been illustrated in multi-modal learning [52] and transfer learning [53] tasks.

5. Conclusions

In this paper, we apply the local information geometric analysis and provide an information-theoretic interpretation to the feature extraction scheme in DNNs. We first establish an information metric for features in inference tasks by formalizing the information-theoretic feature selection problem. In addition, we demonstrate that the features extracted by DNNs coincide with the information-theoretically optimal feature, with the same metric measuring the performance of features, called H-score. Furthermore, we discuss the usage of the H-score for measuring the effectiveness of DNNs. Our framework demonstrates a connection between the practical deep learning implementations and information-theoretic characterizations, which can provide theoretical insights for DNN analysis and learning algorithm designs.

Author Contributions: X.X., S.-L.H., L.Z. and G.W.W. contributed to the conceptualization, methodology, and writing of this paper. All authors have read and agreed to the published version of the manuscript.

Funding: The work of S.-L. Huang was supported in part by the National Natural Science Foundation of China under Grant 61807021 and the Shenzhen Science and Technology Program under Grant KQTD20170810150821146. The work of L. Zheng was supported in part by the National Science

Foundation (NSF) under Award CNS-2002908 and the Office of Naval Research (ONR) under Grant N00014-19-1-2621.

Conflicts of Interest: The authors declare no conflict of interest. The funders had no role in the design of the study; in the collection, analyses, or interpretation of data; in the writing of the manuscript, or in the decision to publish the results.

Appendix A. Proof of Theorem 1

We commence with the characterization of the error exponent.

Lemma A1. *Given a reference distribution $P_X \in \operatorname{relint}(\mathcal{P}^{\mathcal{X}})$, a constant $\epsilon > 0$ and integers n and k, let $x_1, \dots, x_n$ denote i.i.d. samples from one of P_1 or P_2, where $P_1, P_2 \in \mathcal{N}_\epsilon^{\mathcal{X}}(P_X)$. To decide whether P_1 or P_2 is the generating distribution, a sequence of k-dimensional statistics $h^k = (h_1, \dots, h_k)$ is constructed as*

$$h_i = \frac{1}{n}\sum_{l=1}^{n} f_i(x_l), \quad i = 1, \dots, k, \tag{A1}$$

where $(f_1(X), \dots, f_k(X))$ are zero mean, unit-variance, and uncorrelated with respect to P_X, i.e.,

$$\mathbb{E}_{P_X}[f_i(X)] = 0, \quad i \in \{1, \dots, k\} \tag{A2}$$

$$\mathbb{E}_{P_X}\left[f_i(X) f_j(X)\right] = \delta_{ij}, \quad i,j \in \{1, \dots, k\}. \tag{A3}$$

Then, the error probability of the decision based on h^k decays exponentially in n as $n \to \infty$, with (Chernoff) exponent

$$\lim_{n\to\infty} \frac{-\log p_e}{n} \triangleq E_{h^k} = \sum_{i=1}^{k} E_{h_i}, \tag{A4}$$

where

$$E_{h_i} = \frac{1}{8}\langle \phi_1 - \phi_2, \xi_i \rangle^2 + o(\epsilon^2), \tag{A5}$$

and $\phi_1 \leftrightarrow P_1, \phi_2 \leftrightarrow P_2, \xi_i \leftrightarrow f_i(X), i \in \{1, \dots, k\}$ are the corresponding information vectors.

Proof of Lemma A1. Since the rule is to decide based on comparing the projection

$$\sum_{i=1}^{k} h_i \left(\mathbb{E}_{P_1}[f_i(X)] - \mathbb{E}_{P_2}[f_i(X)]\right)$$

to a threshold, via Cramér's theorem [54], the error exponent under P_j $(j = 1, 2)$ is

$$E_j(\lambda) = \min_{P \in \mathcal{S}(\lambda)} D(P \| P_j), \tag{A6}$$

where

$$\mathcal{S}(\lambda) \triangleq \left\{P \in \mathcal{P}^{\mathcal{X}} : \mathbb{E}_P\left[f^k(X)\right] = \lambda\, \mathbb{E}_{P_1}\left[f^k(X)\right] + (1-\lambda)\, \mathbb{E}_{P_2}\left[f^k(X)\right]\right\}. \tag{A7}$$

Now, since (A2) holds, we obtain

$$\begin{aligned}
\mathbb{E}_{P_j}[f_i(X)] &= \sum_{x\in\mathcal{X}} P_j(x)\, f_i(x) \\
&= \sum_{x\in\mathcal{X}} P_X(x)\, f_i(x) + \sum_{x\in\mathcal{X}} (P_j(x) - P_X(x)) f_i(x) \\
&= \mathbb{E}_{P_X}[f_i(X)] + \sum_{x\in\mathcal{X}} \sqrt{P_X(x)}\, \phi_j(x) \cdot \frac{\xi_i(x)}{\sqrt{P_X(x)}} \\
&= \sum_{x\in\mathcal{X}} \phi_j(x)\, \xi_i(x)
\end{aligned}$$

$$= \langle \phi_j, \xi_i \rangle, \quad j = 1,2 \text{ and } i = 1,\ldots,k, \tag{A8}$$

which we express compactly as

$$\mathbb{E}_{P_j}[f^k(X)] = \langle \phi_j, \xi^k \rangle, \quad j = 1,2$$

with $\xi^k \triangleq (\xi_1,\ldots,\xi_k)$.

Hence, the constraint (A7) is expressed in information vectors as

$$\langle \phi, \xi_i \rangle = \langle \lambda\,\phi_1 + (1-\lambda)\,\phi_2, \xi_i \rangle, \quad i = 1,\cdots,k,$$

i.e.,

$$\langle \phi, \xi^k \rangle = \langle \lambda\,\phi_1 + (1-\lambda)\,\phi_2, \xi^k \rangle. \tag{A9}$$

In turn, the optimal P in (A6), which we denoted by P^*, lies in the exponential family through P_j with natural statistic $f^k(x)$, i.e., the k-dimensional family whose members are of the form

$$\log \tilde{P}_{\theta^k}(x) = \sum_{i=1}^{k} \theta_i f_i(x) + \log P_j(x) - \alpha(\theta^k),$$

for which the associated information vector is

$$\tilde{\phi}_{\theta^k}(x) = \sum_{i=1}^{k} \theta_i \xi_i(x) + \phi_j(x) - \alpha(\theta^k)\sqrt{P_X(x)} + o(\epsilon), \tag{A10}$$

where we have used the fact that

$$\begin{aligned}
\log Q_X(x) &= \log P_X(x) + \log \frac{Q_X(x)}{P_X(x)} \\
&= \log P_X(x) + \log\left(1 + \frac{1}{\sqrt{P_X(x)}}\phi(x)\right) \\
&= \log P_X(x) + \frac{1}{\sqrt{P_X(x)}}\phi(x) + o(\epsilon)
\end{aligned}$$

for all $Q_X \in \mathcal{N}_\epsilon^{\mathcal{X}}(P_X)$ with the information vector $\phi \leftrightarrow Q_X$. As a result,

$$\langle \tilde{\phi}_{\theta^k}, \xi_i \rangle = \theta_i + \langle \phi_j, \xi_i \rangle + o(\epsilon),$$

where we have used (A3). Hence, via (A9), we obtain that the intersection with the linear family (A7) is at $P^* = P_{\theta^{k*}}$ with

$$\theta_i^* = \langle \lambda\phi_1 + (1-\lambda)\phi_2 - \phi_j, \xi_i \rangle + o(\epsilon)$$

and thus

$$\begin{aligned}
E_j(\lambda) &= D(P^* \| P_j) \\
&= \frac{1}{2}\left\|\tilde{\phi}_{\theta^k} - \phi_j\right\|^2 + o(\epsilon^2) &\text{(A11)} \\
&= \frac{1}{2}\left\|\sum_{i=1}^{k} \theta_i^* \xi_i\right\|^2 + \frac{1}{2}\alpha(\theta^{k*})^2 + o(\epsilon^2) &\text{(A12)} \\
&= \frac{1}{2}\sum_{i=1}^{k}(\theta_i^*)^2 + \frac{1}{2}\alpha(\theta^{k*})^2 + o(\epsilon^2) &\text{(A13)} \\
&= \frac{1}{2}\sum_{i=1}^{k}\langle \lambda\phi_1 + (1-\lambda)\phi_2 - \phi_j, \xi_i \rangle^2 + o(\epsilon^2), &\text{(A14)}
\end{aligned}$$

where to obtain (A11) we have exploited the local approximation of KL divergence [18], to obtain (A12) we have exploited (A10), to obtain (A13) we have again exploited (A3), and to obtain (A14) we have used that

$$\alpha(\theta^{k^*}) = o(\epsilon^2)$$

since $\theta^{k^*} = O(\epsilon)$ and

$$\alpha(0) = 0, \quad \text{and} \quad \nabla\alpha(0) = \mathbb{E}_{P_j}\left[f^k(X)\right] = \langle \phi_j, \xi^k \rangle = O(\epsilon).$$

Finally, $E_1(\lambda) = E_2(\lambda)$ when $\lambda = 1/2$, so the overall error probability has exponent (A5). □

Then, the following lemma demonstrates a property of information vectors in a Markov chain.

Lemma A2. *Given the Markov relation $X \leftrightarrow Y \leftrightarrow V$ and any $v \in \mathcal{V}$, let $\boldsymbol{\phi}_v^{X|V}$ and $\boldsymbol{\phi}_v^{Y|V}$ denote the associated information vectors for $P_{X|V}(\cdot|v)$ and $P_{Y|V}(\cdot|v)$, then we have*

$$\boldsymbol{\phi}_v^{X|V} = \tilde{\mathbf{B}}^{\mathrm{T}} \boldsymbol{\phi}_v^{Y|V}. \tag{A15}$$

Proof of Lemma A2. From the Markov relation we have

$$P_X(x) = \sum_{y \in \mathcal{Y}} P_{X|Y}(x|y) P_Y(y)$$

and

$$P_{X|V}(x|v) = \sum_{y \in \mathcal{Y}} P_{X|Y,V}(x|y,v) P_{Y|V}(y|v) = \sum_{y \in \mathcal{Y}} P_{X|Y}(x|y) P_{Y|V}(y|v).$$

As a result,

$$P_{X|V}(x|v) - P_X(x) = \sum_{y \in \mathcal{Y}} P_{X|Y}(x|y) [P_{Y|V}(y|v) - P_Y(y)],$$

from which we obtain the corresponding information vector

$$\begin{aligned} \phi_v^{X|V}(x) &= \frac{1}{\sqrt{P_X(x)}} \sum_{y \in \mathcal{Y}} P_{X|Y}(x|y) \sqrt{P_Y(y)} \phi_v^{Y|V}(y) \\ &= \sum_{y \in \mathcal{Y}} \left[\tilde{\mathbf{B}}(y,x) + \sqrt{P_X(x) P_Y(y)} \right] \phi_v^{Y|V}(y) \\ &= \sum_{y \in \mathcal{Y}} \tilde{\mathbf{B}}(y,x) \phi_v^{Y|V}(y), \end{aligned} \tag{A16}$$

where the last equality follows from the fact that

$$\sum_{y \in \mathcal{Y}} \sqrt{P_Y(y)} \phi_v^{Y|V}(y) = \sum_{y \in \mathcal{Y}} [P_{Y|V}(y|v) - P_Y(y)] = 0.$$

Finally, rewrite (A16) in the matrix form and we obtain (A15). □

In addition, the following lemma is useful for dealing with the expectation over an RIE.

Lemma A3. *Let* $\mathbf{z}$ *be a spherically symmetric random vector of dimension* M*, i.e., for any orthogonal* $\mathbf{Q}$ *we have* $\mathbf{z} \overset{\mathrm{d}}{=} \mathbf{Q}\mathbf{z}$*. If* $\mathbf{A}$ *is a fixed matrix of compatible dimensions, then*

$$\mathbb{E}\left[\|\mathbf{z}^{\mathrm{T}}\mathbf{A}\|^2\right] = \frac{1}{M}\mathbb{E}\left[\|\mathbf{z}\|^2\right]\|\mathbf{A}\|_{\mathrm{F}}^2. \tag{A17}$$

Proof of Lemma A3. By definition we have $\mathbf{\Lambda_z} = \mathbf{Q}\mathbf{\Lambda_z}\mathbf{Q}^{\mathrm{T}}$ for any orthogonal $\mathbf{Q}$; hence, $\mathbf{\Lambda_z}$ is diagonal. Suppose $\mathbf{\Lambda_z} = \lambda\,\mathbf{I}$, then from

$$\mathrm{tr}(\mathbf{\Lambda_z}) = \mathbb{E}\left[\|\mathbf{z}\|^2\right] = \lambda M$$

we obtain

$$\lambda = \frac{1}{M}\,\mathrm{tr}(\mathbf{\Lambda_z}).$$

As a result, we have

$$\mathbb{E}\left[\|\mathbf{z}^{\mathrm{T}}\mathbf{A}\|^2\right] = \mathrm{tr}\left(\mathbf{A}^{\mathrm{T}}\mathbf{\Lambda_z}\mathbf{A}\right) = \lambda\,\mathrm{tr}\left(\mathbf{A}^{\mathrm{T}}\mathbf{A}\right) = \frac{1}{M}\mathbb{E}\left[\|\mathbf{z}\|^2\right]\|\mathbf{A}\|_{\mathrm{F}}^2.$$

□

Proceeding to our proof of Theorem 1, by definition of feature functions, we have $\mathbb{E}_{P_X}[f_i(X)] = 0, i = 1,\ldots,k$. Suppose $\boldsymbol{f}$ is the vector representation of f^k and denote by $\tilde{\boldsymbol{f}} \triangleq \mathbf{\Lambda}_{\boldsymbol{f}}^{-1/2}\boldsymbol{f}$ the normalized $\boldsymbol{f}$, with $\mathbf{\Lambda}_{\boldsymbol{f}}^{1/2}$ denoting any square root matrix of $\mathbf{\Lambda}_{\boldsymbol{f}}$. Then, the corresponding statistics $\tilde{f}^k = (\tilde{f}_1,\ldots,\tilde{f}_k)$ satisfy the constraints (A2) and (A3). In addition, we construct the statistic $\tilde{h}^k = (\tilde{h}_1,\ldots,\tilde{h}_k)$ as [cf. (A1)]

$$\tilde{h}_i = \frac{1}{n}\sum_{l=1}^{n}\tilde{f}_i(x_l), \quad i = 1,\ldots,k. \tag{A18}$$

Then, from Lemma A1, the error exponent of distinguishing v and v' based on $\tilde{h}^k$ is

$$\begin{aligned} E_{\tilde{h}^k}(v,v') &= \frac{1}{8}\sum_{i=1}^{k}\left[(\boldsymbol{\phi}_v^{X|V} - \boldsymbol{\phi}_{v'}^{X|V})^{\mathrm{T}}\tilde{\boldsymbol{\xi}}_i^X\right]^2 + o(\epsilon^2) \\ &= \frac{1}{8}\left\|(\boldsymbol{\phi}_v^{X|V} - \boldsymbol{\phi}_{v'}^{X|V})^{\mathrm{T}}\tilde{\mathbf{\Xi}}^X\right\|^2 + o(\epsilon^2), \end{aligned}$$

where $\boldsymbol{\phi}_v^{X|V}$ denotes the associated information vector for $P_{X|V}(\cdot|v)$, $\tilde{\boldsymbol{\xi}}_i^X$ denotes the information vectors of $\tilde{f}_i$, and $\tilde{\mathbf{\Xi}}^X \triangleq [\tilde{\boldsymbol{\xi}}_1^X,\ldots,\tilde{\boldsymbol{\xi}}_k^X]$. Since the optimal decision rule is linear, the error exponent is invariant with linear transformations of statistics, i.e.,

$$\begin{aligned} E_{h^k}(v,v') = E_{\tilde{h}^k}(v,v') &= \frac{1}{8}\left\|(\boldsymbol{\phi}_v^{X|V} - \boldsymbol{\phi}_{v'}^{X|V})^{\mathrm{T}}\tilde{\mathbf{\Xi}}^X\right\|^2 + o(\epsilon^2) \\ &= \frac{1}{8}\left\|(\boldsymbol{\phi}_v^{Y|V} - \boldsymbol{\phi}_{v'}^{Y|V})^{\mathrm{T}}\tilde{\mathbf{B}}\tilde{\mathbf{\Xi}}^X\right\|^2 + o(\epsilon^2), \end{aligned} \tag{A19}$$

where the last equality follows from Lemma A2.

As a result, taking the expectation of (A19) over a given RIE yields

$$\begin{aligned} \mathbb{E}\left[E_{h^k}(v,v')\right] &= \frac{1}{8}\mathbb{E}\left[\left\|(\boldsymbol{\phi}_v^{Y|V} - \boldsymbol{\phi}_{v'}^{Y|V})^{\mathrm{T}}\tilde{\mathbf{B}}\tilde{\mathbf{\Xi}}^X\right\|^2\right] + o(\epsilon^2) \\ &= \frac{\mathbb{E}\left[\|\boldsymbol{\phi}_v^{Y|V} - \boldsymbol{\phi}_{v'}^{Y|V}\|^2\right]}{8|\mathcal{Y}|}\|\tilde{\mathbf{B}}\tilde{\mathbf{\Xi}}^X\|_{\mathrm{F}}^2 + o(\epsilon^2), \end{aligned}$$

where we have exploited Lemma A3. Finally, the error exponent (7) can be obtained via noting from the definition of $\tilde{f}^k$ that

$$\tilde{\Xi}^X = \Xi^X\left(\left(\Xi^X\right)^{\mathrm{T}}\Xi^X\right)^{-\frac{1}{2}}.$$

Appendix B. Proof of Lemma 2

We first prove two useful lemmas.

Lemma A4. *For distributions $P \in \mathrm{relint}(\mathcal{P}^{\mathcal{X}})$, $Q, R \in \mathcal{P}^{\mathcal{X}}$, and sufficiently small ϵ, if $D(P\|Q) \leq \epsilon^2$ and $D(P\|R) \leq \epsilon^2$, then there exists a constant $C > 0$ independent of ϵ, such that $D(Q\|R) \leq C\epsilon^2$.*

Proof of Lemma A4. Denote by $\|\cdot\|_1$ the ℓ_1-distance between distributions, i.e., $\|P - Q\|_1 \triangleq \sum_{x\in\mathcal{X}} |P(x) - Q(x)|$, then from Pinsker's inequality [14], we have

$$\|P - Q\|_1 \leq \sqrt{2D(P\|Q)} < \sqrt{2}\epsilon, \tag{A20}$$

$$\|P - R\|_1 \leq \sqrt{2D(P\|R)} < \sqrt{2}\epsilon, \tag{A21}$$

which implies

$$\|Q - R\|_1 \leq \|P - Q\|_1 + \|P - R\|_1 \leq 2\sqrt{2}\epsilon. \tag{A22}$$

In addition, with $p_{\min} \triangleq \min_{x\in\mathcal{X}} P(x)$, for all $x \in \mathcal{X}$ we have

$$\begin{aligned}
R(x) &> P(x) - |P(x) - R(x)| &\text{(A23)}\\
&> \min_{x\in\mathcal{X}} P(x) - \sqrt{2}\epsilon &\text{(A24)}\\
&= p_{\min} - \sqrt{2}\epsilon, &\text{(A25)}
\end{aligned}$$

where to obtain (A24) we have used (A21). Note that since $P \in \mathrm{relint}(\mathcal{P}^{\mathcal{X}})$ we have $p_{\min} > 0$, and thus $R(x) > p_{\min}/2$ for sufficiently small ϵ. As a result,

$$\begin{aligned}
D(Q\|R) &\leq \sum_{x\in\mathcal{X}} \frac{(Q(x) - R(x))^2}{R(x)} &\text{(A26)}\\
&\leq \frac{2}{p_{\min}} \sum_{x\in\mathcal{X}} [Q(x) - R(x)]^2 &\text{(A27)}\\
&\leq \frac{2\|Q - R\|_1^2}{p_{\min}} &\text{(A28)}\\
&\leq \frac{16}{p_{\min}}\epsilon^2, &\text{(A29)}
\end{aligned}$$

where to obtain (A26) we have used the fact that KL divergence is upper bounded by corresponding χ^2-divergence [55], and to obtain (A29) we have used (A22). □

Lemma A5. *For all $(x, y) \in \mathcal{X} \times \mathcal{Y}$, we have*

$$D(P_X P_Y \| P_X \tilde{P}_{Y|X}^{(s,v,b)}) \geq P_X(x) \log\left[P_Y(y)e^{\tau(x,y)} + (1 - P_Y(y))e^{-\frac{P_Y(y)}{1-P_Y(y)}\tau(x,y)}\right]$$

where $\tilde{P}_{Y|X}^{(s,v,b)}$ is as defined in (4), and where we have defined $\tau(x,y) \triangleq \tilde{v}^{\mathrm{T}}(y)s(x) + \tilde{d}(y)$.

Proof of Lemma A5. First, we can rewrite the conditional distribution $\tilde{P}^{(s,v,b)}_{Y|X}(y|x)$ as

$$\tilde{P}^{(s,v,b)}_{Y|X}(y|x) = \frac{e^{v^{\mathrm{T}}(y)s(x)+b(y)}}{\sum_{y'\in\mathcal{Y}} e^{v^{\mathrm{T}}(y')s(x)+b(y')}} = \frac{P_Y(y)e^{v^{\mathrm{T}}(y)s(x)+d(y)}}{\sum_{y'\in\mathcal{Y}} P_Y(y')e^{v^{\mathrm{T}}(y')s(x)+d(y')}}$$
$$= \frac{P_Y(y)e^{\tilde{v}^{\mathrm{T}}(y)s(x)+\tilde{d}(y)}}{\sum_{y'\in\mathcal{Y}} P_Y(y')e^{\tilde{v}^{\mathrm{T}}(y')s(x)+\tilde{d}(y')}}$$
$$= \frac{P_Y(y)e^{\tau(x,y)}}{\sum_{y'\in\mathcal{Y}} P_Y(y')e^{\tau(x,y')}}. \tag{A30}$$

Then, the KL divergence $D(P_XP_Y\|P_X\,\tilde{P}^{(s,v,b)}_{Y|X})$ can be expressed as

$$D(P_XP_Y\|P_X\,\tilde{P}^{(s,v,b)}_{Y|X}) = \sum_{(x,y)\in\mathcal{X}\times\mathcal{Y}} P_X(x)P_Y(y)\log\frac{\sum_{y'\in\mathcal{Y}} P_Y(y')e^{\tau(x,y')}}{e^{\tau(x,y)}}$$
$$= \sum_{x\in\mathcal{X}} P_X(x)\log\left[\sum_{y'\in\mathcal{Y}} P_Y(y')e^{\tau(x,y')}\right] - \mathbb{E}_{P_XP_Y}[\tau(X,Y)]$$
$$= \sum_{x\in\mathcal{X}} P_X(x)\log\left[\sum_{y'\in\mathcal{Y}} P_Y(y')e^{\tau(x,y')}\right], \tag{A31}$$

where to obtain the last equality we have used the fact $\mathbb{E}_{P_XP_Y}[\tau(X,Y)] = 0$. As a result, we have

$$D(P_XP_Y\|P_X\,\tilde{P}^{(s,v,b)}_{Y|X}) \geq P_X(x)\log\left[\sum_{y'\in\mathcal{Y}} P_Y(y')e^{\tau(x,y')}\right] \tag{A32}$$
$$\geq P_X(x)\log\left[P_Y(y)e^{\tau(x,y)} + (1-P_Y(y))e^{-\frac{P_Y(y)}{1-P_Y(y)}\tau(x,y)}\right], \tag{A33}$$

where the last inequality follows from Jensen's inequality:

$$\sum_{y'\in\mathcal{Y}} P_Y(y')e^{\tau(x,y')} = P_Y(y)e^{\tau(x,y)} + (1-P_Y(y))\sum_{y'\neq y}\frac{P_Y(y')}{1-P_Y(y)}e^{\tau(x,y')}$$
$$\geq P_Y(y)e^{\tau(x,y)} + (1-P_Y(y))\exp\left(\frac{1}{1-P_Y(y)}\sum_{y'\neq y} P_Y(y')\tau(x,y')\right)$$
$$= P_Y(y)e^{\tau(x,y)} + (1-P_Y(y))e^{-\frac{P_Y(y)}{1-P_Y(y)}\tau(x,y)}.$$

□

Proceeding to our proof of Lemma 2, first note that when $v = d = 0$, we have $\tilde{P}^{(s,v,b)}_{Y|X} = P_Y$. As a result, the optimal v, d for (8) satisfy

$$D(P_{XY}\|P_X\,\tilde{P}^{(s,v,b)}_{Y|X}) \leq D(P_{XY}\|P_XP_Y)$$
$$\leq \sum_{(x,y)\in\mathcal{X}\times\mathcal{Y}}\frac{[P_{X,Y}(x,y) - P_X(x)P_Y(y)]^2}{P_X(x)P_Y(y)} \tag{A34}$$
$$\leq \epsilon^2,$$

where to obtain the second inequality we have again exploited χ^2-divergence as an upper bound of KL divergence [55], and to obtain the last inequality we have used the definition of ϵ-dependency.

As $P_{XY} \in \text{relint}(\mathcal{P}^{\mathcal{X}\times\mathcal{Y}})$, from Lemma A4, there exist $C > 0$ and $\epsilon_1 > 0$ such that $D(P_X P_Y \| P_X \tilde{P}^{(s,v,b)}_{Y|X}) < C\epsilon^2$ for all $\epsilon < \epsilon_1$. Furthermore, from Lemma A5, for all $(x,y) \in \mathcal{X}\times\mathcal{Y}$ and $\epsilon \in (0,\epsilon_1)$, we have

$$C\epsilon^2 \geq P_X(x)\log\left[P_Y(y)e^{\tau(x,y)} + (1-P_Y(y))e^{-\frac{P_Y(y)}{1-P_Y(y)}\tau(x,y)}\right]. \tag{A35}$$

Note that the right-hand side of (A35) satisfies

$$\log\left[P_Y(y)e^{\tau(x,y)} + (1-P_Y(y))e^{-\frac{P_Y(y)}{1-P_Y(y)}\tau(x,y)}\right] = \frac{P_Y(y)}{2(1-P_Y(y))}\tau^2(x,y) + o(\tau^2(x,y)).$$

Therefore, there exists $\delta > 0$ independent of ϵ_1, such that for all $|\tau(x,y)| \leq \delta$, we have

$$\log\left[P_Y(y)e^{\tau(x,y)} + (1-P_Y(y))e^{-\frac{P_Y(y)}{1-P_Y(y)}\tau(x,y)}\right] > \frac{P_Y(y)}{2}\tau^2(x,y). \tag{A36}$$

In addition, if $|\tau(x,y)| > \delta$, we have

$$\begin{aligned}
&\log\left[P_Y(y)e^{\tau(x,y)} + (1-P_Y(y))e^{-\frac{P_Y(y)}{1-P_Y(y)}\tau(x,y)}\right]\\
&\geq \min\left\{\log\left[P_Y(y)e^{\delta} + (1-P_Y(y))e^{-\frac{P_Y(y)}{1-P_Y(y)}\delta}\right], \log\left[P_Y(y)e^{-\delta} + (1-P_Y(y))e^{\frac{P_Y(y)}{1-P_Y(y)}\delta}\right]\right\}\\
&\geq \frac{P_Y(y)}{2}\delta^2,
\end{aligned}$$

where to obtain the second inequality we have exploited the monotonicity of function $t \mapsto P_Y(y)e^t + (1-P_Y(y))e^{-\frac{P_Y(y)}{1-P_Y(y)}t}$, and to obtain the third inequality we have exploited (A36).

As a result, we have

$$\log\left[P_Y(y)e^{\tau(x,y)} + (1-P_Y(y))e^{-\frac{P_Y(y)}{1-P_Y(y)}\tau(x,y)}\right] > \frac{P_Y(y)}{2}\cdot\min\{\delta^2, \tau^2(x,y)\}. \tag{A37}$$

Hence, (A35) becomes

$$C\epsilon^2 \geq \frac{P_X(x)P_Y(y)}{2}\cdot\min\{\delta^2, \tau^2(x,y)\}, \tag{A38}$$

from which we can obtain $\tau(x,y) = O(\epsilon)$. To see this, let

$$\epsilon_2 \triangleq \frac{\delta}{\sqrt{2C}}\cdot\min_{(x,y)\in\mathcal{X}\times\mathcal{Y}}\sqrt{P_X(x)P_Y(y)}, \quad \epsilon_0 \triangleq \min\{\epsilon_1, \epsilon_2\}.$$

Then, for all $\epsilon < \epsilon_0$, we have

$$C\epsilon^2 < \frac{P_X(x)P_Y(y)}{2}\cdot\delta^2,$$

and (A38) implies $|\tau(x,y)| < C'\epsilon$ with $C' = \sqrt{\frac{2C}{P_X(x)P_Y(y)}}$.

Appendix C. Proof of Lemma 3

Proof. From Lemma 2, there exists $C' > 0$ such that for all $(x,y) \in \mathcal{X}\times\mathcal{Y}$, we have

$$|\tilde{v}^{\mathrm{T}}(y)s(x) + \tilde{d}(y)| < C'\epsilon, \tag{A39}$$

which implies

$$|\mu_s^{\mathrm{T}}\tilde{v}(y) + \tilde{d}(y)| < C\epsilon, \tag{A40}$$

$$|\tilde{v}^{\mathrm{T}}(y)\tilde{s}(x)| < 2C\epsilon, \tag{A41}$$

with $C = \max\{C', 1\}$.

From (A30), we can assume $\mathbb{E}_{P_Y}[v(Y)] = \mathbb{E}_{P_Y}[d(Y)] = 0$ without loss of generality. Then, (4) can be rewritten as

$$\tilde{P}_{Y|X}^{(s,v,b)}(y|x) = \frac{P_Y(y)e^{\tilde{v}^{\mathrm{T}}(y)s(x)+\tilde{d}(y)}}{\sum_{y'\in\mathcal{Y}} P_Y(y')e^{\tilde{v}^{\mathrm{T}}(y')s(x)+\tilde{d}(y')}}, \tag{A42}$$

and the numerator can be written as

$$\begin{aligned} P_Y(y)e^{\tilde{v}^{\mathrm{T}}(y)s(x)+\tilde{d}(y)} &= P_Y(y)\left(1 + \tilde{v}^{\mathrm{T}}(y)s(x) + \tilde{d}(y) + o(\epsilon)\right) \\ &= P_Y(y)\left(1 + \tilde{v}^{\mathrm{T}}(y)s(x) + \tilde{d}(y)\right) + o(\epsilon), \end{aligned}$$

where we have used (A39). Similarly, from

$$\begin{aligned} \sum_{y'\in\mathcal{Y}} P_Y(y)e^{\tilde{v}^{\mathrm{T}}(y)s(x)+\tilde{d}(y)} &= \sum_{y'\in\mathcal{Y}} P_Y(y)\left(1 + \tilde{v}^{\mathrm{T}}(y)s(x) + \tilde{d}(y)\right) + o(\epsilon) \\ &= 1 + \mathbb{E}_{P_Y}\left[\tilde{v}^{\mathrm{T}}(Y)s(x)\right] + \mathbb{E}_{P_Y}\left[\tilde{d}(y)\right] + o(\epsilon) \\ &= 1 + o(\epsilon) \end{aligned}$$

we obtain

$$\frac{1}{\sum_{y'\in\mathcal{Y}} P_Y(y)e^{\tilde{v}^{\mathrm{T}}(y)s(x)+\tilde{d}(y)}} = \frac{1}{1+o(\epsilon)} = 1 + o(\epsilon).$$

As a result, (A42) can be written as

$$\begin{aligned} \tilde{P}_{Y|X}^{(s,v,b)}(y|x) &= \left[P_Y(y)\left(1 + \tilde{v}^{\mathrm{T}}(y)s(x) + \tilde{d}(y)\right) + o(\epsilon)\right][1 + o(\epsilon)] \\ &= P_Y(y)\left(1 + \tilde{v}^{\mathrm{T}}(y)s(x) + \tilde{d}(y)\right) + o(\epsilon), \end{aligned} \tag{A43}$$

which implies $P_X\,\tilde{P}_{Y|X}^{(v,b)} \in \mathcal{N}_{C\epsilon}^{\mathcal{X}\times\mathcal{Y}}(P_X P_Y)$ for sufficiently small ϵ. In addition, the local assumption of distributions implies that $P_{XY} \in \mathcal{N}_{\epsilon}^{\mathcal{X}\times\mathcal{Y}}(P_X P_Y) \subset \mathcal{N}_{C\epsilon}^{\mathcal{X}\times\mathcal{Y}}(P_X P_Y)$. Again, from the local approximation of KL divergence [18]

$$D(P_1\|P_2) = \frac{1}{2}\|\phi_1 - \phi_2\|^2 + o(\epsilon^2), \tag{A44}$$

we have

$$\begin{aligned} &D(P_{Y,X}\|P_X\,\tilde{P}_{Y|X}^{(s,v,b)}) \\ &= \frac{1}{2}\sum_{x\in\mathcal{X},y\in\mathcal{Y}} \frac{\left[P_{Y,X}(y,x) - \tilde{P}_{Y|X}^{(s,v,b)}(y|x)P_X(x)\right]^2}{P_Y(y)P_X(x)} + o(\epsilon^2) \\ &= \frac{1}{2}\sum_{x\in\mathcal{X},y\in\mathcal{Y}} \left[\frac{P_{Y,X}(y,x)}{\sqrt{P_Y(y)P_X(x)}} - \sqrt{P_Y(y)P_X(x)}\right. \\ &\quad \left. - \sqrt{P_Y(y)P_X(x)}\left(\tilde{v}^{\mathrm{T}}(y)s(x) + \tilde{d}(y) + o(\epsilon)\right)\right]^2 + o(\epsilon^2) \end{aligned}$$

$$= \frac{1}{2} \sum_{x\in\mathcal{X},y\in\mathcal{Y}} \Bigg[\tilde{\mathbf{B}}(y,x) - \sqrt{P_Y(y)P_X(x)}\tilde{v}^{\mathrm{T}}(y)\tilde{s}(x) - \sqrt{P_Y(y)P_X(x)}\left(\tilde{d}(y) + \mu_s^{\mathrm{T}}\tilde{v}(y)\right) - \sqrt{P_Y(y)P_X(x)}o(\epsilon)\Bigg]^2 + o(\epsilon^2)$$

$$\overset{(*)}{=} \frac{1}{2} \sum_{x\in\mathcal{X},y\in\mathcal{Y}} \left[\tilde{\mathbf{B}}(y,x) - \sqrt{P_Y(y)P_X(x)}\tilde{v}^{\mathrm{T}}(y)\tilde{s}(x)\right]^2 + \frac{1}{2} \sum_{x\in\mathcal{X},y\in\mathcal{Y}} \left[\sqrt{P_Y(y)P_X(x)}\left(\tilde{d}(y) + \mu_s^{\mathrm{T}}\tilde{v}(y)\right)\right]^2 + o(\epsilon^2)$$

$$= \frac{1}{2} \sum_{x\in\mathcal{X},y\in\mathcal{Y}} \left[\tilde{\mathbf{B}}(y,x) - \left(\xi^Y(y)\right)^{\mathrm{T}}\xi^X(x)\right]^2 + \frac{1}{2}\mathbb{E}_{P_Y}\left[(\tilde{d}(y) + \mu_s^{\mathrm{T}}\tilde{v}(y))^2\right] + o(\epsilon^2)$$

$$= \frac{1}{2}\|\tilde{\mathbf{B}} - \Xi^Y(\Xi^X)^{\mathrm{T}}\|_{\mathrm{F}}^2 + \frac{1}{2}\eta^{(v,b)}(s) + o(\epsilon^2),$$

where to obtain $(*)$, we have used (A40) and (A41) together with the fact $|\tilde{\mathbf{B}}(y,x)| < \epsilon$, and that

$$\sum_{x\in\mathcal{X},y\in\mathcal{Y}} \tilde{\mathbf{B}}(y,x)\sqrt{P_Y(y)P_X(x)}\left(\tilde{d}(y) + \mu_s^{\mathrm{T}}\tilde{v}(y)\right) = 0,$$

$$\sum_{x\in\mathcal{X},y\in\mathcal{Y}} P_Y(y)P_X(x)\tilde{v}^{\mathrm{T}}(y)\tilde{s}(x)\left(\tilde{d}(y) + \mu_s^{\mathrm{T}}\tilde{v}(y)\right) = 0,$$

since $\mathbb{E}\left[\tilde{d}(Y)\right] = 0, \mathbb{E}[\tilde{s}(X)] = \mathbb{E}[\tilde{v}(Y)] = 0.$ □

Appendix D. Proofs of Theorems 2 and 3

Theorems 2 and 3 can be proved based on Lemma 3.

Proofs of Theorems 2 and 3. Note that the value of $d(\cdot)$ only affects the second term of the KL divergence; hence, we can always choose $d(\cdot)$ such that $\tilde{d}(y) + \mu_s^{\mathrm{T}}\tilde{v}(y) = 0$. Then, the (Ξ^Y, Ξ^X) pair should be chosen as

$$(\Xi^Y, \Xi^X)^* = \underset{(\Xi^Y,\Xi^X)}{\arg\min}\|\tilde{\mathbf{B}} - \Xi^Y(\Xi^X)^{\mathrm{T}}\|_{\mathrm{F}}^2. \tag{A45}$$

Set the derivative (we use the denominator-layout notation of matrix calculus where the scalar-by-matrix derivative will have the same dimension as the matrix)

$$\frac{\partial}{\partial \Xi^Y}\|\tilde{\mathbf{B}} - \Xi^Y(\Xi^X)^{\mathrm{T}}\|_{\mathrm{F}}^2 = 2(\Xi^Y(\Xi^X)^{\mathrm{T}}\Xi^X - \tilde{\mathbf{B}}\Xi^X) \tag{A46}$$

to zero, and the optimal Ξ^Y for fixed Ξ^X is (here, we assume the matrix $(\Xi^X)^{\mathrm{T}}\Xi^X = \mathbf{\Lambda}_{\tilde{s}(X)}$ is invertible; for the case where $(\Xi^X)^{\mathrm{T}}\Xi^X$ is singular, we can obtain a similar result with ordinary matrix inverse replaced by the Moore–Penrose inverse)

$$\Xi^{Y*} = \tilde{\mathbf{B}}\Xi^X((\Xi^X)^{\mathrm{T}}\Xi^X)^{-1}. \tag{A47}$$

As $\mathbf{1}^{\mathrm{T}}\sqrt{\mathbf{P}_Y}\,\tilde{\mathbf{B}} = 0$, we have $\mathbf{1}^{\mathrm{T}}\sqrt{\mathbf{P}_Y}\,\Xi^{Y*} = 0$, which demonstrates that Ξ^{Y*} is a valid matrix for a zero-mean feature vector.

To express Ξ^{Y*} of (A47) in the form of s and v, we can make use of the correspondence between feature and information vectors. We can show that, for a zero-mean feature function $f(X)$ with corresponding information vector ϕ, we have the correspondence

$\mathbb{E}_{P_{X|Y}}[f(X)|Y] \leftrightarrow \tilde{\mathbf{B}}\phi$. To see this, note that the y-th element of information vector $\tilde{\mathbf{B}}\phi$ is given by

$$\begin{aligned}\sum_{x\in\mathcal{X}} \tilde{\mathbf{B}}(y,x)\phi(x) &= \sum_{x\in\mathcal{X}} \frac{P_{XY}(x,y) - P_X(x)P_Y(y)}{\sqrt{P_X(x)P_Y(y)}} f(x)\sqrt{P_X(x)} \\ &= \frac{1}{\sqrt{P_Y(y)}} \sum_{x\in\mathcal{X}} P_{XY}(x,y)f(x) \\ &= \frac{1}{\sqrt{P_Y(y)}} \mathbb{E}_{P_{X|Y}}[f(X)|Y=y].\end{aligned}$$

Using similar methods, we can verify that $\mathbf{\Lambda}_{\tilde{s}(X)} = \left(\Xi^X\right)^{\mathrm{T}}\Xi^X$. As a result, (A47) is equivalent to

$$\tilde{v}^*(y) = \mathbb{E}_{P_{X|Y}}\left[\mathbf{\Lambda}_{\tilde{s}(X)}^{-1}\tilde{s}(X) \,\middle|\, Y=y\right]. \tag{A48}$$

By a symmetry argument, we can also obtain the first two equations of Theorem 3. To obtain the third equations of these two theorems, we need to minimize $\eta^{(v,b)}(s) = \mathbb{E}_{P_Y}\left[(\mu_s^{\mathrm{T}}\tilde{v}(Y) + \tilde{d}(Y))^2\right]$. For given $\tilde{v}$ and μ_s, the optimal $\tilde{d}$ is

$$\tilde{d}^*(y) = -\mu_s^{\mathrm{T}}\tilde{v}(Y), \tag{A49}$$

and the corresponding $\eta^{(v,b)}(s) = 0$.

In addition, for given $\tilde{d}$ and $\tilde{v}$, we have

$$\begin{aligned}\eta^{(v,b)}(s) &= \mathbb{E}_{P_Y}\left[(\mu_s^{\mathrm{T}}\tilde{v}(Y) + \tilde{d}(Y))^2\right] \\ &= \mu_s^{\mathrm{T}}\mathbf{\Lambda}_{\tilde{v}(Y)}\,\mu_s + 2\mu_s^{\mathrm{T}}\mathbb{E}_{P_Y}\left[\tilde{v}(Y)\tilde{d}(Y)\right] + \mathrm{var}(\tilde{d}(Y)).\end{aligned} \tag{A50}$$

Set $\frac{\partial}{\partial \mu_s}\eta^{(v,b)}(s) = 0$ and we obtain

$$\mu_s^* = -\mathbf{\Lambda}_{\tilde{v}(Y)}^{-1}\mathbb{E}_{P_Y}\left[\tilde{v}(Y)\tilde{d}(Y)\right]. \tag{A51}$$

□

Appendix E. Proof of Theorem 4

Proof. From Lemma 3, choosing the optimal (Ξ^Y, Ξ^X) is equivalent to solving the matrix factorization problem of $\tilde{\mathbf{B}}$. Since both Ξ^Y and Ξ^X have rank no greater than k, from the Eckart–Young–Mirsky theorem [56], the optimal choice of $\Xi^Y\left(\Xi^X\right)^{\mathrm{T}}$ should be the truncated singular value decomposition of $\tilde{\mathbf{B}}$ with top k singular values. As a result, $(\Xi^Y, \Xi^X)^*$ are the left and right singular vectors of $\tilde{\mathbf{B}}$ corresponding to the largest k singular values.

The optimality of bias $\tilde{d}(y) = -\mu_s^{\mathrm{T}}\tilde{v}(y)$ has already been shown in Appendix D. □

Appendix F. Proof of Theorem 5

The following lemma is useful to prove Theorem 5.

Lemma A6 (Pythagorean theorem). *Let Ξ^{X*} be the optimal matrix for given Ξ^Y as defined in (13). Then,*

$$\|\tilde{\mathbf{B}} - \Xi^Y\left(\Xi^X\right)^{\mathrm{T}}\|_{\mathrm{F}}^2 - \|\tilde{\mathbf{B}} - \Xi^Y\left(\Xi^{X*}\right)^{\mathrm{T}}\|_{\mathrm{F}}^2 = \|\Xi^Y\left(\Xi^{X*}\right)^{\mathrm{T}} - \Xi^Y\left(\Xi^X\right)^{\mathrm{T}}\|_{\mathrm{F}}^2. \tag{A52}$$

Proof of Lemma A6. Denote by $\langle \mathbf{U}, \mathbf{V} \rangle$ the Frobenius inner product of matrices $\mathbf{U}$ and $\mathbf{V}$, i.e., $\langle \mathbf{U}, \mathbf{V} \rangle \triangleq \mathrm{tr}(\mathbf{U}^{\mathrm{T}}\mathbf{V})$, and we have

$$\begin{aligned}\left\langle \tilde{\mathbf{B}} - \Xi^{Y}(\Xi^{X*})^{\mathrm{T}}, \Xi^{Y}(\Xi^{X})^{\mathrm{T}} \right\rangle &= \mathrm{tr}\left(\tilde{\mathbf{B}}\Xi^{X}(\Xi^{Y})^{\mathrm{T}}\right) - \mathrm{tr}\left(\Xi^{X*}(\Xi^{Y})^{\mathrm{T}}\Xi^{Y}(\Xi^{X})^{\mathrm{T}}\right) \\ &= \mathrm{tr}\left(\tilde{\mathbf{B}}\Xi^{X}(\Xi^{Y})^{\mathrm{T}}\right) - \mathrm{tr}\left(\tilde{\mathbf{B}}^{\mathrm{T}}\Xi^{Y}(\Xi^{X})^{\mathrm{T}}\right) \\ &= 0.\end{aligned}$$

As a result, we obtain

$$\begin{aligned}\|\tilde{\mathbf{B}} - \Xi^{Y}(\Xi^{X})^{\mathrm{T}}\|_{\mathrm{F}}^{2} &= \|\tilde{\mathbf{B}} - \Xi^{Y}(\Xi^{X*})^{\mathrm{T}} + (\Xi^{Y}(\Xi^{X*})^{\mathrm{T}} - \Xi^{Y}(\Xi^{X})^{\mathrm{T}})\|_{\mathrm{F}}^{2} \\ &= \|\tilde{\mathbf{B}} - \Xi^{Y}(\Xi^{X*})^{\mathrm{T}}\|_{\mathrm{F}} + \|\Xi^{Y}(\Xi^{X*})^{\mathrm{T}} - \Xi^{Y}(\Xi^{X})^{\mathrm{T}}\|_{\mathrm{F}}^{2} \\ &\quad + 2\left\langle \tilde{\mathbf{B}} - \Xi^{Y}(\Xi^{X*})^{\mathrm{T}}, \Xi^{Y}((\Xi^{X*})^{\mathrm{T}} - (\Xi^{X})^{\mathrm{T}}) \right\rangle \\ &= \|\tilde{\mathbf{B}} - \Xi^{Y}(\Xi^{X*})^{\mathrm{T}}\|_{\mathrm{F}} + \|\Xi^{Y}(\Xi^{X*})^{\mathrm{T}} - \Xi^{Y}(\Xi^{X})^{\mathrm{T}}\|_{\mathrm{F}}^{2},\end{aligned}$$

which finishes the proof. □

Proceeding to our proof of Theorem 5, from Lemma A6 we have

$$\begin{aligned}&\mathcal{L}(s) - \mathcal{L}(s^{*}) \\ &\quad = \frac{1}{2}\left[\|\tilde{\mathbf{B}} - \Xi^{Y}(\Xi^{X})^{\mathrm{T}}\|_{\mathrm{F}}^{2} - \|\tilde{\mathbf{B}} - \Xi^{Y}(\Xi^{X*})^{\mathrm{T}}\|_{\mathrm{F}}^{2}\right] + \frac{1}{2}\left[\eta^{(v,b)}(s) - \eta^{(v,b)}(s^{*})\right] + o(\epsilon^{2}) \\ &\quad = \frac{1}{2}\|\Xi^{Y}(\Xi^{X*})^{\mathrm{T}} - \Xi^{Y}(\Xi^{X})^{\mathrm{T}}\|_{\mathrm{F}}^{2} + \frac{1}{2}\kappa^{(v,b)}(s, s^{*}) + o(\epsilon^{2}),\end{aligned}$$

where $\kappa^{(v,b)}(s, s^{*}) \triangleq \eta^{(v,b)}(s) - \eta^{(v,b)}(s^{*})$. We then optimize $\|\Xi^{Y}(\Xi^{X*})^{\mathrm{T}} - \Xi^{Y}(\Xi^{X})^{\mathrm{T}}\|_{\mathrm{F}}^{2}$ and $\kappa^{(v,b)}(s, s^{*})$ separately.

For the first term, we need to express Ξ^{X} in terms of $\mathbf{W}$ and Ξ_{1}^{X}. From (17), we obtain

$$\mathbb{E}[s_{z}(X)] = \sigma(c(z)) + o(\epsilon), \tag{A53}$$

$$\tilde{s}_{z}(x) = w^{\mathrm{T}}(z)\tilde{f}(x) \cdot \sigma'(c(z)) + o(\epsilon), \tag{A54}$$

which can be expressed in information vectors as

$$\Xi^{X} = \Xi_{1}^{X}\mathbf{W}^{\mathrm{T}}\mathbf{J} + o(\epsilon). \tag{A55}$$

From Theorem 3, we have

$$\Xi^{X*} = \tilde{\mathbf{B}}^{\mathrm{T}}\,\Xi^{Y}\left((\Xi^{Y})^{\mathrm{T}}\Xi^{Y}\right)^{-1}. \tag{A56}$$

As a result, we have

$$\begin{aligned}\|\Xi^{Y}(\Xi^{X*})^{\mathrm{T}} - \Xi^{Y}(\Xi^{X})^{\mathrm{T}}\|_{\mathrm{F}}^{2} &= \|((\Xi^{Y})^{\mathrm{T}}\Xi^{Y})^{1/2}((\Xi^{X*})^{\mathrm{T}} - (\Xi^{X})^{\mathrm{T}})\|_{\mathrm{F}}^{2} \\ &= \left\|((\Xi^{Y})^{\mathrm{T}}\Xi^{Y})^{1/2}\cdot\left((\Xi^{X*})^{\mathrm{T}} - \mathbf{J}\mathbf{W}(\Xi_{1}^{X})^{\mathrm{T}} - o(\epsilon)\right)\right\|_{\mathrm{F}}^{2} \\ &= \left\|((\Xi^{Y})^{\mathrm{T}}\Xi^{Y})^{1/2}\cdot\left((\Xi^{X*})^{\mathrm{T}} - \mathbf{J}\mathbf{W}(\Xi_{1}^{X})^{\mathrm{T}}\right)\right\|_{\mathrm{F}}^{2} + o(\epsilon^{2}) \\ &= \left\|((\Xi^{Y})^{\mathrm{T}}\Xi^{Y})^{1/2}\mathbf{J}\cdot\left(\mathbf{J}^{-1}(\Xi^{X*})^{\mathrm{T}} - \mathbf{W}(\Xi_{1}^{X})^{\mathrm{T}}\right)\right\|_{\mathrm{F}}^{2} + o(\epsilon^{2}) \\ &= \|\Theta\tilde{\mathbf{B}}_{1} - \Theta\mathbf{W}(\Xi_{1}^{X})^{\mathrm{T}}\|_{\mathrm{F}}^{2} + o(\epsilon^{2}),\end{aligned} \tag{A57}$$

where the third equality follows from the fact that [cf. (A41)] $\tilde{s}(x) = O(\epsilon)$ and $\tilde{v}(y) = O(1)$, and the last equality follows from the definitions $\tilde{\mathbf{B}}_{1} \triangleq \mathbf{J}^{-1}(\Xi^{X*})^{\mathrm{T}}$ and $\Theta \triangleq ((\Xi^{Y})^{\mathrm{T}}\Xi^{Y})^{1/2}\mathbf{J}$.

For the second term, from (A50) and (A51), we have

$$\begin{aligned}\kappa^{(v,b)}(s,s^*) &= [(\mu_s - \mu_{s^*}) + \mu_{s^*}]^{\mathrm{T}}\mathbf{\Lambda}_{\tilde{v}(Y)}[(\mu_s - \mu_{s^*}) + \mu_{s^*}] \\ &\quad - \mu_{s^*}^{\mathrm{T}}\mathbf{\Lambda}_{\tilde{v}(Y)}\mu_{s^*} + 2(\mu_s - \mu_{s^*})^{\mathrm{T}}\mathbb{E}_{P_Y}[\tilde{v}(Y)\tilde{d}(Y)] \\ &= (\mu_s - \mu_{s^*})^{\mathrm{T}}\mathbf{\Lambda}_{\tilde{v}(Y)}(\mu_s - \mu_{s^*}) + 2(\mu_s - \mu_{s^*})^{\mathrm{T}}\left(\mathbf{\Lambda}_{\tilde{v}(Y)}\mu_{s^*} + \mathbb{E}_{P_Y}[\tilde{v}(Y)\tilde{d}(Y)]\right) \\ &= (\mu_s - \mu_{s^*})^{\mathrm{T}}\mathbf{\Lambda}_{\tilde{v}(Y)}(\mu_s - \mu_{s^*}).\end{aligned} \tag{A58}$$

Combining (A57) and (A58) finishes the proof.

Appendix G. Analyses of Hidden Layer Parameters

First, from (A53), the bias $c(z)$ of hidden layer is (when $\mu_t \neq 0$, the formula should be modified as $c(z) = \sigma^{-1}(\mu_s^*(z)) - \mu_t^{\mathrm{T}} w + o(\epsilon)$.)

$$c(z) = \sigma^{-1}(\mu_s^*(z)) + o(\epsilon).$$

To obtain μ_s^*, let us define $\sigma_{\min} \triangleq \inf_x \sigma(x), \sigma_{\max} \triangleq \sup_x \sigma(x)$. Then, the optimal μ_s is the solution of

$$\begin{aligned}&\underset{\mu_s}{\text{minimize}} && (\mu_s - \mu_{s^*})^{\mathrm{T}}\mathbf{\Lambda}_{\tilde{v}(Y)}(\mu_s - \mu_{s^*}) \\ &\text{subject to} && \sigma_{\min} \preceq \mu_s \preceq \sigma_{\max}.\end{aligned} \tag{A59}$$

If μ_{s^*} satisfies the constraint of (A59), then it is the optimal solution. Otherwise, some elements of μ_s^* will become either $\sigma_{\min}$ or $\sigma_{\max}$, known as the saturation phenomenon [21].

To obtain $\mathbf{W}^*$, let

$$\begin{aligned}\tilde{\mathbf{B}}_1' &\triangleq \Theta\tilde{\mathbf{B}}_1 = \left((\Xi^Y)^{\mathrm{T}}\Xi^Y\right)^{-1/2}(\Xi^Y)^{\mathrm{T}}\tilde{\mathbf{B}}, \\ \mathbf{W}' &\triangleq \Theta\mathbf{W} = \left((\Xi^Y)^{\mathrm{T}}\Xi^Y\right)^{1/2}\mathbf{J}\mathbf{W}.\end{aligned}$$

Then, the optimal $\mathbf{W}'$ is given by

$$\mathbf{W}'^* = \underset{\mathbf{W}'}{\arg\min}\ \|\tilde{\mathbf{B}}_1' - \mathbf{W}'(\Xi_1^X)^{\mathrm{T}}\|_{\mathrm{F}}^2 = \tilde{\mathbf{B}}_1'\Xi_1^X((\Xi_1^X)^{\mathrm{T}}\Xi_1^X)^{-1}. \tag{A60}$$

Hence, $\mathbf{W}^*$ is given by

$$\begin{aligned}\mathbf{W}^* = \Theta^{-1}\mathbf{W}'^* &= \Theta^{-1}\tilde{\mathbf{B}}_1'\Xi_1^X((\Xi_1^X)^{\mathrm{T}}\Xi_1^X)^{-1} \\ &= \tilde{\mathbf{B}}_1\Xi_1^X((\Xi_1^X)^{\mathrm{T}}\Xi_1^X)^{-1} \\ &= \mathbf{J}^{-1}\cdot[\Xi^Y((\Xi^Y)^{\mathrm{T}}\Xi^Y)^{-1}]^{\mathrm{T}}\tilde{\mathbf{B}}\,\Xi_1^X((\Xi_1^X)^{\mathrm{T}}\Xi_1^X)^{-1},\end{aligned}$$

where the term $\tilde{\mathbf{B}}\,\Xi_1^X((\Xi_1^X)^{\mathrm{T}}\Xi_1^X)^{-1}$ corresponds to a feature projection of $\tilde{t}(X)$:

$$\tilde{\mathbf{B}}\,\Xi_1^X\left((\Xi_1^X)^{\mathrm{T}}\Xi_1^X\right)^{-1} \leftrightarrow \mathbb{E}_{P_{X|Y}}\left[\mathbf{\Lambda}_{\tilde{t}(X)}^{-1}\tilde{t}(X)\,\middle|\,Y\right]. \tag{A61}$$

As a consequence, this multi-layer neural network conducts a generalized feature projection between features extracted from different layers. Note that the projected feature $\mathbb{E}_{P_{\tilde{t}|Y}}\left[\mathbf{\Lambda}_{\tilde{t}}^{-1}\tilde{t}\middle|Y\right]$ depends only on the distribution $P_{\tilde{t}|Y}$ and does not depend on the distribution $P_{X|Y}$. Therefore, the above computations can be accomplished without knowing the hidden random variable X and can be applied to general cases.

References

1. He, K.; Zhang, X.; Ren, S.; Sun, J. Deep residual learning for image recognition. In Proceedings of the IEEE Conference on Computer Vision and Pattern Recognition, Las Vegas, NV, USA, 26 June–1 July 2016; pp. 770–778.
2. Devlin, J.; Chang, M.W.; Lee, K.; Toutanova, K. BERT: Pre-training of Deep Bidirectional Transformers for Language Understanding. In Proceedings of the 2019 Conference of the North American Chapter of the Association for Computational Linguistics: Human Language Technologies, Minneapolis, MN, USA, 3–5 June 2019; Volume 1 (Long and Short Papers); Association for Computational Linguistics: Minneapolis, MN, USA, 2019; pp. 4171–4186.
3. Brown, T.; Mann, B.; Ryder, N.; Subbiah, M.; Kaplan, J.D.; Dhariwal, P.; Neelakantan, A.; Shyam, P.; Sastry, G.; Askell, A.; et al. Language Models are Few-Shot Learners. In *Advances in Neural Information Processing Systems*; Larochelle, H., Ranzato, M., Hadsell, R., Balcan, M.F., Lin, H., Eds.; Curran Associates, Inc.: Red Hook, NY, USA, 2020; Volume 33, pp. 1877–1901.
4. Silver, D.; Huang, A.; Maddison, C.J.; Guez, A.; Sifre, L.; Van Den Driessche, G.; Schrittwieser, J.; Antonoglou, I.; Panneershelvam, V.; Lanctot, M.; et al. Mastering the game of Go with deep neural networks and tree search. *Nature* **2016**, *529*, 484–489. [CrossRef]
5. Arulkumaran, K.; Cully, A.; Togelius, J. Alphastar: An evolutionary computation perspective. In Proceedings of the Genetic and Evolutionary Computation Conference Companion, Prague, Czech Republic, 13–17 July 2019; pp. 314–315.
6. MacKay, D.J.C. *Information Theory, Inference, and Learning Algorithms*; Cambridge University Press: Cambridge, UK, 2003; ISBN 9780521642989.
7. Zintgraf, L.M.; Cohen, T.S.; Adel, T.; Welling, M. Visualizing Deep Neural Network Decisions: Prediction Difference Analysis. In Proceedings of the 5th International Conference on Learning Representations, ICLR 2017, Toulon, France, 24–26 April 2017.
8. Papyan, V.; Han, X.; Donoho, D.L. Prevalence of neural collapse during the terminal phase of deep learning training. *Proc. Natl. Acad. Sci. USA* **2020**, *117*, 24652–24663. [CrossRef]
9. Adebayo, J.; Gilmer, J.; Muelly, M.; Goodfellow, I.; Hardt, M.; Kim, B. Sanity Checks for Saliency Maps. In *Advances in Neural Information Processing Systems*; Bengio, S., Wallach, H., Larochelle, H., Grauman, K., Cesa-Bianchi, N., Garnett, R., Eds.; Curran Associates, Inc.: Red Hook, NY, USA, 2018; Volume 31.
10. Guidotti, R.; Monreale, A.; Ruggieri, S.; Turini, F.; Giannotti, F.; Pedreschi, D. A survey of methods for explaining black box models. *ACM Comput. Surv. (CSUR)* **2018**, *51*, 1–42. [CrossRef]
11. Jacot, A.; Gabriel, F.; Hongler, C. Neural Tangent Kernel: Convergence and Generalization in Neural Networks. In *Advances in Neural Information Processing Systems*; Bengio, S., Wallach, H., Larochelle, H., Grauman, K., Cesa-Bianchi, N., Garnett, R., Eds.; Curran Associates, Inc.: Red Hook, NY, USA, 2018; Volume 31.
12. Mei, S.; Montanari, A.; Nguyen, P.M. A mean field view of the landscape of two-layer neural networks. *Proc. Natl. Acad. Sci. USA* **2018**, *115*, E7665–E7671. [CrossRef]
13. Arora, S.; Du, S.; Hu, W.; Li, Z.; Wang, R. Fine-grained analysis of optimization and generalization for overparameterized two-layer neural networks. In Proceedings of the International Conference on Machine Learning, PMLR, Long Beach, CA, USA, 9–15 June 2019; pp. 322–332.
14. Cover, T.M.; Thomas, J.A. *Elements of Information Theory*; John Wiley & Sons: Hoboken, NJ, USA, 2012.
15. Huang, S.L.; Xu, X.; Zheng, L.; Wornell, G.W. An information theoretic interpretation to deep neural networks. In Proceedings of the 2019 IEEE International Symposium on Information Theory (ISIT), Paris, France, 7–12 July 2019; pp. 1984–1988.
16. Tishby, N.; Zaslavsky, N. Deep learning and the information bottleneck principle. In Proceedings of the Information Theory Workshop (ITW), Jerusalem, Israel, 26 April–1 May 2015; pp. 1–5.
17. Goldfeld, Z.; Polyanskiy, Y. The information bottleneck problem and its applications in machine learning. *IEEE J. Sel. Areas Inf. Theory* **2020**, *1*, 19–38. [CrossRef]
18. Huang, S.L.; Makur, A.; Zheng, L.; Wornell, G.W. An information-theoretic approach to universal feature selection in high-dimensional inference. In Proceedings of the 2017 IEEE International Symposium on Information Theory (ISIT), Aachen, Germany, 25–30 June 2017; pp. 1336–1340.
19. Arjovsky, M.; Chintala, S.; Bottou, L. Wasserstein generative adversarial networks. In Proceedings of the International Conference on Machine Learning, PMLR, Sydney, Australia, 6–11 August 2017; pp. 214–223.
20. Saxe, A.M.; Bansal, Y.; Dapello, J.; Advani, M.; Kolchinsky, A.; Tracey, B.D.; Cox, D.D. On the information bottleneck theory of deep learning. *J. Stat. Mech. Theory Exp.* **2019**, *2019*, 124020. [CrossRef]
21. Goodfellow, I.; Bengio, J.; Courville, A. *Deep Learning*; MIT Press: Cambridge, MA, USA, 2017.
22. Olga, R.; Jia, D.; Hao, S.; Jonathan, K.; Sanjeev, S.; Sean, M.; Zhiheng, H.; Andrej, K.; Aditya, K.; Michael, B.; et al. ImageNet Large Scale Visual Recognition Challenge. *Int. J. Comput. Vis.* **2015**, *115*, 211–252. [CrossRef]
23. Huang, S.L.; Zheng, L. Linear information coupling problems. In Proceedings of the 2012 IEEE International Symposium on Information Theory Proceedings, Cambridge, MA, USA, 1–6 July 2012; pp. 1029–1033.
24. Huang, S.L.; Makur, A.; Wornell, G.W.; Zheng, L. On universal features for high-dimensional learning and inference. *arXiv* **2019**, arXiv:1911.09105.
25. Hirschfeld, H.O. A connection between correlation and contingency. *Proc. Camb. Phil. Soc.* **1935**, *31*, 520–524. [CrossRef]
26. Gebelein, H. Das statistische problem der Korrelation als variations-und Eigenwertproblem und sein Zusammenhang mit der Ausgleichungsrechnung. *Z. Angew. Math. Mech.* **1941**, *21*, 364–379. [CrossRef]
27. Rényi, A. On Measures of Dependence. *Acta Math. Acad. Sci. Hung.* **1959**, *10*, 441–451. [CrossRef]

28. du Pin Calmon, F.; Makhdoumi, A.; Médard, M.; Varia, M.; Christiansen, M.; Duffy, K.R. Principal inertia components and applications. *IEEE Trans. Inf. Theory* **2017**, *63*, 5011–5038. [CrossRef]
29. Hsu, H.; Asoodeh, S.; Salamatian, S.; Calmon, F.P. Generalizing bottleneck problems. In Proceedings of the 2018 IEEE International Symposium on Information Theory (ISIT), Vail, CO, USA, 17–22 June 2018; pp. 531–535.
30. Hsu, H.; Salamatian, S.; Calmon, F.P. Correspondence analysis using neural networks. In Proceedings of the 22nd International Conference on Artificial Intelligence and Statistics, PMLR, Okinawa, Japan, 16–18 April 2019; pp. 2671–2680.
31. Anantharam, V.; Gohari, A.; Kamath, S.; Nair, C. On hypercontractivity and a data processing inequality. In Proceedings of the 2014 IEEE International Symposium on Information Theory, Honolulu, HI, USA, 29 June–4 July 2014; pp. 3022–3026.
32. Raginsky, M. Strong data processing inequalities and Φ-Sobolev inequalities for discrete channels. *IEEE Trans. Inf. Theory* **2016**, *62*, 3355–3389. [CrossRef]
33. Polyanskiy, Y.; Wu, Y. Strong data-processing inequalities for channels and Bayesian networks. In *Convexity and Concentration*; Springer: Berlin/Heidelberg, Germany, 2017; pp. 211–249.
34. Greenacre, M.J. *Theory and Applications Of Correspondence Analysis*; Academic Press: London, UK, 1984.
35. Wang, H.; Vo, L.; Calmon, F.P.; Médard, M.; Duffy, K.R.; Varia, M. Privacy with estimation guarantees. *IEEE Trans. Inf. Theory* **2019**, *65*, 8025–8042. [CrossRef]
36. Breiman, L.; Friedman, J.H. Estimating Optimal Transformations for Multiple Regression and Correlation. *J. Am. Stat. Assoc.* **1985**, *80*, 614–619.
37. Sutskever, I.; Vinyals, O.; Le, Q.V. Sequence to sequence learning with neural networks. In Proceedings of the Advances in Neural Information Processing Systems, Montreal, QC, Canada, 8–13 December 2014; pp. 3104–3112.
38. Hochreiter, S.; Schmidhuber, J. Long short-term memory. *Neural Comput.* **1997**, *9*, 1735–1780. [CrossRef]
39. Hastie, T.; Tibshirani, R.; Friedman, J. Neural Networks. In *The Elements of Statistical Learning: Data Mining, Inference, and Prediction*; Springer: New York, NY, USA, 2009; pp. 389–416. [CrossRef]
40. Cybenko, G. Approximation by superpositions of a sigmoidal function. *Math. Control. Signals Syst.* **1989**, *2*, 303–314. [CrossRef]
41. Stoer, J.; Bulirsch, R. *Introduction to Numerical Analysis*; Springer Science & Business Media: Berlin/Heidelberg, Germany, 2013; Volume 12.
42. Alain, G.; Bengio, Y. Understanding intermediate layers using linear classifier probes. In Proceedings of the 5th International Conference on Learning Representations, ICLR 2017, Toulon, France, 24–26 April 2017.
43. Sutskever, I.; Martens, J.; Dahl, G.; Hinton, G. On the importance of initialization and momentum in deep learning. In Proceedings of the International Conference on Machine Learning, PMLR, Atlanta, GA, USA, 17–19 June 2013; pp. 1139–1147.
44. Simonyan, K.; Zisserman, A. Very Deep Convolutional Networks for Large-Scale Image Recognition. In Proceedings of the 3rd International Conference on Learning Representations, ICLR 2015, San Diego, CA, USA, 7–9 May 2015; Bengio, Y., LeCun, Y., Eds.; Conference Track Proceedings.
45. Howard, A.G.; Zhu, M.; Chen, B.; Kalenichenko, D.; Wang, W.; Weyand, T.; Andreetto, M.; Adam, H. Mobilenets: Efficient convolutional neural networks for mobile vision applications. *arXiv* **2017**, arXiv:1704.04861.
46. Huang, G.; Liu, Z.; Weinberger, K.Q.; van der Maaten, L. Densely connected convolutional networks. In Proceedings of the IEEE Conference on Computer Vision and Pattern Recognition, Honolulu, HI, USA, 21–29 July 2017; Volume 1, p. 3.
47. Chollet, F. Xception: Deep learning with depthwise separable convolutions. In Proceedings of the IEEE Conference on Computer Vision and Pattern Recognition, Honolulu, HI, USA, 21–29 July 2017; pp. 1251–1258.
48. Szegedy, C.; Vanhoucke, V.; Ioffe, S.; Shlens, J.; Wojna, Z. Rethinking the inception architecture for computer vision. In Proceedings of the IEEE Conference on Computer Vision and Pattern Recognition, Las Vegas, NV, USA, 26 June–1 July 2016; pp. 2818–2826.
49. Szegedy, C.; Ioffe, S.; Vanhoucke, V.; Alemi, A.A. Inception-v4, Inception-ResNet and the Impact of Residual Connections on Learning. In Proceedings of the AAAI, San Francisco, CA, USA, 4–9 February 2017; pp. 4278–4284.
50. Xu, X.; Huang, S.L.; Zheng, L.; Zhang, L. The geometric structure of generalized softmax learning. In Proceedings of the 2018 IEEE Information Theory Workshop (ITW), Guangzhou, China, 25–29 November 2018; pp. 1–5.
51. Wen, W.; Wu, C.; Wang, Y.; Chen, Y.; Li, H. Learning structured sparsity in deep neural networks. *Adv. Neural Inf. Process. Syst.* **2016**, *29*, 2074–2082.
52. Wang, L.; Wu, J.; Huang, S.L.; Zheng, L.; Xu, X.; Zhang, L.; Huang, J. An efficient approach to informative feature extraction from multimodal data. In Proceedings of the AAAI Conference on Artificial Intelligence, Honolulu, HI, USA, 27 January–1 February 2019; Volume 33, pp. 5281–5288.
53. Lee, J.; Sattigeri, P.; Wornell, G. Learning new tricks from old dogs: Multi-source transfer learning from pre-trained networks. *Adv. Neural Inf. Process. Syst.* **2019**, *32*, 4370–4380.
54. Dembo, A.; Zeitouni, O. *Large Deviations Techniques and Applications*; Corrected Reprint of the Second (1998) Edition; Stochastic Modelling and Applied Probability; Springer: Berlin/Heidelberg, Germany, 2010; p. 38.
55. Sason, I.; Verdú, S. f-divergence Inequalities. *IEEE Trans. Inf. Theory* **2016**, *62*, 5973–6006. [CrossRef]
56. Eckart, C.; Young, G. The approximation of one matrix by another of lower rank. *Psychometrika* **1936**, *1*, 211–218. [CrossRef]

entropy MDPI

Article

Population Risk Improvement with Model Compression: An Information-Theoretic Approach †

Yuheng Bu [1,‡], Weihao Gao [2], Shaofeng Zou [3] and Venugopal V. Veeravalli [1,*]

1 Department of Electrical and Computer Engineering, University of Illinois at Urbana-Champaign, Urbana, IL 61820, USA; buyuheng@mit.edu
2 Bytedance Inc., Bellevue, WA 98004, USA; weihao.gao@bytedance.com
3 Department of Electrical Engineering, University at Buffalo, The State University of New York, Buffalo, NY 14221, USA; szou3@buffalo.edu
* Correspondence: vvv@illinois.edu
† This paper is an extended version of our paper published in AAAI Conference on Artificial Intelligence, New York, NY, USA, 7–12 February 2020.
‡ Current address: Department of Electrical Engineering and Computer Science, Massachusetts Institute of Technology, Cambridge, MA 02142, USA.

Abstract: It has been reported in many recent works on deep model compression that the population risk of a compressed model can be even better than that of the original model. In this paper, an information-theoretic explanation for this population risk improvement phenomenon is provided by jointly studying the decrease in the generalization error and the increase in the empirical risk that results from model compression. It is first shown that model compression reduces an information-theoretic bound on the generalization error, which suggests that model compression can be interpreted as a regularization technique to avoid overfitting. The increase in empirical risk caused by model compression is then characterized using rate distortion theory. These results imply that the overall population risk could be improved by model compression if the decrease in generalization error exceeds the increase in empirical risk. A linear regression example is presented to demonstrate that such a decrease in population risk due to model compression is indeed possible. Our theoretical results further suggest a way to improve a widely used model compression algorithm, i.e., Hessian-weighted K-means clustering, by regularizing the distance between the clustering centers. Experiments with neural networks are provided to validate our theoretical assertions.

Keywords: empirical risk; generalization error; K-means clustering; model compression; population risk; rate distortion theory; vector quantization

Citation: Bu, Y.; Gao, W.; Zou, S.; Veeravalli, V.V. Population Risk Improvement with Model Compression: An Information-Theoretic Approach. *Entropy* **2021**, *23*, 1255. https://doi.org/10.3390/e23101255

Academic Editors: Lizhong Zheng and Chao Tian

Received: 13 August 2021
Accepted: 23 September 2021
Published: 27 September 2021

1. Introduction

Although deep neural networks have achieved remarkable success in various domains [1], e.g., computer vision [2], playing games like Go [3], and autonomous driving [4], the improvement of the performance of deep models often comes with deeper layers and more complex network structures, which usually have a large number of parameters. For example, in the application of image classification, it takes over 200 MB to save the parameters of AlexNet [2] and more than 500 MB for VGG-16 net [5]. Hence, it is difficult to port such large models to resource-limited devices such as mobile devices and embedded systems, due to their limited storage, bandwidth, energy, and computational resources.

Due to this reason there has been a flurry of work on compressing deep neural networks (see [6–8] for recent surveys). Existing studies mainly focus on designing compression algorithms to reduce the memory and computational cost, while keeping the same level of population risk. In some recent papers [9–12], aggressive model compression algorithms have been proposed, which require 10% or fewer bits to store the compressed model compared to the storage required by the original model. Surprisingly, it has been observed empirically in these works that the population risk of the compressed model can often be even *better* than that of the original model. This phenomenon is counter-intuitive at first glance, since more compression generally leads to more information loss.

Indeed, a compressed model would usually have a larger empirical risk than the original one, since machine learning methods are usually trained by minimizing the empirical risk. On the other hand, model compression could possibly decrease the generalization error, since it can be interpreted as a regularization technique to avoid overfitting. As the population risk is the sum of the empirical risk and the generalization error, it is possible for the population risk to be reduced by model compression.

1.1. Contributions

In this paper, we provide an information-theoretic explanation for the population risk improvement with model compression by jointly characterizing the decrease in generalization error and the increase in empirical risk. Specifically, we focus on the case where the model is compressed based on a pre-trained model.

We first prove that model compression leads to a tightening of the information-theoretic generalization error bound in [13], and it can therefore be interpreted as a regularization method to reduce overfitting. Furthermore, by defining a distortion metric based on the difference in the empirical risk between the original model obtained by empirical risk minimization (ERM) and compressed models, we use rate distortion theory to characterize the increase in empirical risk as a function of the number of bits R used to describe the model. If the decrease in generalization error exceeds the increase in empirical risk, the population risk can be improved. An empirical illustration of this result for the MNIST dataset is provided in Figure 1, where model compression can lead to population risk improvement (details are given in Section 7). To better demonstrate our theoretical results, we investigate the example of linear regression comprehensively, where we develop explicit bounds on the generalization error and the increase in empirical risk.

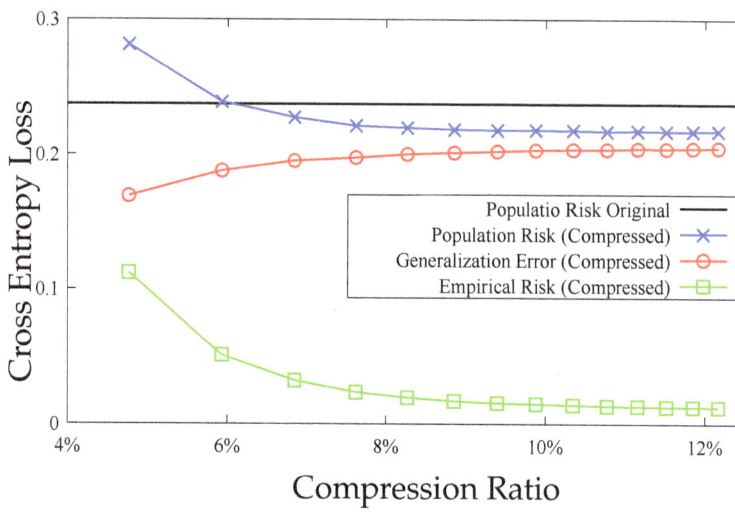

Figure 1. Population risk of the compressed model $\hat{W}$ and the original model W vs. compression ratio (ratio of the number of bits used for compressed model to the number of bits used for original model). The generalization error of $\hat{W}$ decreases and the empirical risk of $\hat{W}$ increases with more compression (smaller compression ratio). The population risk of $\hat{W}$ is less than that of W for compression ratios larger than 6% in this figure. As the compression ratio goes to 100% (no compression), the population risk of $\hat{W}$ will converge to that of the original model W.

Our results also suggest a way to improve a method for compression based on Hessian-weighted K-means clustering [11] in both scalar and vector case, by regularizing the distance between the clustering centers. Our experiments with neural networks validate our theoretical assertions and demonstrate the effectiveness of the proposed regularizer.

1.2. Related Works

There have been many studies on model compression for deep neural networks. The compression could be achieved by varying the training process, e.g., network structure optimization [14], low precision neural networks [15], and neural networks with binary weights [16,17]. Here we mainly discuss compression approaches that are applied on a pre-trained model.

Pruning, quantization, and matrix factorization are the most popular approaches to compressing pre-trained deep neural networks. The study of pruning algorithms for model compression which remove redundant parameters from neural networks dates back to the 1980s and 1990s [18–20]. More recently, an iterative pruning and retraining algorithm to further reduce the size of deep models was proposed in [9,21]. The method of network quantization or weight sharing, i.e., employing a clustering algorithm to group the weights in a neural network, and its variants, including vector quantization [22], soft quantization [23,24], fixed point quantization [25], transform quantization [26], and Hessian weighted quantization [11], have been extensively investigated. Matrix factorization, where low-rank approximation of the weights in neural networks is used instead of the original weight matrix, has also been widely studied in [27–29].

All of the aforementioned works demonstrate the effectiveness of their compression methods via comprehensive numerical experiments. Little research has been done to develop a theoretical understanding of how model compression affects performance. In work [30], an information-theoretic view of model compression via rate-distortion theory is provided, with the focus on characterizing the tradeoff between model compression and only the *empirical risk* of the compressed model. In [31–33], using a PAC-Bayesian framework, a non-vacuous generalization error bound for compressed model is derived based on its smaller model complexity.

In contrast to these works, instead of focusing on minimizing only the empirical risk as in [30], or minimizing only the generalization error as in [33], we use the mutual information based generalization error bound developed in [13,34] jointly with rate distortion theory to connect analyses of generalization error and empirical risk. This way, we are able to characterize the tradeoff between decrease in generalization error and the increase in empirical risk that results from model compression, and thus provide an understanding as to why model compression can improve the population risk. More importantly, our theoretical studies offer insights on designing practical model compression algorithms.

The rest of the paper is organized as follows. In Section 2, we provide relevant definitions and review relevant results from rate distortion theory. In Section 3, we prove that model compression results in the tightening of an information-theoretic generalization error upper bound. In Section 4, we use rate distortion theory to characterize the tradeoff between the increase in empirical risk and the decrease in generalization error that results from model compression. In Section 5, we quantify this tradeoff for a linear regression model. In Section 6, we discuss how the Hessian-weighted K-means clustering compression approach can be improved by using a regularizer motivated by our theoretical results. In Section 7, we provide some experiments with neural network models to validate our theoretical results and demonstrate the effectiveness of the proposed regularizer.

Notation 1. *For a random variable X generated from a distribution μ, we use $\mathbb{E}_{X\sim\mu}$ to denote the expectation taken over X with distribution μ. We use I_d to denote the d-dimensional identity matrix and $\|A\|$ to denote the spectral norm of a matrix A. The cumulant generating function (CGF) of a random variable X is defined as $\Lambda_X(\lambda) \triangleq \ln \mathbb{E}[e^{\lambda(X-\mathbb{E}X)}]$. All logarithms are the natural ones.*

2. Preliminaries

2.1. Review of Rate Distortion Theory

Rate distortion theory, introduced by Shannon [35], is a major branch of information theory that studies the fundamental limits of lossy data compression. It addresses the minimal number of bits per symbol, as measured by the rate R, to transmit a random variable W such that the receiver can reconstruct W without exceeding distortion D.

Specifically, let $W^m = \{W_1, W_2, \cdots, W_m\}$ denote a sequence of m i.i.d. random variables $W_i \in \mathcal{W}$ generated from a source distribution P_W. An encoder $f_m : \mathcal{W}^m \to \{1, 2, \cdots, M\}$ maps the message W^m into a codeword, and a decoder $g_m : \{1, 2, \cdots, M\} \to \hat{\mathcal{W}}^m$ reconstructs the message by an estimate $\hat{W}^m$ from the codeword, where $\hat{\mathcal{W}} \subseteq \mathcal{W}$ denotes the range of $\hat{W}$. A distortion metric $d : \mathcal{W} \times \mathcal{W} \to \mathbb{R}^+$ quantifies the difference between the original and reconstructed messages. The distortion between sequences w^m and $\hat{w}^m$ is defined to be

$$d(w^m, \hat{w}^m) \triangleq \frac{1}{m} \sum_{i=1}^{m} d(w_i, \hat{w}_i). \tag{1}$$

A commonly used distortion metric is the square distortion: $d(w, \hat{w}) = (w - \hat{w})^2$.

Definition 1. *An (m, M, D)-triple is achievable, if there exists a (probabilistic) encoder-decoder pair (f_m, g_m) such that the alphabet of codeword has size M and the expected distortion $\mathbb{E}[d(W^m; g_m(f_m(W^m)))] \le D$.*

Now we define the following rate-distortion and distortion-rate function for lossy data compression.

Definition 2. *The rate-distortion function and the distortion-rate function are defined as*

$$R(D) \triangleq \lim_{m \to \infty} \frac{1}{m} \log_2 M^*(m, D), \tag{2}$$

$$D(R) \triangleq \lim_{m \to \infty} D^*(m, R), \tag{3}$$

where $M^*(m, D) \triangleq \min\{M : (m, M, D)$ is achievable$\}$ and $D^*(m, R) \triangleq \min\{D : (m, 2^{mR}, D)$ is achievable$\}$.

The main theorem of rate distortion theory is as follows.

Lemma 1 ([36]). *For an i.i.d. source W with distribution P_W and distortion function $d(w, \hat{w})$:*

$$R(D) = \min_{P_{\hat{W}|W} : \mathbb{E}[d(W, \hat{W})] \le D} I(W; \hat{W}), \tag{4}$$

$$D(R) = \min_{P_{\hat{W}|W} : I(W; \hat{W}) \le R} \mathbb{E}[d(W, \hat{W})], \tag{5}$$

where $I(W; \hat{W}) \triangleq \mathbb{E}_{W, \hat{W}}[\ln \frac{P_{W, \hat{W}}}{P_W P_{\hat{W}}}]$ denotes the mutual information between W and $\hat{W}$.

The rate-distortion function quantifies the smallest number of bits required to compress the data given the distortion, and the distortion-rate function quantifies the minimal distortion that can be achieved under the rate constraint.

2.2. Generalization Error

Consider an instance space $\mathcal{Z}$, a hypothesis space $\mathcal{W}$, and a non-negative loss function $\ell : \mathcal{W} \times \mathcal{Z} \to \mathbb{R}^+$. A training dataset $S = \{Z_1, \cdots, Z_n\}$ consists of n i.i.d samples $Z_i \in \mathcal{Z}$

drawn from an unknown distribution μ. The goal of a supervised learning algorithm is to find an output hypothesis $w \in \mathcal{W}$ that minimizes the population risk:

$$L_\mu(w) \triangleq \mathbb{E}_{Z\sim\mu}[\ell(w, Z)]. \tag{6}$$

In practice, μ is unknown, and therefore $L_\mu(w)$ cannot be computed directly. Instead, the empirical risk of w on the training dataset S is studied, which is defined as

$$L_S(w) \triangleq \frac{1}{n}\sum_{i=1}^{n} \ell(w, Z_i). \tag{7}$$

A learning algorithm can be characterized by a randomized mapping from the training dataset S to a hypothesis W according to a conditional distribution $P_{W|S}$. The (expected) generalization error of a supervised learning algorithm is the expected difference between the population risk of the output hypothesis and its empirical risk on the training dataset:

$$\text{gen}(\mu, P_{W|S}) \triangleq \mathbb{E}_{W,S}[L_\mu(W) - L_S(W)], \tag{8}$$

where the expectation is taken over the joint distribution $P_{S,W} = P_S \otimes P_{W|S}$. The generalization error is used to measure the extent to which the learning algorithm overfits the training data.

3. Compression Can Improve Generalization

In this section, we show that lossy compression can lead to a tighter mutual information based generalization error upper bound, which potentially reduces the generalization error of a supervised learning algorithm.

We start from the following lemma which provides an upper bound on the generalization error using the mutual information $I(S;W)$ between training dataset S and the output of the learning algorithm W.

Lemma 2 ([13]). *Suppose $\ell(w, Z)$ is σ-sub-Gaussian (A random variable X is σ-sub-Gaussian if $\Lambda_X(\lambda) \le \frac{\sigma^2\lambda^2}{2}, \forall \lambda \in \mathbb{R}$.) under $Z \sim \mu$ for all $w \in \mathcal{W}$, then*

$$|\text{gen}(\mu, P_{W|S})| \le \sqrt{\frac{2\sigma^2}{n} I(S;W)}. \tag{9}$$

Compression can be viewed as a post-processing of the output of a learning algorithm. The output model W generated by a learning algorithm can be quantized, pruned, factorized, or even perturbed by noise, which results in a compressed model $\hat{W}$. Assume that the compression algorithm is only based on W and can be described by a conditional distribution $P_{\hat{W}|W}$. Then the following Markov chain holds: $S \to W \to \hat{W}$. By the data processing inequality,

$$I(S;\hat{W}) \le \min\{I(W;\hat{W}), I(S,W)\}.$$

Thus, we have the following theorem characterizing the generalization error of the compressed model.

Theorem 1. *Consider a learning algorithm $P_{W|S}$, a compression algorithm $P_{\hat{W}|W}$, and suppose $\ell(\hat{w}, Z)$ is σ-sub-Gaussian under $Z \sim \mu$ for all $\hat{w} \in \hat{\mathcal{W}}$. Then*

$$|\text{gen}(\mu, P_{\hat{W}|S})| \le \sqrt{\frac{2\sigma^2}{n} \min\{I(W;\hat{W}), I(S,W)\}}. \tag{10}$$

Note that the generalization error upper bound in Theorem 1 for the compressed model is always no greater than the one in Lemma 2. This allows for the interpretation of compression as a regularization technique to reduce the generalization error.

4. Generalization Error and Model Distortion

In this section, we define a distortion metric in model compression that allows us to relate the distortion (the increase in empirical risk) due to compression with the reduction in the generalization error bound discussed in Section 3.

4.1. Distortion Metric in Model Compression

The expected population risk of a model W can be written as

$$\mathbb{E}_W[L_\mu(W)] = \mathbb{E}[L_S(W)] + \text{gen}(\mu, P_{W|S}), \tag{11}$$

where the first term, which is the expected empirical risk, reflects how well the model W fits the training data, while the second term demonstrates how well the model generalizes. In the empirical risk minimization framework, we control both terms by (1) minimizing the empirical risk of W directly or using other stochastic optimization algorithms, and (2) using regularization methods to control the generalization error, e.g., early stopping and dropout [1].

Now, consider the expected population risk of the compressed model $\hat{W}$:

$$\begin{aligned}\mathbb{E}_{\hat{W}}[L_\mu(\hat{W})] &= \mathbb{E}[L_\mu(\hat{W}) - L_S(\hat{W}) + L_S(\hat{W}) - L_S(W) + L_S(W)] \\ &= \mathbb{E}[L_S(W)] + \text{gen}(\mu, P_{\hat{W}|S}) + \mathbb{E}[L_S(\hat{W}) - L_S(W)].\end{aligned} \tag{12}$$

Compared with (11), we note that the first empirical risk term is independent of the compression algorithm, the second generalization error term can be upper bounded by Theorem 1, and the third term $\mathbb{E}[L_S(\hat{W}) - L_S(W)]$ quantifies the increase in the empirical risk if we use the compressed model $\hat{W}$ instead of the original model W. We then define the following distortion metric for model compression:

$$d_S(w, \hat{w}) \triangleq L_S(\hat{w}) - L_S(w), \tag{13}$$

which is the difference in the empirical risk between the compressed model $\hat{W}$ and the original model W. In general, function $d_S(w, \hat{w})$ is not always non-negative. However, for ERM solution W, which is obtained by minimizing the empirical risk $L_S(W)$, $d_S(w, \hat{w}) \geq 0$, which ensures that $d_S(w, \hat{w})$ is a valid distortion metric. By Theorem 1, it follows that

$$\mathbb{E}_{S,W,\hat{W}}[L_\mu(\hat{W}) - L_S(W)] \leq \sqrt{\frac{2\sigma^2}{n} I(W;\hat{W})} + \mathbb{E}_{S,W,\hat{W}}[d_S(\hat{W}, W)] \triangleq \mathcal{L}_{S,W}(P_{\hat{W}|W}), \tag{14}$$

where $\mathcal{L}_{S,W}(P_{\hat{W}|W})$ is an upper bound on the expected difference between the population risk of $\hat{W}$ and the empirical risk of the original model W on training dataset S. Note that $L_S(W)$ is independent of the compression algorithm. Therefore, the bound in (14) can be viewed as an upper bound of the population risk of the compressed model $\hat{W}$.

4.2. Population Risk Improvement

By Lemma 1, the smallest distortion that can be achieved at rate R is $D(R) = \min_{I(W;\hat{W}) \leq R} \mathbb{E}_{S,W,\hat{W}}[d_S(\hat{W}, W)]$. Thus, the tightest bound in (14) that can be achieved at rate R is given in the following theorem.

Theorem 2. *Suppose the assumptions in Theorem 1 hold, $P_{W|S}$ minimizes the empirical risk $L_S(W)$, and $I(W;\hat{W}) = R$, then*

$$\min_{P_{\hat{W}|W}: I(W;\hat{W}) = R} \mathbb{E}_{S,W,\hat{W}}[L_\mu(\hat{W}) - L_S(W)] \leq \sqrt{\frac{2\sigma^2}{n} R} + D(R). \tag{15}$$

From the properties of the distortion-rate function [36], we know that $D(R)$ is a decreasing function of R. Thus, we see that as R decreases the first term in (15), which corresponds to the generalization error, decreases, while the second term, which corresponds to the empirical risk, increases. Due to this tradeoff, it may be possible for the bound in (15) to be smaller due to compression, i.e., using a smaller rate R. This indicates that the population risk could improve with compression algorithm, which minimizes the upper bound $\mathcal{L}_{S,W}(P_{\hat{W}|W})$.

Remark 1. *In order to conclude definitively that the population risk can be improved with compression, we need to find a lower bound (as a function of R) to match (at least in the order sense) the upper bound in Theorem 2. This appears to be difficult to construct in general. One approach might be to use the same decomposition as in (12) and develop lower bounds for* $\min_{I(W;\hat{W})=R} \mathrm{gen}(\mu, P_{\hat{W}|S})$ *and* $\min_{I(W;\hat{W})=R} \mathbb{E}_{S,W,\hat{W}}[d_S(\hat{W}, W)]$ *independently. However, such an approach runs into the following issues: (1) such a lower bound would be loose since the compression algorithm* $P_{\hat{W}|W}$ *that minimizes generalization error, the one that minimizes the distortion, and the one that minimizes the sum of the two can be quite different; and (2) a lower bound for generalization error needs to be developed, which appears to be difficult, with existing literature mainly focusing on lower bounding the excess risk, e.g., [37].*

As will be shown in Section 7, we can actually improve the population risk with a well designed compression algorithm in practical applications.

5. Example: Linear Regression

In this section, we comprehensively explore the example of linear regression to get a better understanding of the results in Section 4. To this end, we develop explicit upper bounds for generalization error and distortion-rate function $D(R)$. All the proofs of the lemmas and theorems are provided in the Appendixes A–D.

Suppose that the dataset $S = \{Z_1, \cdots, Z_n\} = \{(X_1, Y_1), \cdots, (X_n, Y_n)\}$ is generated from the following linear model with weight vector $w^* = (w^{*(1)}, \cdots, w^{*(d)}) \in \mathbb{R}^d$,

$$Y_i = X_i^\top w^* + \varepsilon_i, \quad i = 1, \cdots, n, \tag{16}$$

where X_i's are i.i.d. d-dimensional random vectors with distribution $\mathcal{N}(0, \Sigma_X)$, and $\varepsilon_i \sim \mathcal{N}(0, \sigma'^2)$ denotes i.i.d. Gaussian noise. We adopt the mean squared error as the loss function, and the corresponding empirical risk on S is

$$L_S(w) = \frac{1}{n}\sum_{i=1}^{n}(Y_i - X_i^\top w)^2 = \frac{1}{n}\|Y - X^\top w\|_2^2, \tag{17}$$

for $w \in \mathcal{W} = \mathbb{R}^d$, where $X \in \mathbb{R}^{d\times n}$ denotes all the input samples, and $Y \in \mathbb{R}^n$ denotes the responses. If $n > d$, the ERM solution is

$$W = (XX^\top)^{-1}XY, \tag{18}$$

which is deterministic given S. Its generalization error can be computed exactly as in the following lemma (see Appendix A for detailed proof).

Lemma 3. *If* $n > d + 1$, *then*

$$\mathrm{gen}(\mu, P_{W|S}) = \frac{\sigma'^2 d}{n}(2 + \frac{d+1}{n-d-1}). \tag{19}$$

5.1. Information-Theoretic Generalization Bounds for Compressed Linear Model

We note that the mutual information based bound in Lemma 2 is not applicable for this linear regression model, since W is a deterministic function of S, and $I(S; W) = \infty$. However, this issue can be resolved if we post-process the ERM solution W by a compression algorithm and upper bound the generalization error by $I(\hat{W};W)$ as shown in Theorem 1.

Consider a compression algorithm, which maps the original weights $W \in \mathbb{R}^d$ to the compressed model $\hat{W} \in \hat{\mathcal{W}} \subseteq \mathbb{R}^d$. For a fixed and compact $\hat{\mathcal{W}}$, we define

$$C(w^*) \triangleq \sup_{\hat{w}\in\hat{\mathcal{W}}} \|\hat{w} - w^*\|_2^2, \tag{20}$$

which measures the largest distance between the reconstruction $\hat{w}$ and the optimal weights w^*. The following proposition provides an upper bound on the generalization error of the compressed model $\hat{W}$, and the detailed proof is provided in Appendix B.

Proposition 1. *Consider the ERM solution $W = (XX^\top)^{-1}XY$, and suppose $\hat{\mathcal{W}}$ is compact, then*

$$\text{gen}(\mu, P_{\hat{W}|S}) \leq 2\sigma_\ell^{*2}\sqrt{\frac{I(W;\hat{W})}{n}}, \tag{21}$$

where $\sigma_\ell^{*2} \triangleq C(w^*)\|\Sigma_X\| + \sigma'^2$.

5.2. Distortion-Rate Function for Linear Model

We now provide an upper bound on the distortion-rate function $D(R)$ for the linear regression model. Note that $\nabla L_S(W) = 0$, since W minimizes the empirical risk. The Hessian matrix of the loss function is

$$H_S(W) = \frac{1}{n}XX^\top, \tag{22}$$

which is not a function of W. Then, the distortion function can be written as:

$$\begin{aligned}\mathbb{E}_{S,W,\hat{W}}[d_S(\hat{W},W)] &= \mathbb{E}_{S,W,\hat{W}}[L_S(\hat{W}) - L_S(W)] \\ &= \mathbb{E}_{S,W,\hat{W}}[(\hat{W}-W)^\top \frac{1}{n}XX^\top(\hat{W}-W)].\end{aligned} \tag{23}$$

The following theorem characterizes upper bounds for $R(D)$ and $D(R)$ for linear regression.

Proposition 2. *For the ERM solution $W = (XX^\top)^{-1}XY$, we have*

$$R(D) \leq \frac{d}{2}\left(\ln\frac{d\sigma'^2}{(n-d-1)D}\right)^+, \quad D \geq 0, \tag{24}$$

$$D(R) \leq \frac{d\sigma'^2}{n-d-1}e^{-\frac{2R}{d}}, \quad R \geq 0, \tag{25}$$

where $(x)^+ = \max\{0, x\}$.

Proof sketch. The proof of the upper bound for $R(D)$ is based on considering a Gaussian random vector which has the same mean and covariance matrix as W. In addition, the upper bound is achieved when $W - \hat{W}$ is independent of the dataset S with the following conditional distribution,

$$P_{\hat{W}|W} = \mathcal{N}((1-\alpha)W + \alpha w^*, (1-\alpha)\frac{D}{d}\Sigma_X^{-1}), \tag{26}$$

where $\alpha \triangleq \frac{nD}{d\sigma'^2} \leq 1$. Note that this "compression algorithm" requires the knowledge of optimal weights w^*, which is unknown in practice.

The details can be found in Appendix C. □

Remark 2. *As shown in [38], if* $n > d/\epsilon^2$, $\|\frac{1}{n}XX^\top - \Sigma_X\| \le \epsilon$ *holds with high probability. Then, the following lower bound on* $R(D)$ *holds if we can approximate* $\frac{1}{n}XX^\top$ *in (23) using* Σ_X,

$$R(D) \gtrsim \frac{d}{2}\left(\ln \frac{d\sigma'^2}{(n-d-1)D}\right)^+ - D(P_W \| P_{W_G}), \tag{27}$$

where W_G *denotes a Gaussian random vector with the same mean and variance as* W*. The details can be found in Appendix D.*

Combing Propositions 1 and 2, we have the following result.

Corollary 1. *Under the same assumptions as in Propositions 1, we have*

$$\min_{P_{\hat{W}|W}: I(W;\hat{W}) = R} \mathbb{E}_{S,W,\hat{W}}[L_\mu(\hat{W}) - L_S(W)] \le 2\sigma_\ell^{*2}\sqrt{\frac{R}{n}} + \frac{d\sigma'^2}{n-d-1}e^{-\frac{2R}{d}}, \quad R \ge 0. \tag{28}$$

In (28) the first term corresponds to the generalization error, which decreases with compression, and the second term corresponds to the empirical risk, which increases with compression.

5.3. Evaluation and Visualization

In the following plots, we generate the training dataset S using the linear model in (16) by letting $d = 50$, $n = 80$, $\Sigma_X = I_d$ and $\sigma'^2 = 1$. We consider the following two compression algorithms. The first one is the conditional distribution $P_{\hat{W}|W}$ in the proof of achievability (26), which requires the knowledge of w^* and is denoted as "Oracle". The second one is the well-known K-means clustering algorithm, where the weights in W are grouped into K clusters and represented by the cluster centers in the reconstruction $\hat{W}$. By changing the number of clusters K, we can control the rate R, i.e., $I(W;\hat{W})$. We average the performance and estimate $I(W;\hat{W})$ of these algorithms with 10,000 Monte-Carlo trials in the simulation.

We note that $I(W;\hat{W})$ is equal to the number of bits used in compression only in the asymptotic regime of large number of samples. In practice, we may have only one sample of the weights W, and therefore $I(W;\hat{W})$ simply measures the extent to which compression is performed by the compression algorithm.

In Figure 2a, we plot the generalization error bound in Proposition 1 as a function of the rate R and compare the generalization errors of the Oracle and K-means algorithms. It can be seen that Proposition 1 provides a valid upper bound for the generalization error, but this bound is tight only when R is small. Moreover, both compression algorithms can achieve smaller generalization errors compared to that of the ERM solution W, which validates the result in Theorem 1.

Figure 2b plots the upper bound on the distortion-rate function in Theorem 2 and the distortions achieved by the Oracle and K-means algorithms. The distortion of the Oracle decreases as we increase the rate R and matches the $D(R)$ function well. However, there is a large gap between the distortion achieved by K-means algorithms and $D(R)$. One possible explanation is that since w^* is unknown, it is impossible for the K-means algorithm to learn the optimal cluster center with only one sample of W. Even if we view $W^{(j)}, j = 1, \cdots, d$ as i.i.d. samples from the same distribution, there is still a gap between the distortion achieved by the K-means algorithm and the optimal quantization as studied in [39].

We plot the population risks of the ERM solution W, the Oracle, and K-means algorithms in Figure 2c. It is not surprising that the Oracle algorithm achieves a small population risk, since $\hat{W}$ is a function of w^* and $\hat{W} = w^*$ when $R = 0$. However, it can be seen that the K-means algorithm achieves a smaller population risk than the original model W, since the decrease in generalization error exceeds the increase in empirical risk, when we use fewer clusters in the K-means algorithm, i.e., a smaller rate R. We note that the minimal population risk is achieved when $K = 2$, since we initialize w^* so that $w^{*(i)}$, $1 \leq i \leq d$, can be well approximated by two cluster centers.

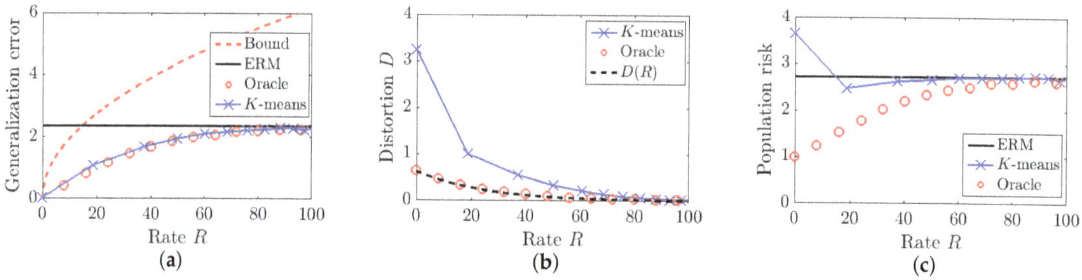

Figure 2. Comparison of three different quantities for linear regression as a function of rate R in bits. (**a**) Generalization error. (**b**) Distortion. (**c**) Population risk.

6. Clustering Algorithm Minimizing $\mathcal{L}_{S,W}$

In this section, we propose an improvement of the Hessian-weighted (HW) K-means clustering algorithm [11] for model compression by regularizing the distance between the cluster centers, which minimizes the upper bound $\mathcal{L}_{S,W}(P_{\hat{W}|W})$, as suggested by our theoretical results in Section 4.

6.1. Hessian-Weighted K-Means Clustering

The goal of HW K-means is to minimize the distortion on the empirical risk $d_S(\hat{W}, W)$, which has the following Taylor series approximation:

$$d_S(\hat{W}, W) \approx (\hat{W} - W)^T \nabla L_S(W) + \frac{1}{2}(\hat{W} - W)^T H_S(W)(\hat{W} - W), \tag{29}$$

where $H_S(W)$ is the Hessian matrix. Assuming that W is a local minimum of $L_S(W)$ (ERM solution) and $\nabla L_S(W) \approx 0$, the first term can be ignored. Furthermore, the Hessian matrix $H_S(W)$ can be approximated by a diagonal matrix, which further simplifies the objective to $d_S(\hat{W}, W) \approx \sum_{j=1}^{d} h^{(j)}(W^{(j)} - \hat{W}^{(j)})^2$, where $h^{(j)}$ is the j-th diagonal element of the Hessian matrix.

Given network parameters $w = \{w^{(1)}, \cdots, w^{(d)}\}$, the HW K-means clustering algorithm [11] partitions them into K disjoint clusters, using a set of cluster centers $c = \{c^{(1)}, \cdots, c^{(K)}\}$, and a cluster assignment $C = \left\{C^{(1)}, \cdots, C^{(K)}\right\}$, while solving the following optimization problem:

$$\min \sum_{k=1}^{K} \sum_{w^{(j)} \in C^{(k)}} h^{(j)} |w^{(j)} - c^{(k)}|^2. \tag{30}$$

6.2. Diameter Regularization

In contrast to HW K-means which only cares about empirical risk, our goal is to obtain as small a population risk as possible by minimizing the upper bound

$$\mathcal{L}_{S,W}(P_{\hat{W}|W}) = \sqrt{\frac{2\sigma^2}{n} I(W; \hat{W})} + \mathbb{E}[d_S(\hat{W}, W)]. \tag{31}$$

Here, we let the number of clusters K to be an input argument of the algorithm, so that $I(W; \hat{W}) \leq \log_2 K$, and we want to minimize $\mathcal{L}_{S,W}(P_{\hat{W}|W})$ by carefully designing the reconstructed weights given K, i.e., by choosing cluster centers $\{c^{(1)}, \cdots, c^{(K)}\}$. Then, minimizing the sub-Gaussian parameter σ is one way to control the generalization error of the compression algorithm. Recall that in Proposition 1, we have

$$\text{gen}(\mu, P_{\hat{W}|S}) \leq 2\big(C(w^*)\|\Sigma_X\| + \sigma'^2\big)\sqrt{\frac{I(W;\hat{W})}{n}}, \tag{32}$$

where the sub-Gaussian parameter is related to $C(w^*) = \sup_{\hat{w}\in\hat{\mathcal{W}}} \|\hat{w} - w^*\|_2^2$ in linear regression. Note that this quantity can be interpreted as the diameter of the set $\mathcal{W}$. Since the ground truth w^* is unknown in practice, we then propose the following diameter regularization by approximating $C(w^*)$ in (32) by

$$\beta \max_{k_1,k_2} |c^{(k_1)} - c^{(k_2)}|^2, \quad \beta \geq 0, \tag{33}$$

where β is a parameter controls the penalty term and can be selected by cross validation in practice. Our diameter-regularized Hessian-weighted (DRHW) K-means algorithm solves the following optimization problem:

$$\min \sum_{k=1}^{K} \sum_{w^{(j)}\in C^{(k)}} h^{(j)}|w^{(j)} - c^{(k)}|^2 + \beta \max_{k_1,k_2} |c^{(k_1)} - c^{(k_2)}|^2. \tag{34}$$

Such an optimization problem can be easily extended to the vector case which leads to a vector quantization algorithm. Suppose that we group the d-dimensional weights $w = \{w^{(1)}, \cdots, w^{(d)}\}$ into $d' = d/m$ vectors with length m, i.e., $\{\mathbf{w}^{(1)}, \cdots, \mathbf{w}^{(d')}\}$, $\mathbf{w}^{(j)} \in \mathbb{R}^m$, then our goal is to find cluster centers $\mathbf{c}^k \in \mathbb{R}^m$ and assignments minimizing the following cost function:

$$\min \sum_{k=1}^{K} \sum_{\mathbf{w}^{(j)}\in C^{(k)}} (\mathbf{w}^{(j)} - \mathbf{c}^{(k)})^\top H^{(j)}(\mathbf{w}^{(j)} - \mathbf{c}^{(k)}) + \beta \max_{k_1,k_2} \|\mathbf{c}^{(k_1)} - \mathbf{c}^{(k_2)}\|_2^2, \tag{35}$$

where $H^{(j)}$ is the diagonal Hessian matrix corresponding to the vector $\mathbf{w}^{(j)}$. An iterative algorithm to solve the above optimization problem for vector quantization is provided in Algorithm 1.

The algorithm alternates between minimizing the objective function over the cluster centers and the assignments. In the Assignment step, we first fix centers and assign each $\mathbf{w}^{(j)}$ to its nearest neighbor. We then fix assignments and update the centers by the weighted mean of each cluster in the Update step. For the farthest pair of centers, the diameter regularizer pushes them toward each other, so that the output centers have potentially smaller diameters than those of regular K-means. We note that the time complexity of the proposed diameter-regularized Hessian weighted K-means algorithm is the same as that of the original K-means algorithm.

Algorithm 1 Diameter-regularized Hessian weighted K-means in vector case

Input: Weights vector $\{\mathbf{w}^{(1)}, \ldots, \mathbf{w}^{(d')}\}$, Hessian matrices $\{H^{(1)}, \ldots, H^{(d')}\}$, diameter regularizer $\beta > 0$, number of clusters K, iterations T

Initialize the K cluster centers $\{\mathbf{c}_0^{(1)}, \ldots, \mathbf{c}_0^{(K)}\}$ randomly

for $t = 1$ to T **do**

Assignment step:

Initialize $C_t^{(k)} = \emptyset$ for all $k \in [K]$.

for $j = 1$ to d' **do**

Assign $\mathbf{w}^{(j)}$ to the nearest cluster center, i.e., find $k_t^{(j)} = \arg\min_{k\in[K]} \|\mathbf{w}^{(j)} - \mathbf{c}_{t-1}^{(k)}\|_2^2$ and let

$$C_t^{(k_t^{(j)})} \leftarrow C_t^{(k_t^{(j)})} \cup \{\mathbf{w}^{(j)}\} \tag{36}$$

end for
Update step:
Find current farthest pair of centers $(k_1, k_2) = \arg\max_{k_1,k_2} \|\mathbf{c}_{t-1}^{(k_1)} - \mathbf{c}_{t-1}^{(k_2)}\|_2^2$.
Update $\mathbf{c}_t^{(k_1)}$ and $\mathbf{c}_t^{(k_2)}$ by

$$\begin{aligned} \mathbf{c}_t^{(k_1)} &= \Big(\sum_{\mathbf{w}^{(j)} \in C_t^{(k_1)}} H^{(j)} + \beta I_m \Big)^{-1} \Big(\sum_{\mathbf{w}^{(j)} \in C_t^{(k_1)}} H^{(j)} \mathbf{w}^{(j)} + \beta \mathbf{c}_t^{(k_2)} \Big) \\ \mathbf{c}_t^{(k_2)} &= \Big(\sum_{\mathbf{w}^{(j)} \in C_t^{(k_2)}} H^{(j)} + \beta I_m \Big)^{-1} \Big(\sum_{\mathbf{w}^{(j)} \in C_t^{(k_2)}} H^{(j)} \mathbf{w}^{(j)} + \beta \mathbf{c}_t^{(k_1)} \Big) \end{aligned} \tag{37}$$

for $k = 1$ to K, $k \notin \{k_1, k_2\}$ **do**
Update the cluster centers by

$$\mathbf{c}_t^{(k)} = \Big(\sum_{\mathbf{w}^{(j)} \in C_t^{(k)}} H^{(j)} \Big)^{-1} \Big(\sum_{\mathbf{w}^{(j)} \in C_t^{(k)}} H^{(j)} \mathbf{w}^{(j)} \Big) \tag{38}$$

end for
end for
Output: centers $\{\mathbf{c}_T^{(1)}, \ldots, \mathbf{c}_T^{(K)}\}$ and assignments $\{C_T^{(1)}, \ldots, C_T^{(K)}\}$.

7. Experiments

In this section, we provide some real-world experiments to validate our theoretical assertions and the DRHW K-means algorithm. (The code for our experiments is available at the following link https://github.com/wgao9/weight-quant (accessed on 13 August 2021)) Our experiments include compression of: (i) a three-layer fully connected network on the MNIST dataset [40]; and (ii) a convolutional neural network with five convolutional layers and three linear layers on the CIFAR10 dataset [41] (We downloaded the pre-trained model in PyTorch from https://github.com/aaron-xichen/pytorch-playground (accessed on 13 August 2021)).

In Theorem 1, an upper bound on the *expected* generalization error is provided, and therefore we independently train 50 different models (with the same structure but different parameter initializations) using different subset of training samples, and average the results. We use 10% of the training data to train the model for MNIST and use 20% of the training data to train the model for CIFAR10. For each experiment, we use the same number of clusters for each convolutional layer and fully connected layer.

In the following experiments, we plot the cross entropy loss as a function of compression ratio. Note that compression ratio can be controlled by changing the number of clusters K in the quantization algorithm. To see this, suppose that the neural networks have total of d parameters that need to be compressed, and each parameter is of b bits. Let $C^{(k)}$ be the set of weights in cluster k and let b_k be the number of bits of the codeword assigned to the network parameters in cluster k for $1 \le k \le K$. For a lookup table to decode quantized values, we need Kb bits to store all the reconstructed weights, i.e., cluster centers $c = \{c^{(1)}, \cdots, c^{(K)}\}$. Then, the compression ratio is given by

$$\text{Compression Ratio} = \frac{\sum_{k=1}^{K} |C^{(k)}| b_k + Kb}{db}, \tag{39}$$

where $|\cdot|$ denotes the number of elements in the set. In our experiments, we use a variable-length code such as the Huffman code to compute the compression ratio under different numbers of clusters K.

In Figures 3 and 4, we compare the scalar DRHW K-means algorithm with the scalar HW K-means algorithm for different compression ratios on the MNIST and CIFAR10 datasets. Both figures demonstrate that the compression algorithm increases the empirical risk but decreases the generalization error, and the net effect is that the both compressed models have smaller population risks than those of the original models. More importantly, the DRHW K-means algorithm produces a compressed model that has a better population risk than that of the HW K-means algorithm.

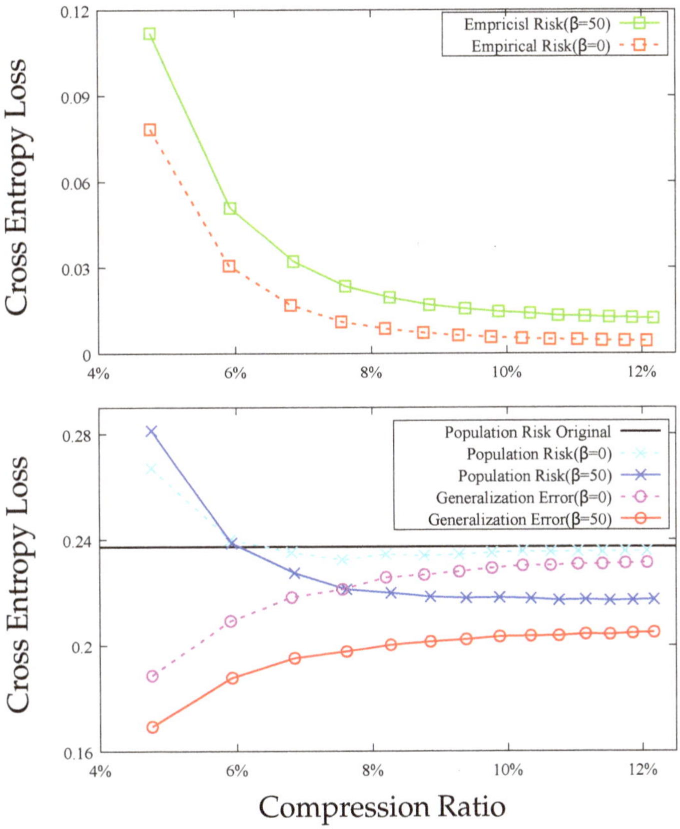

Figure 3. Comparison between DRHW K-means ($\beta = 50$) and HW K-means ($\beta = 0$) on MNIST. **Top**: empirical risks. **Bottom**: population risks and generalization errors.

In Figure 5, we compare the population risk of scalar DRHW K-means algorithm and that of the vector DRHW K-means algorithm with block length $m = 2$ for different compression ratios on the MNIST dataset. It can be seen from the figure that the improvement by using vector quantization ($m = 2$) is quite modest, which implies that the dependence between the weights $W^{(j)}$ is weak. However, we can still observe the improvement of adding the diameter regularizer in vector DRHW K-means algorithm by comparing the curves with $\beta = 50$ and $\beta = 0$.

In Figure 6, we demonstrate how β affects the performance of our diameter-regularized Hessian-weighted K-means algorithm in scalar case. It can be seen that as β increases, the generalization error decreases and the distortion in empirical risk increases, which validates the idea that this proposed diameter regularizer can be used to reduce the generalization error. The value of β that results in the best population risk therefore can be chosen via cross-validation in practice.

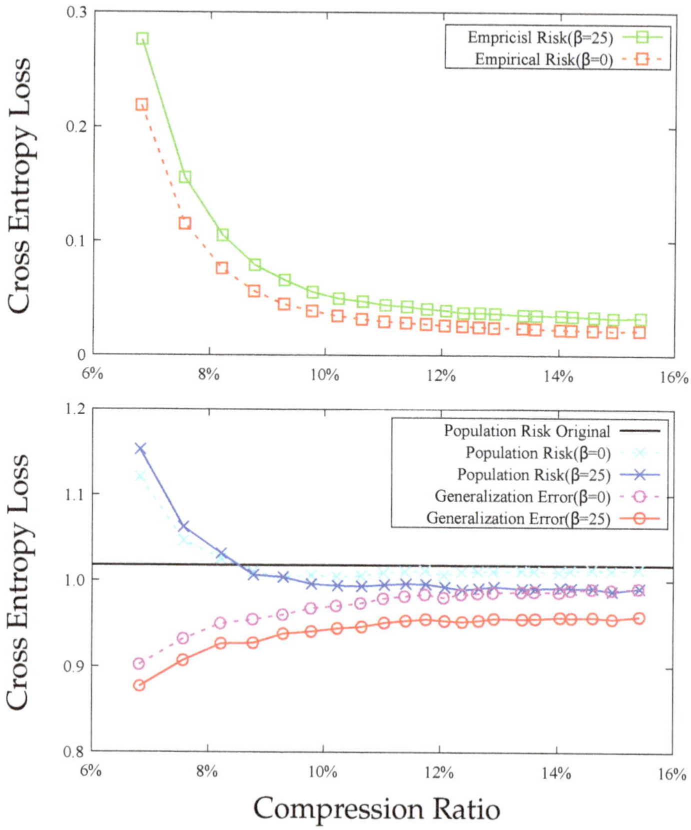

Figure 4. Comparison between DRHW K-means ($\beta = 25$) and HW K-means ($\beta = 0$) on CIFAR10. **Top**: empirical risks. **Bottom**: population risks and generalization errors.

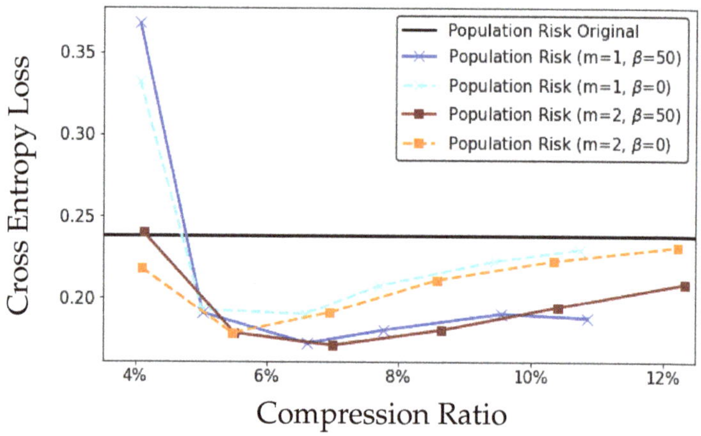

Figure 5. Comparison between scalar DRHW K-means ($m = 1$) and vector DRHW K-means ($m = 2$) on the MNIST dataset.

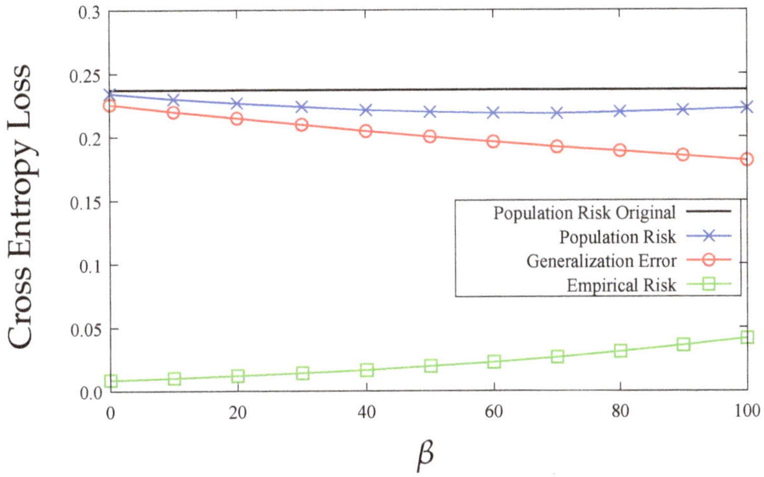

Figure 6. DRHW K-means with different β on the MNIST dataset with $K = 7$.

8. Conclusions

In this paper, we have provided an information-theoretical understanding of how model compression affects the population risk of a compressed model. In particular, our results indicate that model compression may increase the empirical risk but decrease the generalization error. Therefore, it might be possible to achieve a smaller population risk via model compression. Our experiments validate these theoretical findings. Furthermore, we showed how our information-theoretic bound on the population risk can be used to optimize practical compression algorithms.

We note that our results could be applied to improve other compression algorithms, such as pruning and matrix factorization. Moreover, we believe that the information-theoretic analysis adopted here could be generalized to characterize a similar tradeoff between the generalization error and empirical risk in other applications beyond compressing pre-trained models, e.g., distributed optimization [42] and low precision training [15].

Author Contributions: Y.B.: theoretical analysis, methodology, conceptualization, writing—original draft. W.G.: software, methodology and visualization. S.Z.: writing—review and editing. V.V.V.: supervision, funding acquisition, and writing—review and editing. All authors have read and agreed to the published version of the manuscript.

Funding: This work was supported by Army Research Laboratory under Cooperative Agreement W911NF-17-2-0196, through the University of Illinois at Urbana-Champaign.

Data Availability Statement: Data and code can be found in Section 7.

Conflicts of Interest: The authors declare no conflict of interest.

Appendix A. Proof of Lemma 3

Let $\widetilde{Z} = (\widetilde{X}, \widetilde{Y})$, $\widetilde{X} \in \mathbb{R}^d$ and $\widetilde{Y} \in \mathbb{R}$ denote an independent copy of the training sample Z_i. Then, it can be shown that

$$\begin{aligned} \text{gen}(\mu, P_{W|S}) &= \mathbb{E}_{W,S}[L_\mu(W) - L_S(W)] \\ &= \mathbb{E}_{W,S}\Big[\mathbb{E}_{\widetilde{Z}}[(\widetilde{Y} - \widetilde{X}^\top W)^2] - \frac{1}{n}\|Y - X^\top W\|_2^2\Big] \\ &= \mathbb{E}_S\Big[\mathbb{E}_{\widetilde{Z}}[(\widetilde{Y} - \widetilde{X}^\top (XX^\top)^{-1}XY)^2] - \frac{1}{n}\|Y - X^\top (XX^\top)^{-1}XY\|_2^2\Big], \end{aligned} \quad \text{(A1)}$$

where $\widetilde{Y} = \widetilde{X}^\top w^* + \widetilde{\varepsilon}$ and $Y = X^\top w^* + \varepsilon$. Then, we have

$$\begin{aligned}
\text{gen}(\mu, P_{W|S}) &= \mathbb{E}_{\varepsilon,\widetilde{\varepsilon},X,\widetilde{X}}\left[(\widetilde{\varepsilon} - \widetilde{X}^\top (XX^\top)^{-1}X\varepsilon)^2\right] - \frac{1}{n}\mathbb{E}_{\varepsilon,X}\left[\|\varepsilon - X^\top (XX^\top)^{-1}X\varepsilon\|_2^2\right] \\
&= \mathbb{E}_{\varepsilon,X,\widetilde{X}}\left[\varepsilon^\top X^\top (XX^\top)^{-1}\widetilde{X}\widetilde{X}^\top (XX^\top)^{-1}X\varepsilon + \frac{1}{n}\varepsilon^\top X^\top (XX^\top)^{-1}X\varepsilon\right] \\
&= \mathbb{E}_{\varepsilon,X}\left[\text{Tr}(X^\top (XX^\top)^{-1}\Sigma_X (XX^\top)^{-1}X\varepsilon\varepsilon^\top)\right] + \frac{\sigma'^2 d}{n} \\
&= \sigma'^2\mathbb{E}_X\left[\text{Tr}((XX^\top)^{-1}\Sigma_X)\right] + \frac{\sigma'^2 d}{n}.
\end{aligned} \tag{A2}$$

Note that X_i's are i.i.d. samples from $\mathcal{N}(0, \Sigma_X)$, then we have $(XX^\top)^{-1}$ distributed according to $\text{Wishart}^{-1}(\Sigma_X^{-1}, n)$, where Wishart^{-1} denotes the inverse Wishart distribution with n degrees of freedom, and $\mathbb{E}[(XX^\top)^{-1}] = \frac{\Sigma_X^{-1}}{n-d-1}$. It then follows that

$$\text{gen}(\mu, P_{W|S}) = \frac{\sigma'^2}{n-d-1}\left[\text{Tr}(\Sigma_X^{-1}\Sigma_X)\right] + \frac{\sigma'^2 d}{n} = \frac{\sigma'^2 d}{n}(2 + \frac{d+1}{n-d-1}). \tag{A3}$$

Appendix B. Proof of Proposition 1

For all $\hat{w} \in \hat{\mathcal{W}}$, it can be shown that

$$\ell(\hat{w}, \widetilde{Z}) = (\widetilde{Y} - \widetilde{X}^\top \hat{w})^2 = (\widetilde{X}^\top (w^* - \hat{w}) + \widetilde{\varepsilon})^2. \tag{A4}$$

Since $\widetilde{X} \sim \mathcal{N}(0, \Sigma_X)$ and $\widetilde{\varepsilon} \sim \mathcal{N}(0, \sigma'^2)$, then $\ell(\hat{w}, \widetilde{Z}) \sim \sigma_\ell^2 \chi_1^2$, where

$$\sigma_\ell^2 \triangleq (\hat{w} - w^*)^\top \Sigma_X (\hat{w} - w^*) + \sigma'^2,$$

and χ_1^2 denotes the chi-squared distribution with one degree of freedom. Then, the CGF of $\ell(\hat{w}, \widetilde{Z})$ is

$$\Lambda_{\ell(\hat{w},\widetilde{Z})}(\lambda) = -\sigma_\ell^2\lambda - \frac{1}{2}\ln(1 - 2\sigma_\ell^2\lambda),\ \lambda \in (-\infty, \frac{1}{2\sigma_\ell^2}). \tag{A5}$$

Thus, $\ell(\hat{w}, \widetilde{Z})$ is not sub-Gaussian for all $\lambda \in \mathbb{R}$. However, it can be shown that

$$\Lambda_{\ell(\hat{w},\widetilde{Z})}(\lambda) \leq \sigma_\ell^4 \lambda^2, \quad \lambda < 0. \tag{A6}$$

We need the following lemma from the Theorem 1 of [43] to proceed our analysis.

Lemma A1 ([43]). *Assume that for all $\hat{w} \in \hat{\mathcal{W}}$, $\Lambda_{\ell(\hat{w},\widetilde{Z})}(\lambda) \leq \frac{\sigma^2\lambda^2}{2}$ for $\lambda \leq 0$. Then,*

$$\text{gen}(\mu, P_{\hat{W}|S}) \leq \sqrt{\frac{2\sigma^2}{n} I(\hat{W}; S)}. \tag{A7}$$

Recall that $C(w^*) = \sup_{\hat{w}\in\hat{\mathcal{W}}} \|\hat{w} - w^*\|_2^2$. We then have the following bound on the CGF of $\ell(\hat{w}, \widetilde{Z})$,

$$\Lambda_{\ell(\hat{w},\widetilde{Z})}(\lambda) \leq \lambda^2 \max_{\hat{w}\in\hat{\mathcal{W}}} \sigma_\ell^4 \leq \lambda^2 (C(w^*)\|\Sigma_X\| + \sigma'^2)^2, \quad \lambda < 0. \tag{A8}$$

Applying Lemma A1 and data processing inequality, we have

$$\text{gen}(\mu, P_{\hat{W}|S}) \leq 2(C(w^*)\|\Sigma_X\| + \sigma'^2)\sqrt{\frac{I(\hat{W}; W)}{n}}. \tag{A9}$$

Appendix C. Proof of Proposition 2

The constraint on the distortion function can be written as follows:

$$D \geq \mathbb{E}_{S,W,\hat{W}}[d_S(\hat{W},W)] = \frac{1}{n}\mathbb{E}_{S,W,\hat{W}}[(\hat{W}-W)^\top XX^\top(\hat{W}-W)]. \quad \text{(A10)}$$

It follows from Lemma 1 that

$$R(D) = \min_{P_{\hat{W}|W}} I(\hat{W};W), \quad \text{s.t.} \quad \mathbb{E}_{S,W,\hat{W}}[(\hat{W}-W)^\top \frac{1}{n} XX^\top(\hat{W}-W)] \leq D. \quad \text{(A11)}$$

Note that $\mathbb{E}[W] = w^*$ and $\mathrm{Cov}[W] = \frac{\sigma'^2}{n-d-1}\Sigma_X^{-1}$ since W is the ERM solution. In the following proof, we consider a Gaussian random vector with the same mean and covariance matrix $W_G \sim \mathcal{N}(w^*, \frac{\sigma'^2}{n-d-1}\Sigma_X^{-1})$ as W.

For the upper bound of $R(D)$, consider the channel $P^*_{\hat{W}|W} = \mathcal{N}((1-\alpha)W + \alpha w^*, (1-\alpha)\frac{D}{d}\Sigma_X^{-1})$, where $\alpha = \frac{nD}{d\sigma'^2} \leq 1$. It can be verified that this channel satisfies the constraint on the distortion:

$$\begin{aligned}
&\mathbb{E}_{S,W,\hat{W}}[d_S(\hat{W},W)] \\
&= \alpha^2 \mathbb{E}[(W-w^*)^\top \frac{1}{n} XX^\top (W-w^*)] + (1-\alpha)\frac{D}{d}\mathrm{Tr}\Big(\mathbb{E}[\frac{1}{n}XX^\top]\Sigma_X^{-1}\Big) \\
&= \alpha^2 \mathbb{E}[((XX^\top)^{-1}X\varepsilon)^\top \frac{1}{n} XX^\top ((XX^\top)^{-1}X\varepsilon)] + (1-\alpha)D \\
&= \alpha^2 \frac{1}{n}\mathbb{E}[\varepsilon^\top X^\top (XX^\top)^{-1} X\varepsilon] + (1-\alpha)D \\
&= D.
\end{aligned} \quad \text{(A12)}$$

If we let $\xi \sim \mathcal{N}(0, (1-\alpha)\frac{D}{d}\Sigma_X^{-1})$, it follows that

$$\begin{aligned}
R(D) &\leq I(W;(1-\alpha)W + \alpha w^* + \xi) \\
&\overset{(a)}{\leq} I(W_G;(1-\alpha)W_G + \xi) \\
&= \frac{d}{2}\ln\Big(\frac{d\sigma'^2}{(n-d-1)D} - \frac{n}{n-d-1} + 1\Big) \\
&\leq \frac{d}{2}\Big(\ln\frac{d\sigma'^2}{(n-d-1)D}\Big)^+,
\end{aligned} \quad \text{(A13)}$$

where (a) is due to the fact that Gaussian distribution maximizes the mutual information in an additive white Gaussian noise channels.

The upper bound on $D(R)$ follows immediately from the upper bound on $R(D)$.

Appendix D. Discussion of Remark 2

Suppose that $\frac{1}{n}XX^\top$ can be approximated by Σ_X for large n in (A10). It then follows that

$$R(D) = \min_{P_{\hat{W}|W}} I(\hat{W};W), \quad \text{s.t.} \quad \mathbb{E}_{S,W,\hat{W}}[(\hat{W}-W)^\top \Sigma_X (\hat{W}-W)] \leq D. \quad \text{(A14)}$$

It can be easily verified that the channel $P^*_{W|\hat{W}} = \mathcal{N}(\hat{W}, \frac{D}{d}\Sigma_X^{-1})$ satisfies the distortion constraint. For any $P_{W|\hat{W}}$ such that $\mathbb{E}_{S,W,\hat{W}}[d_S(\hat{W},W)] \leq D$, it follows that

$$
\begin{aligned}
I(W;\hat{W}) &= \mathbb{E}_{W,\hat{W}}\left[\ln\frac{P_{W|\hat{W}}}{P_W}\right] \\
&= \mathbb{E}_{W,\hat{W}}\left[\ln\frac{P_{W|\hat{W}}}{P^*_{W|\hat{W}}}\right] + \mathbb{E}_{W,\hat{W}}\left[\ln\frac{P^*_{W|\hat{W}}}{P_{W_G}}\right] - \mathrm{KL}(P_W\|P_{W_G}) \\
&\geq \mathbb{E}_{W,\hat{W}}\left[\ln\frac{P^*_{W|\hat{W}}}{P_{W_G}}\right] - \mathrm{KL}(P_W\|P_{W_G}),
\end{aligned} \tag{A15}
$$

where $\mathrm{KL}(P_W\|P_{W_G})$ is the Kullback–Leibler divergence between the two distributions, and the last step follows from the fact that $\mathrm{KL}(P_{W,\hat{W}}\|P^*_{W,\hat{W}}) \geq 0$. Note that

$$
\begin{aligned}
&\mathbb{E}_{W,\hat{W}}\left[\ln\frac{P^*_{W|\hat{W}}}{P_{W_G}}\right] \\
&= \mathbb{E}_{W,\hat{W}}\left[\frac{(n-d-1)(W-w^*)^\top\Sigma_X(W-w^*)}{2\sigma'^2} - \frac{d(\hat{W}-W)^\top\Sigma_X(\hat{W}-W)}{2D}\right] \\
&\quad + \frac{d}{2}\ln\frac{d\sigma'^2}{(n-d-1)D} \\
&\overset{(a)}{=} \frac{d}{2}\ln\frac{d\sigma'^2}{(n-d-1)D} + \mathbb{E}_{W,\hat{W}}\left[\frac{d}{2} - \frac{d(\hat{W}-W)^\top\Sigma_X(\hat{W}-W)}{2D}\right] \\
&\overset{(b)}{\geq} \frac{d}{2}\ln\frac{d\sigma'^2}{(n-d-1)D},
\end{aligned} \tag{A16}
$$

where (a) follows from the fact that $\mathbb{E}[W] = w^*$ and $\mathrm{Cov}[W] = \frac{\sigma'^2}{n-d-1}\Sigma_X^{-1}$, and (b) is due to the fact that $P_{\hat{W}|W}$ satisfies the distortion constraint. Thus,

$$
R(D) \geq \frac{d}{2}\ln\frac{d\sigma'^2}{(n-d-1)D} - \mathrm{KL}(P_W\|P_{W_G}). \tag{A17}
$$

References

1. Goodfellow, I.; Bengio, Y.; Courville, A.; Bengio, Y. *Deep Learning*; MIT Press: Cambridge, UK, 2016; Volume 1.
2. Krizhevsky, A.; Sutskever, I.; Hinton, G.E. Imagenet classification with deep convolutional neural networks. In Proceedings of the Advances in Neural Information Processing Systems (NIPS), Lake Tahoe, NV, USA, 3–6 December 2012; pp. 1097–1105.
3. Silver, D.; Schrittwieser, J.; Simonyan, K.; Antonoglou, I.; Huang, A.; Guez, A.; Hubert, T.; Baker, L.; Lai, M.; Bolton, A.; et al. Mastering the game of Go without human knowledge. *Nature* **2017**, *550*, 354. [CrossRef]
4. Huval, B.; Wang, T.; Tandon, S.; Kiske, J.; Song, W.; Pazhayampallil, J.; Andriluka, M.; Rajpurkar, P.; Migimatsu, T.; Cheng-Yue, R. An empirical evaluation of deep learning on highway driving. *arXiv* **2015**, arXiv:1504.01716.
5. Simonyan, K.; Zisserman, A. Very deep convolutional networks for large-scale image recognition. *arXiv* **2014**, arXiv:1409.1556.
6. Cheng, Y.; Wang, D.; Zhou, P.; Zhang, T. A survey of model compression and acceleration for deep neural networks. *arXiv* **2017**, arXiv:1710.09282.
7. Krishnamoorthi, R. Quantizing deep convolutional networks for efficient inference: A whitepaper. *arXiv* **2018**, arXiv:1806.08342.
8. Guo, Y. A survey on methods and theories of quantized neural networks. *arXiv* **2018**, arXiv:1808.04752.
9. Han, S.; Mao, H.; Dally, W.J. Deep compression: Compressing deep neural networks with pruning, trained quantization and huffman coding. *arXiv* **2015**, arXiv:1510.00149.
10. Zhu, C.; Han, S.; Mao, H.; Dally, W.J. Trained ternary quantization. *arXiv* **2016**, arXiv:1612.01064.
11. Choi, Y.; El-Khamy, M.; Lee, J. Towards the limit of network quantization. *arXiv* **2016**, arXiv:1612.01543.
12. Lin, Y.; Han, S.; Mao, H.; Wang, Y.; Dally, W.J. Deep gradient compression: Reducing the communication bandwidth for distributed training. *arXiv* **2017**, arXiv:1712.01887.
13. Xu, A.; Raginsky, M. Information-theoretic analysis of generalization capability of learning algorithms. In Proceedings of the Advances in Neural Information Processing Systems (NIPS), Long Beach, CA, USA, 4–9 December 2017; pp. 2524–2533.

14. Howard, A.G.; Zhu, M.; Chen, B.; Kalenichenko, D.; Wang, W.; Weyand, T.; Andreetto, M.; Adam, H. Mobilenets: Efficient convolutional neural networks for mobile vision applications. *arXiv* **2017**, arXiv:1704.04861.
15. Gupta, S.; Agrawal, A.; Gopalakrishnan, K.; Narayanan, P. Deep learning with limited numerical precision. In Proceedings of the International Conference on Machine Learning (ICML), Lille, France, 6–11 July 2015; pp. 1737–1746.
16. Courbariaux, M.; Bengio, Y.; David, J.P. Binaryconnect: Training deep neural networks with binary weights during propagations. In Proceedings of the Advances in Neural Information Processing Systems (NIPS), Montreal, QC, Canada, 7–12 December 2015; pp. 3123–3131.
17. Rastegari, M.; Ordonez, V.; Redmon, J.; Farhadi, A. Xnor-net: Imagenet classification using binary convolutional neural networks. In *European Conference on Computer Vision*; Springer: Berlin/Heidelberg, Germany, 2016; pp. 525–542.
18. Mozer, M.C.; Smolensky, P. Skeletonization: A technique for trimming the fat from a network via relevance assessment. In Proceedings of the Advances in Neural Information Processing Systems (NIPS), Denver, CO, USA, 27–30 November 1989; pp. 107–115.
19. LeCun, Y.; Denker, J.S.; Solla, S.A. Optimal brain damage. In Proceedings of the Advances in Neural Information Processing Systems (NIPS), Denver, CO, USA, 26–29 November 1990; pp. 598–605.
20. Hassibi, B.; Stork, D.G. Second order derivatives for network pruning: Optimal brain surgeon. In Proceedings of the Advances in Neural Information Processing Systems (NIPS), San Francisco, CA, USA, 30 November–3 December 1992; pp. 164–171.
21. Han, S.; Pool, J.; Tran, J.; Dally, W. Learning both weights and connections for efficient neural network. In Proceedings of the Advances in Neural Information Processing Systems (NIPS), Montreal, QC, Canada, 7–12 December 2015; pp. 1135–1143.
22. Gong, Y.; Liu, L.; Yang, M.; Bourdev, L. Compressing deep convolutional networks using vector quantization. *arXiv* **2014**, arXiv:1412.6115.
23. Ullrich, K.; Meeds, E.; Welling, M. Soft weight-sharing for neural network compression. *arXiv* **2017**, arXiv:1702.04008.
24. Louizos, C.; Ullrich, K.; Welling, M. Bayesian compression for deep learning. In Proceedings of the Advances in Neural Information Processing Systems (NIPS), Long Beach, CA, USA, 4–9 December 2017; pp. 3288–3298.
25. Lin, D.; Talathi, S.; Annapureddy, S. Fixed point quantization of deep convolutional networks. In Proceedings of the International Conference on Machine Learning, PMLR, New York, NY, USA, 19–24 June 2016; pp. 2849–2858.
26. Young, S.; Wang, Z.; Taubman, D.; Girod, B. Transform Quantization for CNN Compression. *IEEE Trans. Pattern Anal. Mach. Intell.* **2021**, 1. [CrossRef] [PubMed]
27. Denton, E.L.; Zaremba, W.; Bruna, J.; LeCun, Y.; Fergus, R. Exploiting linear structure within convolutional networks for efficient evaluation. In Proceedings of the Advances in Neural Information Processing Systems (NIPS), Montreal, QC, Canada, 8–13 December 2014; pp. 1269–1277.
28. Tai, C.; Xiao, T.; Zhang, Y.; Wang, X. Convolutional neural networks with low-rank regularization. *arXiv* **2017**, arXiv:1511.06067.
29. Novikov, A.; Podoprikhin, D.; Osokin, A.; Vetrov, D.P. Tensorizing neural networks. In Proceedings of the Advances in Neural Information Processing Systems (NIPS), Montreal, QC, Canada, 7–12 December 2015; pp. 442–450.
30. Gao, W.; Liu, Y.H.; Wang, C.; Oh, S. Rate distortion for model compression: From theory to practice. In Proceedings of the International Conference on Machine Learning (ICML), PMLR, Long Beach, CA, USA, 9–15 June 2019; pp. 2102–2111.
31. Dziugaite, G.K.; Roy, D.M. Computing nonvacuous generalization bounds for deep (stochastic) neural networks with many more parameters than training data. *arXiv* **2017**, arXiv:1703.11008.
32. Neyshabur, B.; Bhojanapalli, S.; McAllester, D.; Srebro, N. Exploring generalization in deep learning. *arXiv* **2017**, arXiv:1706.08947.
33. Zhou, W.; Veitch, V.; Austern, M.; Adams, R.P.; Orbanz, P. Non-vacuous generalization bounds at the imagenet scale: A PAC-bayesian compression approach. *arXiv* **2018**, arXiv:1804.05862.
34. Russo, D.; Zou, J. Controlling bias in adaptive data analysis using information theory. In Proceedings of the International Conference on Artifical Intelligence and Statistics (AISTATS), Cadiz, Spain, 9–11 May 2016; pp. 1232–1240.
35. Shannon, C.E. Coding theorems for a discrete source with a fidelity criterion. *IRE Nat. Conv. Rec* **1959**, *4*, 142–163.
36. Cover, T.M.; Thomas, J.A. *Elements of Information Theory*; John Wiley & Sons: Hoboken, NJ, USA, 2012.
37. Grønlund, A.; Kamma, L.; Larsen, K.G.; Mathiasen, A.; Nelson, J. Margin-Based Generalization Lower Bounds for Boosted Classifiers. In Proceedings of the Advances in Neural Information Processing Systems (NeurIPS), Vancouver, BC, Canada, 8–14 December 2019; pp. 11940–11949.
38. Vershynin, R. Introduction to the non-asymptotic analysis of random matrices. *arXiv* **2010**, arXiv:1011.3027.
39. Linder, T.; Lugosi, G.; Zeger, K. Rates of convergence in the source coding theorem, in empirical quantizer design, and in universal lossy source coding. *IEEE Trans. Inf. Theory* **1994**, *40*, 1728–1740. [CrossRef]
40. LeCun, Y.; Bottou, L.; Bengio, Y.; Haffner, P. Gradient-based learning applied to document recognition. *Proc. IEEE* **1998**, *86*, 2278–2324. [CrossRef]
41. Krizhevsky, A.; Hinton, G. *Learning Multiple Layers of Features from Tiny Images*; Technical Report; University of Toronto: Toronto, ON, Canada, 2009.

42. Basu, D.; Data, D.; Karakus, C.; Diggavi, S. Qsparse-local-SGD: Distributed SGD with Quantization, Sparsification and Local Computations. In Proceedings of the Advances in Neural Information Processing Systems (NeurIPS), Vancouver, BC, Canada, 8–14 December 2019; pp. 14668–14679.
43. Bu, Y.; Zou, S.; Veeravalli, V.V. Tightening Mutual Information-Based Bounds on Generalization Error. *IEEE J. Sel. Areas Inf. Theory* **2020**, *1*, 121–130. [CrossRef]

entropy

MDPI

Article

Robust Spike-Based Continual Meta-Learning Improved by Restricted Minimum Error Entropy Criterion

Shuangming Yang [1], Jiangtong Tan [1] and Badong Chen [2,*]

[1] School of Electrical and Information Engineering, Tianjin University, Tianjin 300072, China; yangshuangming@tju.edu.cn (S.Y.); shuangmingyang@tju.edu.cn (J.T.)
[2] Institute of Artificial Intelligence and Robotics, Xi'an Jiaotong University, Xi'an 710049, China
* Correspondence: chenbd@mail.xjtu.edu.cn

Abstract: The spiking neural network (SNN) is regarded as a promising candidate to deal with the great challenges presented by current machine learning techniques, including the high energy consumption induced by deep neural networks. However, there is still a great gap between SNNs and the online meta-learning performance of artificial neural networks. Importantly, existing spike-based online meta-learning models do not target the robust learning based on spatio-temporal dynamics and superior machine learning theory. In this invited article, we propose a novel spike-based framework with minimum error entropy, called MeMEE, using the entropy theory to establish the gradient-based online meta-learning scheme in a recurrent SNN architecture. We examine the performance based on various types of tasks, including autonomous navigation and the working memory test. The experimental results show that the proposed MeMEE model can effectively improve the accuracy and the robustness of the spike-based meta-learning performance. More importantly, the proposed MeMEE model emphasizes the application of the modern information theoretic learning approach on the state-of-the-art spike-based learning algorithms. Therefore, in this invited paper, we provide new perspectives for further integration of advanced information theory in machine learning to improve the learning performance of SNNs, which could be of great merit to applied developments with spike-based neuromorphic systems.

Keywords: spiking neural network; meta-learning; information theoretic learning; minimum error entropy; artificial general intelligence

Citation: Yang, S.; Tan, J.; Chen, B. Robust Spike-Based Continual Meta-Learning Improved by Restricted Minimum Error Entropy Criterion. *Entropy* **2022**, *24*, 455. https://doi.org/10.3390/e24040455

Academic Editors: Lizhong Zheng and Chao Tian

Received: 25 February 2022
Accepted: 23 March 2022
Published: 25 March 2022

1. Introduction

In recent years, deep learning has shown a superior performance that exceeds the human-level performance in various types of individual narrow tasks [1]. However, in comparison with human intelligence that can learn to learn continually in order to execute unlimited tasks, the current successful deep learning methods still have a lot of drawbacks and limitations. In fact, humans can learn to learn by accumulating knowledge across their life time, which is a great challenge for artificial neural networks (ANNs) [2]. From this point of view, continual meta-learning aims at realizing machine intelligence at a higher level by providing machines with the meta-learning capability of learning to learn continually [3].

The human brain can realize meta-learning continually and avoid the catastrophic forgetting problem based on a combination of neural mechanisms [4]. The catastrophic forgetting problem is the critical challenge for developing the capability of continual meta-learning [5]. The human brain has implemented an efficient and scalable mechanism for continual learning based on neuronal activity patterns that represent previous experiences [6]. Neurons communicate with each other and process the neural information by using neural spikes, which is one of the most critical fundamental mechanism in the brain. Based on this mechanism, the human brain can realize superior performance in different aspects, such as low power consumption and high spatio-temporal processing capability [7]. Therefore, implementing a brain-inspired continual meta-learning algorithm based on spike patterns and the brain's mechanisms is a promising technique.

The spiking neural network (SNN) uses the biologically plausible neuron model based on spiking dynamics, while the conventional ANN only uses the neurons based on a static rate [8]. SNNs are applied to reproduce the brain's mechanisms and to deal with the cognitive tasks [9]. In addition, the neuromorphic hardware based on SNNs can realize high performance in artificial intelligence tasks, including low power consumption, high noise tolerance, and low computation latency [10]. Previous neuromorphic hardware researches have proven these advantages by using various types of tasks, such as Tianjic, Loihi, BiCoSS, CerebelluMorphic, and LaCSNN [11–15]. Researchers have proposed SNN models to realize the short-term memory capability in a spike-based framework [16]. However, the current SNN models still suffer from the continual meta-learning problem under the non-Gaussian noise, and no previous study has solved this problem. Therefore, this is the focus of this study.

Information theoretic learning (ITL) has attracted increasing attention in the field of machine learning in recent years to improve the learning robustness and enhance the explainable capability [17–19]. Previously, Chen et al. proposed researches focusing on maximum correntropy theory and minimum error entropy criteria to improve the robustness of machine learning theory [20–22]. In addition, a series of entropy-based learning algorithms have been presented to deal with the robustness improvement of machine learning models, including guided complement entropy and fuzzy entropy [23–25]. Nevertheless, there is no application of the ITL-based approach in the spike-based continual meta-learning to improve its learning robustness. Therefore, in this invited article, we aim to propose a novel approach to deal with this challenging problem. A novel model is presented, which is called meta-learning with minimum error entropy (MeMEE). We test the meta-learning capability of the proposed SNN model. Then, we investigate the robust working memory capability in non-Gaussian noise. Finally, the robust transfer learning performance is explored under a non-Gaussian noisy condition. Experimental results strongly suggest the robust meta-learning capability of the SNN model with a working memory feature in a non-Gaussian noisy environment.

2. Materials and Methods

2.1. SNN Model

Previous studies have shown that the firing timing and activity space of dendrites can significantly affect neural function. Excitability of dendrites can excite the membrane to fire, whereas inhibitory dendrites can have the opposite effect [26–29]. Inspired by this morphological structure and function of the neuron model, we propose a spiking neuron model, which has three compartments, including a somatic compartment and two dendritic compartments. The model utilizes distinct dendritic compartments to receive excitatory and inhibitory inputs, while using dendrites and somatic cells to receive and send spiking activities, respectively. The formulation for calculating the membrane potential of dendrites and soma are as follows

$$\begin{cases} \tau_m \frac{dU_m(t)}{dt} = -U_m(t) + R_m I_m(t) + g_i(U_i(t) - \theta_i) + g_e(U_e(t) - \theta_e) - \Gamma_j(t) z_j(t) \\ \tau_i \frac{dU_i(t)}{dt} = -U_i(t) + R_i I_i(t) \\ \tau_e \frac{dU_e(t)}{dt} = -U_e(t) + R_e I_e(t) \end{cases} \tag{1}$$

where τ_v represents the time constant of membrane. The variables $U(t)$, $U_i(t)$, and $U_e(t)$ represent the somatic membrane potentials, inhibitory dendritic membrane potentials, and excitatory dendritic membrane potentials, respectively. The parameters θ_e and θ_i represent the reversal membrane potential of excitatory dendrite and inhibitory dendrite, respectively. R_m, R_e, and R_i represent the membrane resistance of the soma, excitatory dendrite, and inhibitory dendrite, respectively. The parameters g_e and g_i represent the synaptic conductance of excitatory dendrites and inhibitory dendrites, respectively. Neuron emits a spike at time t when it is currently not in a refractory period. The soma of neurons uses the spike adaptation mechanism. The threshold size can be changed by analyzing the firing pattern of neurons. Variable $z_j(t)$ represents the spike train of neuron j and assumes value in $\{0, 1/\Delta t\}$. The dynamics of $\Gamma_j(t)$ is changed with each spike, representing the firing rate of neuron j, which is defined as

$$\Gamma_j(t) = \tau_j^0 + \alpha \cdot \tau_j(t) \tag{2}$$

where α represents a constant that scales the deviation $\tau_j(t)$ from the baseline $\tau_j{}^0$. The variable $\tau_j(t)$ can be defined as

$$\tau_j(t + \Delta t) = \beta_j \tau_j(t) + (1 - \beta_j) z_j(t) \tag{3}$$

where $\beta_j = \exp(-\Delta t/\tau_{a,j})$. The constant $\tau_{a,j}$ represents the adaptation time constant. Variable $z_j(t)$ represents the spike train of neuron j and assumes value in $\{0, 1/\Delta t\}$. The parameter values of the spiking neuron model that we proposed are listed in Table 1. The input current $I_j(t)$ of a neuron is defined as the weighted sum of the pulses, which come from external neurons or other neurons. Its mathematical formula is as follows

$$\begin{cases} I_m^j(t) = \sum\limits_{j=1}^{n} W_{ij}\chi_i(t - \kappa_{ij}) + \sum\limits_{j=1}^{n} W_{ij}^{rec}\varepsilon_i(t - \kappa_{ij}^{rec}) \\ I_i^j(t) = \sum\limits_{j=1}^{n} W_{ij}^{i}\chi_i(t - \kappa_{ij}^{i}) + \sum\limits_{j=1}^{n} W_{ij}^{irec}\varepsilon_i(t - \kappa_{ij}^{irec}) \\ I_e^j(t) = \sum\limits_{j=1}^{n} W_{ij}^{e}\chi_i(t - \kappa_{ij}^{e}) + \sum\limits_{j=1}^{n} W_{ij}^{erec}\varepsilon_i(t - \kappa_{ij}^{erec}) \end{cases} \tag{4}$$

where W_{ij}^{rec}, W_{ij}^{erec}, and W_{ij}^{irec} represent the recurrent synaptic weights of soma, excitatory dendrites, and inhibitory dendrites, respectively. In addition,W_{ij}, W_{ij}^{e}, and W_{ij}^{i} represent the synaptic weights of soma, excitatory dendrite, and inhibitory dendrite, respectively. The constants κ_{ij}, κ_{ij}^{e}, and κ_{ij}^{i} represent the delays of input synapses for soma, excitatory dendrite, and inhibitory dendrite, respectively. The constants κ_{ij}^{rec}, κ_{ij}^{erec}, and κ_{ij}^{irec} represent the delays of recurrent synapses for soma, excitatory dendrite, and inhibitory dendrite, respectively. The spike trains $\chi_i(t)$ and $\varepsilon_i(t)$ are modeled as sums of Dirac pulses, representing the spike trains from input neurons and recurrent neurons with recurrent connections, respectively. The dynamics of the proposed spiking neuron model are shown in Figure 1 accordingly.

Table 1. Parameter settings of the spiking neuron model.

Parameter	Value	Parameter	Value
R_m	1 Ω	R_i, R_e	1 Ω
τ_m	20 ms	θ_i, θ_e	0 mV
κ, κ^i, κ^e	5 ms	κ^{rec}, κ^{irec}, κ^{erec}	5 ms
α	1.8	τ^0	0.01
τ_a	700 ms	g_i, g_e	1 nS

We integrate the spiking neuron model into an SNN framework and test the accuracy of this new model on different types of learning tasks. The structure of the SNN model is shown in Figure 2. The model is divided into three layers: input layer, hidden layer, and output layer. According to different tasks, we choose different encoding methods of the input layer and decoding methods of the output layer. In Figure 2, the solid blue lines represent feed-forward inhibitory synaptic connections, while the red dashed lines represent lateral inhibitory synaptic connections. The dendrites and soma of different neurons in the hidden layer are connected by lateral inhibitory synapses that are random and sparse at the same time. Information is transmitted from the input layer to the dendrites, and the soma transmits impulse signals to the output layer. The initial network weights in the proposed SNN model are set via a Gaussian distribution $W_{ij} \sim \frac{w_0}{\sqrt{n_{in}}} N(0,1)$, where n_{in} represents the number of input neurons in the spiking neural network in the weight matrix. $N(0, 1)$ represents the Gaussian distribution with zero mean and unit variance, while $w_0 = \Delta t / R_m$ represents a weight-scaling factor depending on the time step Δt and membrane resistance R_m. This scaling factor is significant as it is used to initialize the spiking neural network with a practical firing rate needed for efficient training.

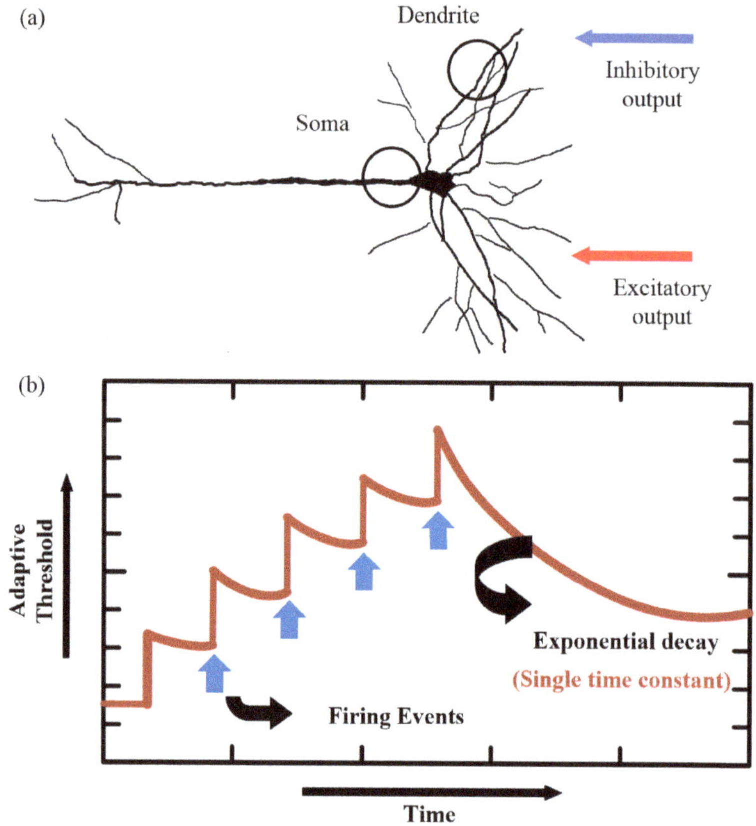

Figure 1. Dynamics of the proposed spiking neuron. (**a**) The biological structure that inspires the proposed neuron model. (**b**) The adaptive dynamics of the threshold along with the firing events.

We use a deep rewiring algorithm because it is able to maintain the sign of each synapse during the learning process [30]. Hence, this sign is inherited from the initial weights of the network. In consideration of this, the model needs efficient and reasonable initialization weights for both excitatory and inhibitory neurons. To achieve this, we sample neurons from a Bernoulli distribution, generating the symbol sign $k_i \in \{-1, 1\}$ randomly. At the same time, to avoid the problem of exploding gradients, we scale the weights so that the largest eigenvalue is less than 1. A large square matrix is generated with the number of rows selected, ultimately with uniform probability. This square matrix is then multiplied by a binary mask, resulting in a sparse matrix, as a part of the depth rewiring algorithm that we mentioned before. This algorithm achieves the goal of maintaining the level of sparse connectivity in the network by dynamically disconnecting some synapses while reconnecting others. In this algorithm, we set the temperature parameter to 0 and the L1-norm regularization parameter to 0.01.

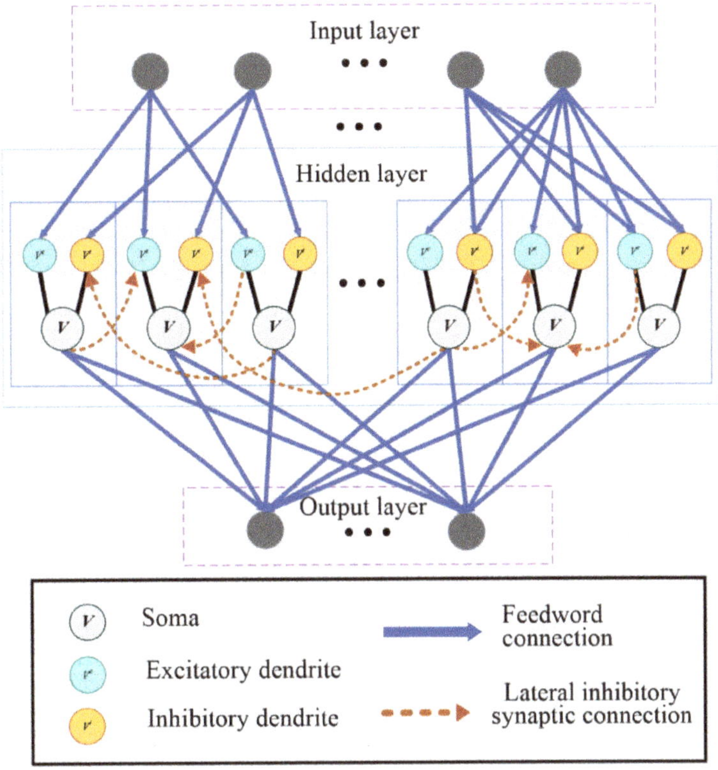

Figure 2. Network architecture for learning and memory integrated with the proposed SAM model. This network architecture is comparable to a 2-layer network of point neurons. The soma and dendrites of different neurons in the hidden layer are connected to lateral inhibitory synapses randomly. The gray circles in the input layer and output layer are not SAM neurons, representing the input spiking neuron and output spiking neuron, respectively. The input and output encodings are determined for different tasks, which will be described in the section of experimental results.

2.2. BPTT Training Algorithm

In common ANN models, the gradients of the loss function are obtained with respect to the weights in the network using back propagation. Nevertheless, the training method of back propagation cannot be directly applied to SNNs due to the non-differentiability of spikes from SNNs. Providing that time is discretized, the gradient needs to be propagated through continuous time or multiple time steps. To enable the SNN model to learn in the training process, we use a pseudo-derivative technique as shown below

$$\frac{dz_j(t)}{dv_j(t)} = k\max\{0, 1 - |v_j(t)|\} \tag{5}$$

where $k = 0.3$ (typically less than 1) is a constant value that can dampen the increase in back propagated errors through spikes by using a pseudo-derivative of amplitude to achieve the goal of stable performance. The variable $z_j(t)$ represents the spike train of neuron j that assumes values in {0, 1}. The variable $v_j(t)$ represents the normalized membrane potential, which is defined as follows

$$v_j(t) = \frac{V_j(t) - \Gamma_j(t)}{\Gamma_j(t)} \quad (6)$$

where Γ_j represents the firing rate of neuron j. With the purpose of providing the self-learning capability required for reinforcement learning for the proposed SAM model, we utilize a proximal policy optimization algorithm [31]. This algorithm is easy to implement and allows the model to have self-learning capabilities. The clipped surrogate objective of this algorithm is defined as $O^{PPO}(\vartheta_{old}, \vartheta, t, k)$. Therefore, the loss function with respect to ϑ is formulated as

$$L_P(\theta) = -\frac{\sum\limits_{k<K}\sum\limits_{t<T} O^{PPO}(\vartheta_{old}, \vartheta, t, k)}{KT} + \mu_f \frac{1}{n}\sum_j \left\| \frac{\sum\limits_{k,t} z_j(t,k) - f^0}{KT} \right\|^2 \quad (7)$$

where f^0 represents a target firing rate of 10 Hz and μ_f represents a regularization hyper-parameter. Variables t and k represent the simulation time step and the total number of epochs. The variable ϑ represents the current policy parameter, which is defined in the previous research [31]. In each iteration of training, $K = 10$ episodes of $T = 2000$ time steps are generated with a fixed parameter ϑ_{old}, which is the vector of policy parameters before the update as expressed in [31]. At the same time, the loss function $L(\vartheta)$ is minimized by the ADAM optimizer [32].

2.3. Minimum Error Entropy Criterion (MEEC)

The minimum error entropy (MEE) can minimize the entropy of the estimation error, so that decreases the uncertainty in the learning process. The α-order Renyi's entropy is used assuming a random variable e with probability density function $f^{\alpha}(e)$, which is defined as

$$H(e) \triangleq \frac{1}{1-\alpha} \log \int f^{\alpha}(e) de \quad (8)$$

where α is set to 2 for 2-order Renyi's entropy in this study. The kernel density estimation (KDE) is used to estimate the PDF of the error samples, which has three advantages. First, it is a non-parameter approach, which does not require the prior knowledge of the error distribution. Second, it does not require the integration calculation. Third, it can be smooth and differentiable, which is vital for the gradient computation. Considering a set of i.i.d data $\{e_i\}_{i=1}^{N}$ drawn from the distribution, the KDE of the PDF can be formulated as

$$\hat{f}_E(e) = \frac{1}{N}\sum_{i=1}^{N} G_{\Sigma}(e - e_i) \quad (9)$$

where $G_{\Sigma}(e - e_i)$ represents the Gaussian function with the following expression as

$$G_{\Sigma}(e - e_i) = \frac{1}{\sqrt{2\pi(\det \Sigma)}} \cdot \exp\left(-\frac{1}{2}(e - e_i)^T \Sigma^{-1} (e - e_i)\right) \quad (10)$$

where N and Σ represent the number of the data points and the kernel parameter, respectively. In this research, Σ represents a diagonal matrix with the s-th diagonal element with the variance δ_s^2 for e_s in e, where $s = 1, 2, \ldots, S$. The kernel parameter represents a free parameter. Thus, the Renyi's quadratic entropy can be expressed as

$$\begin{aligned} H_2(e) &= -\log \int \left(\frac{1}{N} \sum_{i=1}^{N} G_{\Sigma}(e - e_i) \right)^2 de \\ &= -\log \frac{1}{N^2} \int \left(\sum_{i=1}^{N} \sum_{j=1}^{N} G_{\Sigma}(e - e_i) G_{\Sigma}(e - e_j) \right) de \\ &= -\log \frac{1}{N^2} \int \left(\sum_{i=1}^{N} \sum_{j=1}^{N} G_{\Sigma}(e - e_i) G_{\Sigma}(e - e_j) \right) de \\ &= -\log \frac{1}{N^2} \left(\sum_{i=1}^{N} \sum_{j=1}^{N} G_{\sqrt{2}\Sigma}(e_i - e_j) \right) \\ &= -\log \frac{1}{N^2} \left(\sum_{i=1}^{N} \sum_{j=1}^{N} G_{\Sigma_2}(e_i - e_j) \right) \end{aligned} \tag{11}$$

Based on the Formula (11), we define a function $V(e)$ to represent the information potential of variable e, which is formulated as

$$V(e) = \frac{1}{N^2} \left(\sum_{i=1}^{N} \sum_{j=1}^{N} G_{\Sigma_2}(e_i - e_j) \right) \tag{12}$$

Therefore, the minimization of the Renyi's entropy $H_2(e)$ means the maximization of the information potential $V(e)$ because of the monotonic increasing feature of the log function. The Parzen window is used to decrease the computational complexity and the instantaneous information potential at time t, which can be formulated as

$$J_1(e) = \frac{1}{W} \sum_{i=k-W+1}^{k} G_{\Sigma_2}(e_k - e_i) \tag{13}$$

where W represents the length of the Parzen window. It should be noted that MEE is a kind of local optimization criterion but suffers from the shift-invariant problem. It can only determine the location of error PDF but cannot know the distribution location. The function $G_{\Sigma 2}(.)$ can be defined as the Gaussian kernel function with bandwidth σ

$$G_{\Sigma 2}(x) = \frac{1}{\sqrt{2\pi}\sigma} \exp\left(-\frac{x^2}{2\sigma^2} \right) \tag{14}$$

In order to reduce the computational complexity, quantization technique is used to realize the quantized MEE (QMEE). Thus, the information potential is expressed as

$$V^Q(e) = \frac{1}{N^2} \left(\sum_{i=1}^{N} \sum_{j=1}^{N} G_{\Sigma_2}(e_i - Q|e_j|) \right) = \frac{1}{N^2} \sum_{i=1}^{N} \sum_{j=1}^{M} \varphi_j G_{\Sigma 2}(e_i - c_j) \tag{15}$$

where $Q[.]$ represents a quantization operator mapping each $\{e_i\}_{i=1}^{N}$ to one of $\{c_j\}_{j=1}^{M}$, resulting in a codebook $C = (c_1, c_2, c_3, \ldots, c_M)$. $\Phi = (\varphi_1, \varphi_2, \ldots, \varphi_M)$ represents the number of the samples quantized to the corresponding set $\{c_j\}_{j=1}^{M}$. It should be noted that $\sum_{j=1}^{M} \varphi_j = N$. Theoretical proof of the robustness has been presented in [22].

2.4. Restricted MEEC

In this study, the fundamental inner product to measure the similarity is used, which is generalized from its vectors' application [33]. The inner product similarity between continuous pdfs $f_X(x)$ and $g_X(x)$ can be expressed as

$$\langle f_X(x), g_X(x)\rangle = \int_X f_X(x) g_X(x) dx \tag{16}$$

The desired distribution $\rho_E(e)$, which is expressed in [33] in detail, can be defined as follows

$$\rho_E(e) = \begin{cases} \zeta_0, & e = 0 \\ \zeta_{-1}, & e = -1 \\ \zeta_1, & e = 1 \\ 0, & \text{otherwise} \end{cases} \tag{17}$$

where ζ_i ($i = 0, -1, 1$) denotes the corresponding density for each peak, which is simplified into a Dirac-δ function.

The maximization of the similarity measure between the error pdf $f_E(e)$ and the desired distribution $\rho_E(e)$ can be formulated as

$$\begin{array}{l} \max\langle f_E(e), \rho_E(e)\rangle \\ \Leftrightarrow \max \int_X f_E(e)\rho_E(e) dx \\ \Leftrightarrow \max \zeta_0 f_E(0) + \zeta_{-1} f_E(-1) + \zeta_1 f_E(1) \end{array} \tag{18}$$

Furthermore, the model parameter can be expressed as

$$\begin{array}{l} w^* = \operatorname{argmax} \zeta_0 \hat{f}_E(0) + \zeta_{-1}\hat{f}_E(-1)\zeta_1 \hat{f}_E(1) \\ = \operatorname{argmax} \begin{pmatrix} \zeta_0 \frac{1}{N}\sum\limits_{i=1}^{N} G_{\Sigma 2}(0 - e_i) \\ +\zeta_{-1}\frac{1}{N}\sum\limits_{i=1}^{N} G_{\Sigma 2}(-1 - e_i) \\ \zeta_1 \frac{1}{N}\sum\limits_{i=1}^{N} G_{\Sigma 2}(1 - e_i) \end{pmatrix} \\ = \operatorname{argmax} \frac{1}{N^2}\sum\limits_{i=1}^{N} \begin{pmatrix} N\zeta_0 G_{\Sigma 2}(e_i) \\ +N\zeta_{-1} G_{\Sigma 2}(e_i + 1) \\ +N\zeta_1 G_{\Sigma 2}(e_i - 1) \end{pmatrix} \end{array} \tag{19}$$

In fact, QMEE converges the prediction errors $\{c_j\}_{j=1}^{M}$ to obtain a compact error distribution. Based on the method in [33], a predetermined codebook $C = (0, -1, 1)$ implements QMEE to restrict errors to three positions and avoid the undesirable double-peak learning consequence. Therefore, the restricted MEE (RMEE) algorithm can be formulated as

$$V^R(e) = \frac{1}{N^2}\sum_{i=1}^{N} \begin{pmatrix} \varphi_0 G_{\Sigma 2}(e_i) \\ +\varphi_{-1} G_{\Sigma 2}(e_i + 1) \\ +\varphi_1 G_{\Sigma 2}(e_i - 1) \end{pmatrix} \tag{20}$$

where $\Phi = (\varphi_0, \varphi_{-1}, \varphi_1) = (N\zeta_0, N\zeta_{-1}, N\zeta_1)$ that represents the corresponding number for each quantization word $C = (0, -1, 1)$. The proposed RMEE algorithm maximizes the inner product similarity between error pdf $f_E(e)$ and the optimal three-peak distribution $\rho_E(e)$. RMEE is a specific formation of QMEE where the codebook is predetermined as $C = (0, -1, 1)$ and converges learning errors on these three locations.

In order to optimize Equation (19), the half-quadratic technique is used to solve optimization issues. A convex function $g(x) = -x\log(-x) + x$ is defined, and the information potential can be expressed as

$$V^R(e) = \sum_{i=1}^{N} \begin{pmatrix} \varphi_0\left\{u_i\frac{e_i^2}{2\sigma^2} - g(u_i)\right\} \\ +\varphi_{-1}\left\{v_i\frac{(e_i+1)^2}{2\sigma^2} - g(v_i)\right\} \\ +\varphi_1\left\{s_i\frac{(e_i-1)^2}{2\sigma^2} - g(s_i)\right\} \end{pmatrix} \triangleq J_{R1}(w, u_i, v_i, s_i) \tag{21}$$

In half-quadratic technique, it has the following relationship

$$\begin{aligned} u_i^k &= -\exp\left(-\frac{e_i^2}{2\sigma^2}\right) < 0 \\ v_i^k &= -\exp\left(-\frac{(e_i+1)^2}{2\sigma^2}\right) < 0 \\ s_i^k &= -\exp\left(-\frac{(e_i-1)^2}{2\sigma^2}\right) < 0 \\ (i &= 1, 2, \ldots, N). \end{aligned} \tag{22}$$

By attaining the optimal (u_i^k, v_i^k, s_i^k) in the kth iteration, the information potential can be formulated as

$$V^R(e) = \sum_{i=1}^{N} \begin{pmatrix} \varphi_0 u_i(t_i - y_i)^2 \\ +\varphi_{-1} v_i(t_i + 1 - y_i)^2 \\ +\varphi_1 s_i(t_i - 1 - y_i)^2 \end{pmatrix} \triangleq J_{R2}(w) \tag{23}$$

The $J_{R2}(w)$ can be optimized based on gradient-based methods because the objective function is differentiable and continuous. For example, the gradient of $J_{R2}(w)$ can be expressed as

$$\frac{\partial}{\partial w}J_{R2}(w) = \sum_{i=1}^{N} \begin{pmatrix} \varphi_0 u_i\frac{\partial(t_i-y_i)^2}{\partial w} \\ +\varphi_{-1} v_i\frac{\partial(t_i+1-y_i)^2}{\partial w} \\ +\varphi_1 s_i\frac{\partial(t_i-1-y_i)^2}{\partial w} \end{pmatrix} = -2\sum_{i=1}^{N} \begin{pmatrix} \varphi_0 u_i e_i \\ +\varphi_{-1} v_i(e_i+1) \\ +\varphi_{-1} s_i(e_i-1) \end{pmatrix} x_i y_i(1-y_i) \tag{24}$$

The detailed algorithm of the HQ-based optimization and its convergence analysis for RMEE are presented in [33].

3. Results

3.1. Proposed Network with RMEE Criterion

Since MEE has the shift-invariant feature, and estimation results based on MEEC will not always converge to the true value. A consideration is to combine the RMEE criterion with CEE for a global optimal solution. The cross-entropy loss function, also regarded as log loss, is the most commonly used loss function for back propagation. The cross-entropy loss function increases as the predicted probability deviates from the actual label, and can be described as follows

$$L_{ce}\left(\hat{y}_i, y_i\right) = -\sum_i y_i \log\left(\hat{y}_i\right) \tag{25}$$

In this paper, the label l^n of each image is used, which is only assumed to be 1 for images belonging to the same class of images during testing, and 0 otherwise. The cross-entropy formula can be expressed as

$$J_2 = \sum_{n=1}^{5} -l^n \log\sigma\left(y^{20+20\cdot n}\right) - (1-l^n)\log\left(1-\sigma^{20+20\cdot n}\right) \tag{26}$$

where the output of the SNN model is only counted after all images are fully rendered. Therefore, for the novel criterion, the performance index can be formulated as

$$J_k(e) = \mu\left[\sum_{i=1}^{N}\begin{pmatrix} \varphi_0 u_i(t_i - y_i)^2 \\ +\varphi_{-1} v_i(t_i + 1 - y_i)^2 \\ +\varphi_1 s_i(t_i - 1 - y_i)^2 \end{pmatrix}\right] + (1-\mu)\left[\sum_{n=1}^{5}\begin{pmatrix} -l^n \log\sigma(y^{20+20\cdot n}) \\ -(1-l^n)\log(1-\sigma^{20+20\cdot n}) \end{pmatrix}\right] \tag{27}$$

where μ represents a weighting constant. In the supervised learning tasks, there only exist cross-entropy and RMEE, which is described in Equation (27).

3.2. Autonomous Navigation

We first apply the proposed SNN model in the agent navigation task, which requires the network to have reinforcement learning capabilities. The agent needs to learn to find objects in a 2D area and eventually be able to navigate to find objects at random locations in the area. This task is interrelated with the neuroscience paradigm of the well-known Morris water maze task, which is designed to study learning in the brain [34]. In this task, a virtual agent is simulated as a point in the 2D simulation arena and is controlled by the proposed SNN model. The position of the agent is configured randomly with a uniform probability in the overall arena at the beginning of an episode. The agent produces a small velocity vector of the Euclidean norm and selects an action at each time step. It receives a reward value '1' after reaching the destination.

In the navigation task, the information $s(t)$ of the current environment state and the reward score $r(t)$ are received as input data by neurons in the input layer at each time step. The coordinate information of the position is encoded by the input neurons through the Gaussian population rate encoding method. Furthermore, each neuron in the input layer is assigned a coordinate value with a firing rate, which is defined as: $r_{\max} = \exp(-100(\xi_i\text{-}\xi)^2)$, where ξ_i and ξ represent the actual coordinate value and the preferred coordinate value, respectively. $r_{\max}$ is supposed to be set as 500 Hz. Moreover, the instantaneous reward $r(t)$ is encoded by two sets of input neurons. In the first group, the neurons generate spikes in sync when a positive reward is received, while in the second group, the neurons generate spikes as long as the proposed SNN model receives a negative reward. The output of the network is represented by five readout neurons in the output layer with membrane potential $\lambda_i(t)$. The action vector $\zeta(t) = (\zeta_x(t), \zeta_y(t))^T$ is used to determine the movement of the agent in the navigation task that we mentioned before. It is calculated from a Gaussian distribution with mean $\mu_x = \tanh(\lambda_1(t))$ and $\mu_y = \tanh(\lambda_2(t))$ as well as variances $\Phi_x = \sigma(\lambda_3(t))$ and $\Phi_y = \sigma(\lambda_4(t))$. In the end, the output of the last readout neuron λ_5 is calculated to predict the value function $\mu_\theta(t)$. This predicts the expected discounted sum of future rewards $\Omega(t) = \Sigma_{t'} > t_{\gamma t'} - t_{\omega(t')}$, where $\omega(t')$ represents the reward at time t' and γ represents the discount factor, whose value is usually 0.99.

The agent based on the proposed SNN model learns to learn in the navigation task towards the correct destination location after the meta-learning process. The overall training process in the reward learning process is described by Algorithm 1. We add other loss functions to support the reinforcement learning framework, maintaining the loss function consistent with Equation (26). Figure 3 shows the successful destination reached number (DRN) per learning iteration. Each iteration contains a batch of ten episodes, and network weights are updated during the navigation task. For each episode, the model is expected to explore until reaching and storing the destination location, and uses the prior knowledge to find the shortest path to the destination. This reveals that the proposed SNN model has meta-learning capability in the autonomous navigation task.

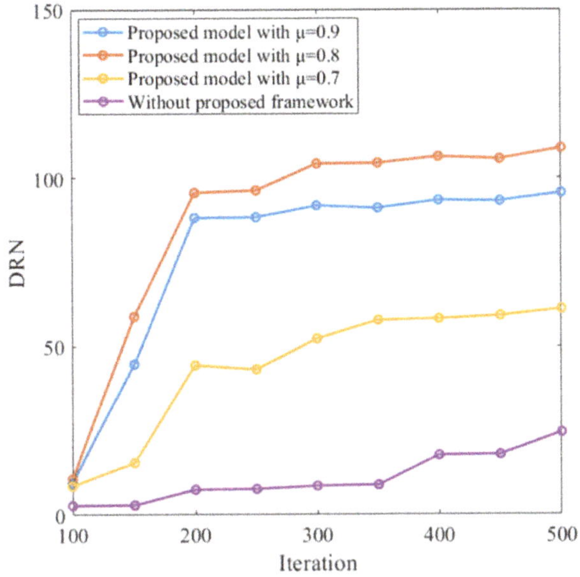

Figure 3. Navigation performance of the proposed model with different settings.

Algorithm 1 Training process in the reward learning process

Input: number of full episodes **K**, timesteps **T**, fixed parameters θ_{old}, target firing rate f^0, regularization hyper-parameters μ_v, μ_e, μ_{firing}, bandwidth σ, predicted value function $\mathbf{V}_\theta(t,k)$ and sum of future rewards $\mathbf{R}(\mathbf{t},\mathbf{k})$
Output: total loss L_θ.

1. Parameters setting: f^0, μ_v, μ_e, μ_{firing} *and* σ.
2. **for** n **in** batch size N:
3. Set $e_n = \mathbf{R}(\mathbf{t},\mathbf{k}) - \mathbf{V}_\theta(t,k)$
4. **if** number of literation is 0:
5. $(\varphi_0,\ \varphi_{-1},\ \varphi_1) = (\mathbf{N},\ \mathbf{0},\ \mathbf{0})$
6. **else**:
   $(\varphi_0,\ \varphi_{-1},\ \varphi_1) = (\#\{e_n \in (-\mathbf{0.5},\ \mathbf{0.5})\},$
   $\#\{e_n \in (-\mathbf{1}, -\mathbf{0.5})\},$
   $\#\{e_n \in (\mathbf{0.5},\ \mathbf{1})\})$
   where $\#\{\cdot\}$ indicates counting the samples that satisfy the condition
7. $(u_n, v_n, s_n) = (-exp(-\frac{e_n{}^2}{2\sigma^2}), -exp(-\frac{(e_n+1)^2}{2\sigma^2}), -exp(-\frac{(e_n-1)^2}{2\sigma^2}))$
8. $L_n^{RMEE}{=}\varphi_0 u_n e_n{}^2 + \varphi_{-1} v_n (e_n+\mathbf{1})^2 + \varphi_1 s_n (e_n - \mathbf{1})^2$
9. **end for**
10. **for** k **in** K:
11. **for** t **in** T:
    $L_{(t,k)}^{PPO} = O^{PPO}(\theta_{old}, \theta,\ t,\ k)$
12. **end for**
13. **end for**
14. Calculate the total loss:
    $L(e) = L_p(e) + J_k(e)$
15. **return** $L(e)$

3.3. Working Memory Performance on Store–Recall Task with Non-Gaussian Noise

To further demonstrate the robust working memory capability of the proposed SNN model, we apply the model in a store–recall task with non-Gaussian noise. The detailed settings of the store–recall task have been previously presented in [35]. The SNN model receives a sequence of frames that are represented by ten spike trains in a period of time. The inputs #1 and #2 are represented by the spiking activities of input neurons from #1 to #10 and from #11 to #20, respectively. As shown in Figure 4, the neurons from #21 to #30 and from #31 to #40 receive the random store and recall commands, respectively. The store command means direct attention is paid to the specific frame of input data flow. Then, this frame will be reproduced when receiving the recall command. Figure 4 shows one test example with the spiking activities after working memory training. The dynamic threshold changes along with the learning procedure, which is shown in Figure 4. This reveals that the proposed SNN model can exhibit the working memory performance and realize the store–recall task successfully. Since working memory is a vital feature and the foundation for meta-learning, this also suggests that the MeMEE model can exhibit the meta-learning tasks based on its working memory mechanisms with a robust performance.

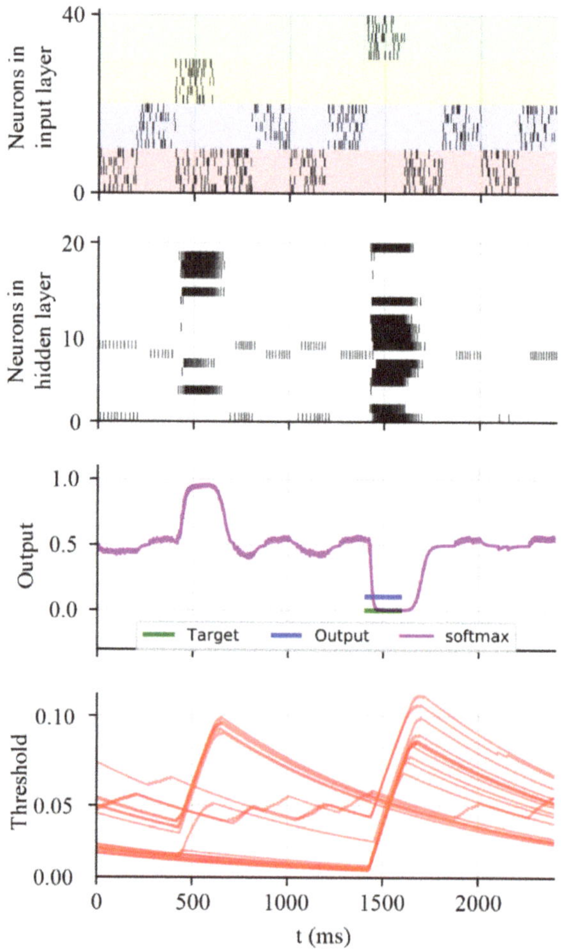

Figure 4. Working memory capability of the proposed SNN model after training.

3.4. *Meta-Learning Performance on Sequential MNIST Data Set with Non-Gaussian Noise*

We further demonstrate the meta-learning capability of the proposed SNN model in a transfer learning task based on the sequential MNIST (sMNIST) data set. We divide the sMNIST data set into two parts. The first part includes 30,000 images for digits '0', '1', '2', '3', and '4', and the second part includes 30,000 patterns for digits '5', '6', '7', '8', and '9'. In the first phase, the first part is employed to train the SNN model, and the second part is then used for training. In the second phase, 10% salt and pepper noise is added to the testing data set as the non-Gaussian noise for the performance evaluation. Figure 5 shows the performance of the MeMEE model and compares it with other counterpart models, including recurrent SNN (RSNN) and the conventional LIF-based SNN model without the RMEE criterion. This shows that the proposed model outperforms the other solutions, and the reasoning behind this includes three points. Firstly, the proposed model has the meta-learning capability, so it can illustrate the transfer learning capability, and its transfer learning performance is superior to the RSNN model accordingly, considering accuracy and convergence speed. Secondly, due to the RMEE criterion being the loss function, its robustness to the non-Gaussian noise is superior to the model without the RMEE criterion in terms of the learning accuracy. The result suggests that the MeMEE model with RMEE criterion has a more powerful robust meta-learning capability in learning sequential spatio-temporal patterns.

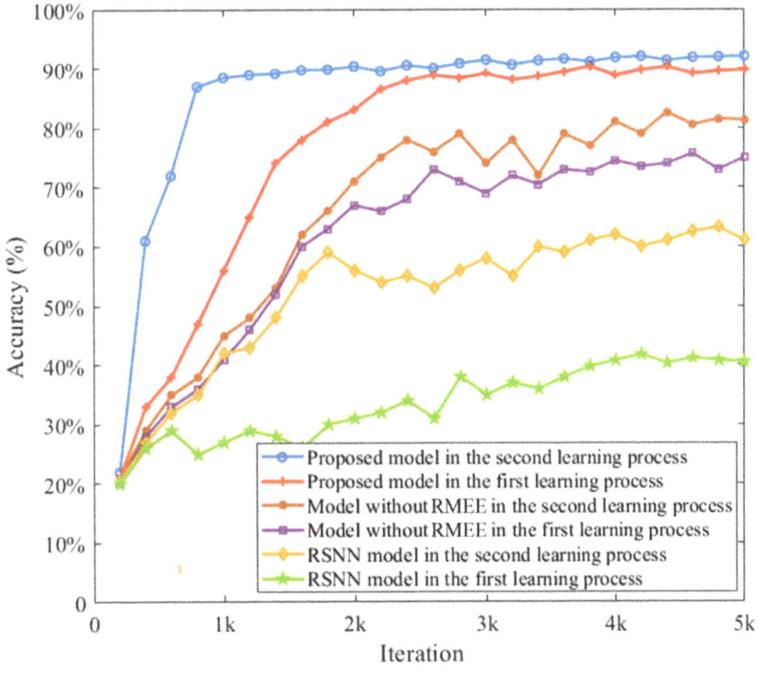

Figure 5. Meta-learning capability of the proposed MeMEE model on sequential MNIST data set.

3.5. *Effects of Loss Parameters on Learning Performance*

In this study, we further investigate how each loss function affects the learning performance of the proposed MeMEE model. We use the sMNIST data set to evaluate and quantify the learning accuracy along with the changing loss parameter. In order to demonstrate the learning robustness based on the proposed MeMEE model, salt and pepper noise is added to the sMNIST data set. Different levels are considered, which are selected from 3.19% to 19.13%. Different values of parameter μ are investigated, which are set from 0.3 to 1.0. As shown in Figure 6, the value of μ with 0.7, 0.8, and 0.9 can induce the higher learning accuracy on sequential visual recognition. This reveals that the RMEE criterion can further enhance the robustness of the proposed MeMEE model without the RMEE criterion, i.e., $\mu = 1$. Since the model without RMEE criterion with 3.19% non-Gaussian noise only reaches 83.6% accuracy, the RMEE criterion can improve the learning accuracy of the proposed MeMEE model with non-Gaussian salt and pepper noise.

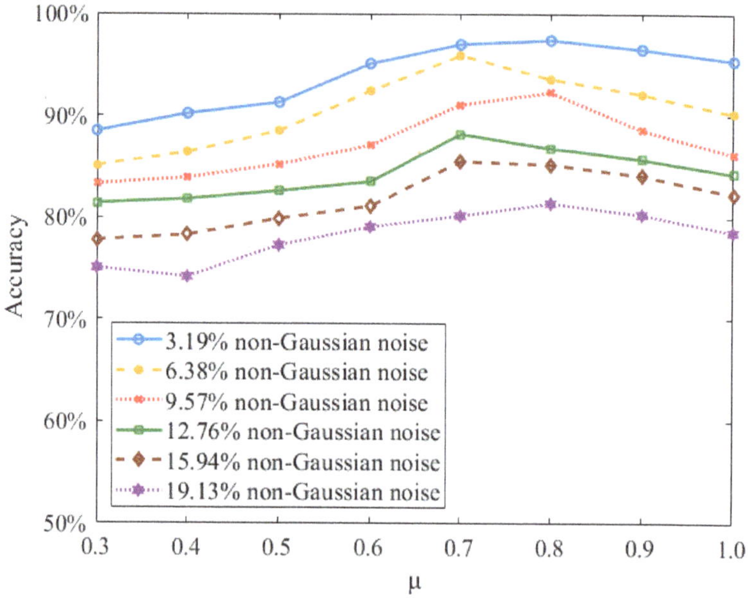

Figure 6. Effects of loss parameters on the learning performance of sequential classification.

4. Discussion

This paper presents an information theoretic learning framework for robust spike-driven continual meta-learning. Different from the previous SNN learning research, we first introduce the RMEE criterion to develop and improve the spike-based learning framework, which is significantly general and can also provide a series of theoretic insights. Moreover, the information theoretic framework allows us to obtain a direct understanding and better interpretation of the robust learning solutions of SNN models, compared with some previous studies focusing on improving the learning robustness of SNNs [36].

As a first step in establishing a rigorous framework for SNN continual meta-learning with RMEE, the presented research can be extended in both theoretical and practical aspects. From the theoretical point of view, one extension is to use the information potential to train the presented SNN model. For example, as shown in [37], Chen et al. presented a survival information potential algorithm for adaptive system training. This does not require computing of the kernel function and has good robustness performance accordingly. The other extension is to apply the proposed framework in other spike-based learning paradigms, including few-shot learning, multitask learning, and unsupervised learning [38].

From a practical point of view, the model is expected to be implemented on neuromorphic platforms to realize low-power and real-time systems for various types of applications. The state-of-the-art digital neuromorphic systems include Loihi [12], Tianjic [11], BiCoSS [13], CerebelluMorphic [14], LaCSNN [15], TrueNorth [39], and SpiNNaker [40]. By implementing embedded neuromorphic systems, it can be applied in different fields such as edge computing devices, brain–machine integration systems, and intelligent systems [41–43].

5. Conclusions

In this invited paper, we first presented an ITL-based scheme for robust spike-based continual meta-learning, which is improved by the RMEE criterion. A gradient descent learning principle is presented in a recurrent SNN architecture. Several tasks are realized to demonstrate the learning performance of the proposed MeMEE model, including autonomous navigation, robust working memory in the store–recall task and robust meta-learning capability for the sMNIST data set. In the first autonomous navigation task, the SNN model learns to find the correct destination by continual meta-learning from the task reward and punishment. This demonstrates that the MeMEE model based on the proposed RMEE criterion realizes the meta-learning capability for navigation and outperforms the conventional RSNN model. In the second task, the proposed MeMEE model improves the working memory performance by recalling the stored noisy patterns. In the third task, the proposed MeMEE model with RMEE criterion can enhance the robustness in the meta-learning task for noisy sMNIST images. This invited paper provides a novel insight into the improvement of the spike-based machine learning performance based on information theoretic learning strategy, which is critical for the further research of artificial general intelligence. In addition, it can be implemented by the low-power neuromorphic system, which can be applied in edge computing of internet of things (IoT) and unmanned systems.

Author Contributions: S.Y. and B.C. contributed to the conceptualization, methodology, and writing of this paper. J.T. helped to conduct the experiment. All authors have read and agreed to the published version of the manuscript.

Funding: This study was funded partly by the National Natural Science Foundation of China with grant numbers (Grant No. 62006170, No. 62088102, No. U21A20485) and partly by China Postdoctoral Science Foundation (Grant Nos. 2020M680885, 2021T140510).

Acknowledgments: We would like to thank the editor and reviewer for their comments on the manuscript.

Conflicts of Interest: The authors declare no conflict of interest. The funders had no role in the design of the study; in the collection, analyses, or interpretation of data; in the writing of the manuscript, or in the decision to publish the results.

References

1. Krizhevsky, A.; Sutskever, I.; Hinton, G. ImageNet classification with deep convolutional neural networks. *Adv. Neural Inf. Process. Syst.* **2012**, *25*, 89–94. [CrossRef]
2. Parisi, G.I.; Kemker, R.; Part, J.L.; Kanan, C.; Wermter, S. Continual lifelong learning with neural networks: A review. *Neural Netw.* **2019**, *113*, 54–71. [CrossRef] [PubMed]
3. Yao, H.; Zhou, Y.; Mahdavi, M.; Li, Z.; Socher, R.; Xiong, C. Online structured meta-learning. *Adv. Neural Inf. Process. Syst.* **2020**, *33*, 6779–6790.
4. Javed, K.; White, M. Meta-learning representations for continual learning. *Adv. Neural Inf. Process. Syst.* **2019**, *32*, 172.
5. Serrà, J.; Surís, D.; Miron, M.; Karatzoglou, A. Overcoming catastrophic forgetting with hard attention to the task. In Proceedings of the International Conference on Machine Learning (PMLR 80), Stockholmsmässan, Stockholm, Sweden, 10–15 July 2018; pp. 4548–4557.
6. Zeng, G.; Chen, Y.; Cui, B.; Yu, S. Continual learning of context-dependent processing in neural networks. *Nat. Mach. Intell.* **2019**, *1*, 364–372. [CrossRef]
7. van de Ven, G.M.; Siegelmann, H.T.; Tolias, A.S. Brain-inspired replay for continual learning with artificial neural networks. *Nat. Commun.* **2020**, *11*, 4069. [CrossRef]
8. Tavanaei, A.; Ghodrati, M.; Kheradpisheh, S.R.; Masquelier, T.; Maida, A. Deep learning in spiking neural networks. *Neural Netw.* **2019**, *111*, 47–63. [CrossRef]
9. Lee, C.; Panda, P.; Srinivasan, G.; Roy, K. Training deep spiking convolutional neural networks with stdp-based unsupervised pre-training followed by supervised fine-tuning. *Front. Neurosci.* **2018**, *12*, 435. [CrossRef]
10. Xia, Q.; Yang, J.J. Memristive crossbar arrays for brain-inspired computing. *Nat. Mat.* **2019**, *18*, 309–323. [CrossRef]
11. Pei, J.; Deng, L.; Song, S.; Zhao, M.; Zhang, Y.; Wu, S.; Wang, G.; Zou, Z.; Wu, Z.; He, W.; et al. Towards artificial general intelligence with hybrid Tianjic chip architecture. *Nature* **2019**, *572*, 106–111. [CrossRef]

12. Davies, M.; Srinivasa, N.; Lin, T.-H.; Chinya, G.; Cao, Y.; Choday, S.H.; Dimou, G.; Joshi, P.; Imam, N.; Jain, S.; et al. Loihi: A neuromorphic manycore processor with on-chip learning. *IEEE Micro* **2018**, *38*, 82–99. [CrossRef]
13. Yang, S.; Wang, J.; Hao, X.; Li, H.; Wei, X.; Deng, B.; Loparo, K.A. BiCoSS: Toward large-scale cognition brain with multigranular neuromorphic architecture. *IEEE Trans. Neural Netw. Learn. Syst.* **2021**, *11*, 1–15. [CrossRef] [PubMed]
14. Yang, S.; Wang, J.; Zhang, N.; Deng, B.; Pang, Y.; Azghadi, M.R. Cerebellumorphic: Large-scale neuromorphic model and architecture for supervised motor learning. *IEEE Trans. Neural Netw. Learn. Syst.* **2021**, *23*, 1–15. [CrossRef] [PubMed]
15. Yang, S.; Wang, J.; Deng, B.; Liu, C.; Li, H.; Fietkiewicz, C.; Loparo, K.A. Real-time neuromorphic system for large-scale conductance-based spiking neural networks. *IEEE Trans. Cybern.* **2019**, *49*, 2490–2503. [CrossRef]
16. Bellec, G.; Salaj, D.; Subramoney, A.; Legenstein, R.; Maass, W. Long short-term memory and learning-to-learn in networks of spiking neurons. *Adv. Neural Inf. Process. Syst.* **2018**, *31*, 247.
17. Li, Y.; Zhou, J.; Tian, J.; Zheng, X.; Tang, Y.Y. Weighted error entropy-based information theoretic learning for robust subspace representation. *IEEE Trans. Neural Netw. Learn. Syst.* **2021**, *19*, 1–15. [CrossRef]
18. Chen, J.; Song, L.; Wainwright, M.; Jordan, M. Learning to explain: An information-theoretic perspective on model interpretation. In Proceedings of the 35th International Conference on Machine Learning (PMLR 80), Stockholmsmässan, Stockholm, Sweden, 10–15 July 2018; pp. 883–892.
19. Xu, Y.; Cao, P.; Kong, Y.; Wang, Y. DMI: A novel information-theoretic loss function for training deep nets robust to label noise. *Adv. Neural Inf. Process. Syst.* **2019**, *32*, 76.
20. Chen, B.; Xing, L.; Zhao, H.; Du, S.; Principe, J.C. Effects of outliers on the maximum correntropy estimation: A robustness analysis. *IEEE Trans. Syst. Man Cybern. Syst.* **2021**, *51*, 4007–4012. [CrossRef]
21. Chen, B.; Li, Y.; Dong, J.; Lu, N.; Qin, J. Common spatial patterns based on the quantized minimum error entropy criterion. *IEEE Trans. Syst. Man Cybern. Syst.* **2020**, *50*, 4557–4568. [CrossRef]
22. Chen, B.; Xing, L.; Xu, B.; Zhao, H.; Principe, J.C. Insights into the robustness of minimum error entropy estimation. *IEEE Trans. Neural Netw. Learn. Syst.* **2018**, *29*, 731–737. [CrossRef]
23. Chen, H.-Y.; Liang, J.-H.; Chang, S.-C.; Pan, J.-Y.; Chen, Y.-T.; Wei, W.; Juan, D.-C. Improving adversarial robustness via guided complement entropy. In Proceedings of the IEEE/CVF International Conference on Computer Vision (ICCV), Seoul, Korea, 27 October–2 November 2019; pp. 4880–4888.
24. Rachdi, M.; Waku, J.; Hazgui, H.; Demongeot, J. Entropy as a robustness marker in genetic regulatory networks. *Entropy* **2020**, *22*, 260. [CrossRef] [PubMed]
25. Borin, J.A.M.S.; Humeau-Heurtier, A.; Virgílio Silva, L.E.; Murta, L.O. Multiscale entropy analysis of short signals: The robustness of fuzzy entropy-based variants compared to full-length long signals. *Entropy* **2021**, *23*, 1620. [CrossRef] [PubMed]
26. Grienberger, C.; Milstein, A.D.; Bittner, K.C.; Romani, S.; Magee, J.C. Inhibitory suppression of heterogeneously tuned excitation enhances spatial coding in CA1 place cells. *Nat. Neurosci.* **2017**, *20*, 417–426. [CrossRef]
27. Muñoz, W.; Tremblay, R.; Levenstein, D.; Rudy, B. Layer-specific modulation of neocortical dendritic inhibition during active wakefulness. *Science* **2017**, *355*, 954–959. [CrossRef]
28. Poleg-Polsky, A.; Ding, H.; Diamond, J.S. Functional compartmentalization within starburst amacrine cell dendrites in the retina. *Cell Rep.* **2018**, *22*, 2898–2908. [CrossRef]
29. Ranganathan, G.N.; Apostolides, P.F.; Harnett, M.T.; Xu, N.L.; Druckmann, S.; Magee, J.C. Active dendritic integration and mixed neocortical network representations during an adaptive sensing behavior. *Nat. Neurosci.* **2018**, *21*, 1583–1590. [CrossRef]
30. Bellec, G.; Kappel, D.; Maass, W.; Legenstein, R. Deep rewiring: Training very sparse deep networks. *arXiv* **2017**, arXiv:1711.05136.
31. Schulman, J.; Wolski, F.; Dhariwal, P.; Radford, A.; Klimov, O. Proximal policy optimization Algorithms. *arXiv* **2017**, arXiv:1707.06347.
32. Kingma, D.P.; Ba, J. Adam: A method for stochastic optimization. *arXiv* **2014**, arXiv:1412.6980.
33. Li, Y.; Chen, B.; Yoshimura, N.; Koike, Y. Restricted minimum error entropy criterion for robust classification. *IEEE Trans. Neural Netw. Learn. Syst.* **2021**, *2*, 1–14. [CrossRef] [PubMed]
34. Vasilaki, E.; Frémaux, N.; Urbanczik, R.; Senn, W.; Gerstner, W. Spike-based reinforcement learning in continuous state and action space: When policy gradient methods fail. *PLoS Comput. Biol.* **2009**, *5*, e1000586. [CrossRef]
35. Wolff, M.J.; Jochim, J.; Akyürek, E.G.; Stokes, M.G. Dynamic hidden states underlying working-memory-guided behavior. *Nat. Neurosci.* **2017**, *20*, 864–871. [CrossRef] [PubMed]
36. Yang, S.; Gao, T.; Wang, J.; Deng, B.; Lansdell, B.; Linares-Barranco, B. Efficient spike-driven learning with dendritic event-based processing. *Front. Neurosci.* **2021**, *15*, 601109. [CrossRef] [PubMed]
37. Chen, B.; Zhu, P.; Principe, J.C. Survival information potential: A new criterion for adaptive system training. *IEEE Trans. Signal Process.* **2012**, *60*, 1184–1194. [CrossRef]
38. Jiang, R.; Zhang, J.; Yan, R.; Tang, H. Few-shot learning in spiking neural networks by multi-timescale optimization. *Neural Comput.* **2021**, *33*, 2439–2472. [CrossRef]
39. DeBole, M.V.; Appuswamy, R.; Carlson, P.J.; Cassidy, A.S.; Datta, P.; Esser, S.K.; Garreau, G.J.; Holland, K.L.; Lekuch, S.; Mastro, M.; et al. Truenorth: Accelerating from zero to 64 million neurons in 10 years. *Computer* **2019**, *52*, 20–29. [CrossRef]
40. Furber, S.B.; Galluppi, F.; Temple, S.; Plana, L.A. The SpiNNaker project. *Proc. IEEE* **2014**, *102*, 652–665. [CrossRef]
41. Krestinskaya, O.; James, A.P.; Chua, L.O. Neuromemristive circuits for edge computing: A review. *IEEE Trans. Neural Netw. Learn. Syst.* **2020**, *31*, 4–23. [CrossRef]

42. Yoo, J.; Shoaran, M. Neural interface systems with on-device computing: Machine learning and neuromorphic architectures. *Curr. Opin. Biotechnol.* **2021**, *72*, 95–101. [CrossRef]
43. Cho, S.W.; Kwon, S.M.; Kim, Y.; Park, S.K. Recent progress in transistor-based optoelectronic synapses: From neuromorphic computing to artificial sensory system. *Adv. Intell. Syst.* **2021**, *3*, 2000162. [CrossRef]

entropy

MDPI

Article

Probabilistic Deterministic Finite Automata and Recurrent Networks, Revisited

Sarah E. Marzen [1,*,†] and James P. Crutchfield [2,*,†]

1 W. M. Keck Science Department, Pitzer, Scripps, and Claremont McKenna College, Claremont, CA 91711, USA
2 Complexity Sciences Center, Physics Department, University of California at Davis, One Shields Avenue, Davis, CA 95616, USA
* Correspondence: marzen.sarah@gmail.com (S.E.M.); chaos@ucdavis.edu (J.P.C.)
† These authors contributed equally to this work.

Abstract: Reservoir computers (RCs) and recurrent neural networks (RNNs) can mimic any finite-state automaton in theory, and some workers demonstrated that this can hold in practice. We test the capability of generalized linear models, RCs, and Long Short-Term Memory (LSTM) RNN architectures to predict the stochastic processes generated by a large suite of probabilistic deterministic finite-state automata (PDFA) in the small-data limit according to two metrics: predictive accuracy and distance to a predictive rate-distortion curve. The latter provides a sense of whether or not the RNN is a lossy predictive feature extractor in the information-theoretic sense. PDFAs provide an excellent performance benchmark in that they can be systematically enumerated, the randomness and correlation structure of their generated processes are exactly known, and their optimal memory-limited predictors are easily computed. With less data than is needed to make a good prediction, LSTMs surprisingly lose at predictive accuracy, but win at lossy predictive feature extraction. These results highlight the utility of causal states in understanding the capabilities of RNNs to predict.

Keywords: time series prediction; finite state machines; hidden Markov models; recurrent neural networks; reservoir computers; long short-term memory

Citation: Marzen, S.E.; Crutchfield, J.P. Probabilistic Deterministic Finite Automata and Recurrent Networks, Revisited. *Entropy* **2022**, *24*, 90. https://doi.org/10.3390/e24010090

Academic Editors: Lizhong Zheng and Chao Tian

Received: 26 November 2021
Accepted: 30 December 2021
Published: 6 January 2022

1. Introduction

Many real-world tasks rely on prediction. Given past stock prices, traders try to predict if a stock price will go up or down, adjusting investment strategies accordingly. Given past weather, farmers endeavor to predict future temperatures, rainfall, and humidity, adapting crop and pesticide choices. Manufacturers try to predict which goods will appeal most to consumers, adjusting raw materials purchases. Self-driving cars must predict the motion of other objects on and off the road. Furthermore, when it comes to biology, evidence suggests that organisms endeavor to predict their environment as a key survival strategy [1–3]. One simple metric often used to evaluate the quality of our predictive algorithms is simply the accuracy of our predictions—how well we can predict what will happen next given what has happened previously.

However, we also care about the cost of formulating and communicating a prediction of the next symbol in some sequence of symbols, either to another person or from one part an organism to another. Costs of formulation might include the time, memory, and/or energy taken to compute a prediction. Once the prediction is made, it is often communicated to some other downstream region that will use the prediction to take an action. This communication requires some amount of channel capacity, and channel capacity can be energetically expensive. All other concerns equal, one is inclined to employ a predictor with a lower transmission rate [4].

Here, we focus solely on communication, ignoring costs in formulating the prediction. As such, note that transmission rate is unrelated to sample complexity or time complexity.

Rather, we allow for an unbounded number of samples in testing (thus avoiding the question of generalization error) and an unbounded time to train and compute predictions, and merely ask: what channel capacity do we need to faithfully communicate the predictions?

Simultaneously optimizing the objectives—high predictive accuracy and low code rate—leads to *predictive rate-distortion* [5–7]. The predictive rate-distortion curve separates combinations of achievable rates and distortions from unachievable rates and distortions. The closer a lossy predictive compressor is to the curve, the better. This diagnosis has been used, for example, to suggest that salamander retinal ganglion cells are near-optimal lossy predictors of visual input [8].

Surprisingly, we do not yet know how well recurrent neural networks perform relative to the predictive rate-distortion curve, though rate-distortion curves have been used to explain and calibrate the performance of artificial feedforward neural networks [9,10]. Note that recurrent neural networks allow us to store information, in principle, about semi-infinite pasts, while feedforward neural networks only allow for storage of finite pasts. The following calibrates the performance of various predictors (generalized linear models, reservoir computers, and recurrent neural networks) using the predictive rate-distortion curve. We stimulate predictors with output of probabilistic deterministic finite automata (PDFA), also called unifilar hidden Markov models in information theory [11]. The PDFAs used in the following are simple, in that their statistical complexity [12] and excess entropy [13,14] are finite and relatively small. The following explores PDFAs since optimal predictors of the time series they generate are easily computed [12], and the tradeoffs between code rate and predictive accuracy (encapsulated by the predictive rate-distortion function) are easily computed as well [7].

This work builds on seminal results establishing that both reservoir computers (RCs) [15,16] and recurrent neural networks (RNNs) [17] can reproduce any dynamical system, when given a sufficient number of nodes. Further work gave example RNNs that faithfully reproduce finite state automata, to the point that RNN nodes mimicked the automata states [18], and established bounds on the required RNN complexity [19]. One would conjecture, then, that Long Short-Term Memory (LSTM) architectures—an easily-trainable RNN variety [20,21]—should easily learn to predict the outputs of PDFAs. The further question we ask is: do these models not only predict, but predict *efficiently*?

We use predictive rate-distortion curves to calibrate the performance of three time series predictors: generalized linear models (GLMs) [22], RCs [15,16], and LSTMs [20]. Unsurprisingly, LSTMs are generally more efficient than reservoirs, which are generally more efficient than GLMs. Perhaps unsurprisingly, LSTMs are less accurate than both methods, seemingly due to overfitting. Surprisingly, despite the simplicity of the generated stochastic time series, we find that all tested prediction methods can fail to attain maximal predictive accuracy (measured by the probability of being correct) by as much as 50% and often need higher rates than necessary to attain maximal predictive performance. However, existing methods for inferring PDFAs [23] can correctly infer the PDFA and generate the optimal predictor with orders-of-magnitude less data. This leads us to conclude that prediction algorithms that first infer *causal states* [6,23–25] can surpass trained RNNs if the time series in question has (approximately) finite causal states, sometimes also called *predictive state representations* [26].

In Section 2, we describe how rate-distortion functions can provide a benchmark for prediction algorithms. In Section 3, we describe PDFAs, GLMs, RCs, and LSTMs. In Section 4, we describe our results. Section 5 summarizes our conclusions.

2. Rate-Distortion Benchmarks for Prediction Algorithms

Typically, when one talks about recurrent neural networks, one considers a setup as in Figure 1 (top). Input is sent to the network, which updates its state based on both the input and its previous state. The network's state is then used to make a prediction. The only metric that characterizes the final performance of the network, post-training, is the prediction accuracy—how well it predicts future symbols given past symbols.

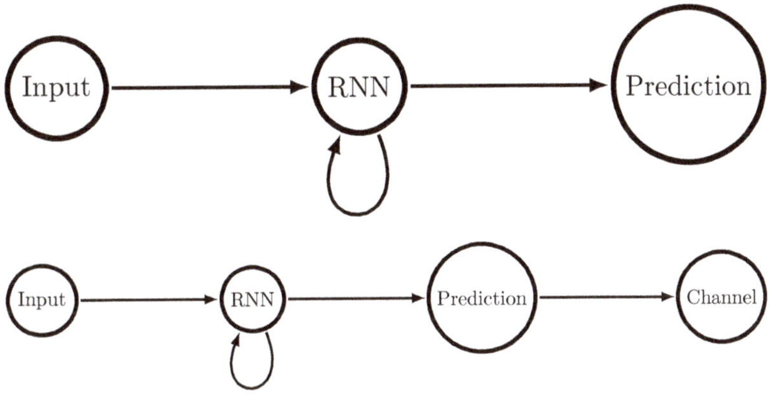

Figure 1. At (**top**), a typical setup for a recurrent neural network (or any other predictor): input is sent to the recurrent neural network, which makes a prediction about future inputs. At (**bottom**), our setup for a recurrent neural network in which predictions must be made and the prediction must be communicated losslessly through the channel.

We now augment that setup slightly. Consider a channel over which the prediction must be communicated, as in Figure 1 (bottom). Now there are two metrics that characterize the network's performance, post-training: the predictive accuracy and the required channel capacity. In the particular setup of Figure 1 (bottom), the required channel capacity must be at least the entropy of the predictions [4]. If one is allowed longer blocklengths, meaning that one can communicate several predictions at once using the channel, the required channel capacity somewhat diminishes.

One can now trace out a plane of the two metrics, prediction accuracy and channel capacity, and ask which combinations of the two are achievable. The curve that separates the achievable combinations from the unachievable combinations is called the predictive rate-accuracy curve, very closely related to the predictive rate-distortion curve. See Figure 2.

Let R be the random variable representing our representation of the past that we use to predict the future, and r be its realization. When the accuracy is the conditional mutual information $I[\overrightarrow{X};\overleftarrow{X}|R]$, the predictive rate-accuracy function is exactly the predictive information curve [5,6]. Finding representations that lie on the information curve motivates slow feature analysis [27], recovers canonical correlation analysis [28], and identifies the minimal sufficient statistics of prediction—the causal states [5]. Predictive information curves have even been used to evaluate the predictive efficiency of salamander retinal neural spiking patterns [8].

Here, however, we work only with binary processes, and we adopt the stance that predictive accuracy could be taken to be the probability that one's prediction is correct. Accordingly, we force our representation $r \in \{0,1\}$ to be a prediction, and calculate *accuracy* via:

$$a(r_t, x_{t+1}) = 1 - \delta_{r_t, x_{t+1}} ,$$

which implies:

$$E[a] = \sum_{\overleftarrow{x}_t} p(\overleftarrow{x}_t) \sum_{r_t = x_{t+1}} p(r_t|\overleftarrow{x}_t) p(x_{t+1}|\overleftarrow{x}_t) .$$

The choice of distortion or accuracy measure is an important one, and determined by one's particular application.

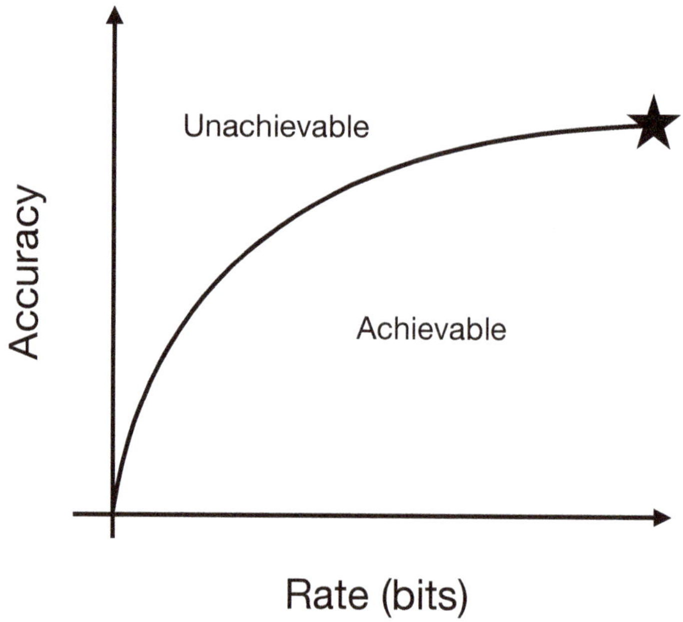

Figure 2. A sample predictive rate-accuracy curve, which is dependent not on how we process the time series but only on intrinsic properties of the time series. It is quite possible, and typical, to have zero rate and a nonzero predictive accuracy, and so the meeting of the x-axis and y-axis is not at the origin. The rate can run between zero and one bit for the binary-valued time series we study here. The starred point, which encodes the rate and accuracy of a minimal optimal predictor, has a rate of the single-symbol Shannon entropy of the time series and a predictive accuracy that depends in a complicated way on the specific time series. (Note the slight difference between this communication setup and that of standard predictive rate-distortion.) It is possible to have rates larger than the rate of the starred point, up to and including one bit.

There is another way to understand predictive rate-accuracy curves. With an eye to making contact with nonpredictive rate-distortion theory, we summarize the setup of predictive rate-accuracy as follows. Semi-infinite pasts are drawn independently from the same process-dependent distribution and sent to an encoder, which then produces a prediction or a probability distribution over possible predictions. A predictive distortion measures how far the estimated predictions differ from correct predictions. Distortion is often taken, for example, to be the Kullback-Leibler divergence between the true distribution $p(\overrightarrow{x}|\overleftarrow{x})$ over futures $\overrightarrow{x}$ conditioned on the past $\overleftarrow{x}$ and the distribution $p(\overrightarrow{x}|r)$ over futures conditioned on our *representation* r [29]. A predictive accuracy might then be some maximal achievable accuracy minus the predictive distortion. The predictive rate-accuracy curve $\mathrm{R}(A)$, the minimal necessary rate at a given expected accuracy, separates the plane of rates and predictive distortions into regions of achievable and unachievable combinations. A slight variant of the rate-distortion theorem gives:

$$\mathrm{R}(A) = \min_{p(\vec{x}|r):E[a]\geq A} I[\overleftarrow{X};R]\,, \quad (1)$$

where $I[\cdot;\cdot]$ is the mutual information.

3. Background

In what follows, we review time-series generation and the widely-used prediction methods we compare. We first discuss PDFAs and then prediction methods.

3.1. PDFAs and Predictive Rate-Distortion

We focus on minimal PDFAs—for a given stochastic process, that with the smallest number of states. A PDFA consists of a set $\mathcal{S}$ of states $\sigma \in \mathcal{S}$, a set $\mathcal{A}$ of emission symbols, and transition probabilities $p(\sigma_{t+1}, x_t|\sigma_t)$, where $\sigma_t, \sigma_{t+1} \in \mathcal{S}$ and $x_t \in \mathcal{A}$. The "deterministic" descriptor comes from the fact that $p(\sigma_{t+1}|x_t, \sigma_t)$ has support on only one state. (This is "determinism" in the sense of formal language theory [30]—an automaton deterministically *recognizes* a string—not in the sense of nonstochastic. It was originally called *unifilarity* in the information theoretic analysis of hidden Markov chains [11]. Thus, PDFAs are also known as *unifilar hidden Markov models* [12].)

Here, we concern ourselves with minimal and binary-alphabet ($\mathcal{A} = \{0,1\}$) PDFAs. In dynamical systems theory minimal unifilar HMMs (minimal PDFAs) are called *ϵ-machines* and their states σ *causal states*. Due to the automaton's determinism, one can uniquely determine the state from the past symbols with probability 1. Each state is therefore a cluster of pasts that have the same conditional probability distribution over futures. As a result, all that one needs to know to optimally predict the future is given by the causal state [12].

For example, the simple two-state PDFA shown in Figure 3 generates the Even Process: only an even number of 1's are seen between two successive 0's. This leads to a simple prediction algorithm: find the parity of the number of 1's since the last 0; if even, we are in state A, so predict 0 and 1 with equal probability; if odd, we are in state B, so predict 1. There is only one past for which our prediction algorithm yields no fruit: given the past of all 1s a single state is never identified. One only knows that the machine is in either state A or B and the best prediction is a mixture of what the states indicate. Even though that past occurs with probability 0, it causes the Even Process to be an infinite-order Markov Process [31]. See Ref. [32] for a measure-theoretic treatment.

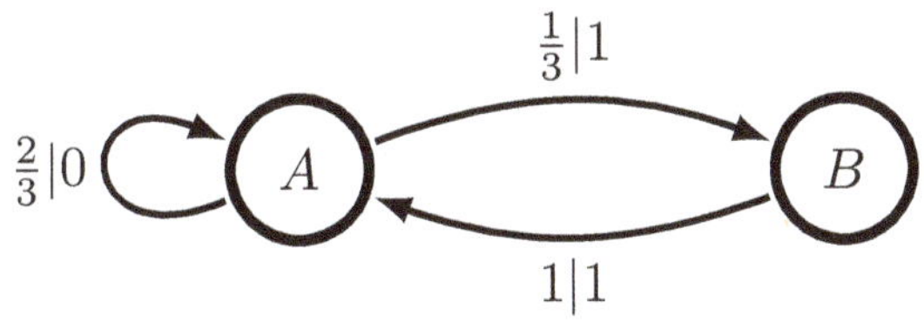

Figure 3. Minimal two-state PDFA that generates the Even Process, so-called since there are always an even number of 1s between 0's. Arrows indicate allowed transitions, while transition labels $p|s$ indicate the transition (and so too emission) probabilities $p \in [0,1]$ for the symbol $s \in \mathcal{A}$. Given a current state and next symbol, one knows the next state—the deterministic or unifilar property of this PDFA.

Causal states and ϵ-machines can be inferred from data in a variety of ways [6,23,25,33].

The causal states are uniquely useful to calculating predictive rate-distortion curves. Under weak assumptions, the predictive rate-accuracy function of Section 2 becomes:

$$R(A) = \min_{p(r|\sigma):E[d]\geq A} I[\mathcal{S};R]$$

with:

$$E[d] = \sum_{\sigma_t} p(\sigma_t) \sum_{x_{t+1}=r_t} p(r_t|\sigma_t) p(x_{t+1}|\sigma_t) .$$

See Ref. [7] for the proof. With this substitution—of a finite object ($\mathcal{S}$) for an infinite one ($\overleftarrow{X}$)—the Blahut-Arimoto algorithm can be used to accurately calculate the predictive rate-accuracy function, in that the algorithm provably converges to the optimal $p(r|\sigma)$ [34]. The same cannot be said of the predictive information curve [7], which converges to a local optimum of the objective function, but may not converge to a global optimum.

In practice, we always augment the predictive rate-accuracy function with the rate and accuracy of the optimal predictor, which is (as described earlier) straightforwardly derived from the ϵ-machine. Simply put, we infer the causal state σ_t from past data and predict the next symbol to be $\arg\max_{x_{t+1}} p(x_{t+1}|\sigma_t)$.

The following tests the various time series predictors on all of the (uniformly sampled) binary-alphabet ϵ-machine topologies [35] with randomly-chosen emission probabilities. Due to the super-exponential explosion of the set of topological ϵ-machines with number of states, we only look at binary-alphabet machines with four or fewer (causal) states. (There are 1338 unique topologies for four states, but over 10^6 for six states.) The analysis discards any ϵ-machine with zero-rate optimal predictor, which can arise depending on the emission probabilities.

3.2. Time Series Methods

We focus on three methods for time series prediction: generalized linear models (GLM), reservoir computers (RCs), and LSTMs.

The GLM we use predicts x_t from a linear combination of the last k symbols $x_{t-k}, x_{t-k+1}, \dots, x_{t-1}$. More precisely, a GLM models the probability of x_t being a 0 via:

$$p_{GLM}(x_t = 0|x_{t-k}, \dots, x_{t-1}) = \frac{e^{w_k x_{t-k} + \dots + w_1 x_{t-1} + w_0}}{1 + e^{w_k x_{t-k} + \dots + w_1 x_{t-1} + w_0}} . \quad (2)$$

The model's estimate of the probability of $x_t = 1$ follows:

$$p_{GLM}(x_t = 1|x_{t-k}, \dots, x_{t-1}) = \frac{1}{1 + e^{w_k x_{t-k} + \dots + w_1 x_{t-1} + w_0}} . \quad (3)$$

We use Scikit-learn logistic regression to find the best weights $w_0, w_1, \dots, w_k$. Predictions are then made via $\arg\max_{x_t} p_{GLM}(x_t|x_{t-k}, \dots, x_{t-1})$.

The RC is more powerful in that it uses logistic regression with features that contain information about symbols arbitrarily far into the past. We employ a *tanh* activation function, so that the reservoir's state advances via:

$$h_{t+1} = \tanh(W h_t + v x_t + b) \quad (4)$$

and initialize W, v, b with i.i.d. normally distributed elements. The matrix W is then scaled so that it is near the "edge of chaos" [36–39], where RCs are conjectured to have maximal memory [40,41]. We then use logistic regression with h_t as features to predict x_t:

$$p_{reservoir}(x_t = 0|h_t) = \frac{e^{w^\top h_t + w_0}}{1 + e^{w^\top h_t + w_0}} ,$$
$$p_{reservoir}(x_t = 1|h_t) = \frac{1}{1 + e^{w^\top h_t + w_0}} .$$

It is straightforward to devise a weight matrix W and bias b so that $p_{reservoir}(x_t|h_t)$ attains the restricted linear form of p_{GLM} of Equations (2) and (3). That is, RCs are more powerful than GLMs, as they use nonlinear functions of semi-infinite pasts for their summary statistics. We use Scikit-learn logistic regression to find the best weights w_0 and w. Note that the weights W, v, and b are not learned, but held constant; we only train w and w_0. Predictions are made via $\arg\max_{x_t} p_{reservoir}(x_t|h_t)$.

Finally, we analyze the LSTM's predictive capabilities. LSTMs are no more powerful than vanilla RNNs; e.g., those as in Equation (4). However, they are far more trainable in that it is possible to achieve good results without extensive hyperparameter tuning [21]. An LSTM has several hidden states f_t, i_t, o_t, c_t, and h_t that update via the following:

$$\begin{aligned} f_t &= \sigma_g(W_f x_t + U_f h_{t-1} + b_f) \\ i_t &= \sigma_g(W_i x_t + U_i h_{t-1} + b_i) \\ o_t &= \sigma_g(W_o x_t + U_o h_{t-1} + b_o) \\ c_t &= f_t \odot c_{t-1} + i_t \odot \sigma_c(W_c x_t + U_c h_{t-1} + b_c) \\ h_t &= o_t \odot c_t \,, \end{aligned}$$

where σ_g is the sigmoid function and σ_c is the hyperbolic tangent. The variable c_t is updated linearly, therefore avoiding issues with vanishing gradients [42]. Meanwhile, the gating function f_t allows us to forget the past selectively. We then predict the probability of x_t given the past using:

$$\begin{aligned} p_{LSTM}(x_t = 0|h_t) &= \frac{e^{w^\top h_t + w_0}}{1 + e^{w^\top h_t + w_0}} \,, \\ p_{LSTM}(x_t = 1|h_t) &= \frac{1}{1 + e^{w^\top h_t + w_0}} \,. \end{aligned} \tag{5}$$

Weights w and w_0 are learned while we estimate parameters W_f, U_f, b_f, W_i, U_i, W_o, U_o, b_o, W_c, U_c, and b_c to maximize the log-likelihood. Predictions are made via $\arg\max_{x_t} p_{LSTM}(x_t|h_t)$.

Predictive accuracy is calculated by comparing the predictions to the actual values of the next symbol and counting the frequency of correct predictions. The code rate is calculated via the prediction entropy [4].

4. Results

An aim here is to thoroughly and systematically analyze the predictive accuracy as measured by the probability of correctly guessing the next symbol and code rate of our three time series predictors of a large swath of PDFAs in the small-data limit, in which only 5000 samples are shown to the RNN. To implement this, we ran through Ref. [35]'s topological ϵ-machine library—binary-alphabet PDFAs with four states or less and randomly chosen emission probabilities, in which transition probabilities were drawn from a uniform distribution. For each PDFA, we generated a length-5000 time series. The first half was presented to a predictor and used to train its weights. We then evaluated each time series predictor based on its predictions for the second half of the time series. Predictive accuracy and code rate were calculated and compared to the predictive rate-distortion function. Predictive accuracy was calculated as the probability of having a correct prediction; code rate was calculated empirically as the single-symbol entropy of the predictions [14].

Note that Bayesian structural inference (BSI) provides a useful comparison [23]. In BSI, we compute the maximum a posteriori (MAP) estimate of the PDFA generating an observed time series, and use this MAP estimate to build an optimal predictor of the process. BSI can correctly infer the PDFA essentially 100% of the time with orders-of-magnitude less data than used to monitor the three prediction methods tested here. Hence, it achieves optimal predictive accuracy with minimal rate. Our aim is to test the ability of GLMs, RCs, and RNNs to equal BSI's previously-published performance.

The time series predictors used have hyperparameters. A variety of orders (k's) were used for the GLMs and reservoirs and LSTMs of different sizes (number of nodes) were tested. Learning rate and optimizer type, including gradient descent and Adam [43], were also varied for the LSTM, with little effect on results. Regularization was necessary and utilized in both L_1 and L_2 forms on all three predictors. As is typical, a validation set was used to select the strength and type of regularization, and results were reported on a separate test set. In total, 5000 steps of the time series were simulated, which was small enough to test how these machine learning methods responded to too little data, but enough data that the machine learning methods could have picked up on patterns.

4.1. The Difference between Theory and Practice: The Even and Neven Process

We first analyze two easily-described PDFAs, deriving RNNs that correctly infer causal states and, therefore, that match the optimal predictor—the ϵ-machine. We then compare the trained GLMs, RCs, and LSTMs to the easily-inferred optimal predictors. In theory, RCs and LSTMs should be able to mimic the derived RNNs, in that it is possible to find weights of an RC and LSTM that yield nodes that mimic the causal states of the PDFA. In practice, surprisingly, RCs and LSTMs have some difficulty.

First, we analyze the Even Process shown in Figure 3. The optimal prediction algorithm is easily seen by inspection of Figure 3. When we determine the machine is in state A, we predict a 0 or a 1 with equal probability; if it is in state B, we predict a 1. We determine whether or not it is in state A or B by the parity of the number of 1s since the last 0. If odd, it is in state B; if even, it is in state A. The inferred state is easily encoded by the following RNN:

$$h_{t+1} = x_t(1 - h_t) \ . \tag{6}$$

If x_t is 0, the hidden state of the RNN "resets" to 0; e.g., state A. If $x_t = 1$, then the hidden state updates by flipping from 0 to 1 or vice versa, mimicking the transitions from A to B and back. One can show that a one-node LSTM hidden state h_t can, with proper weight choices, mimic the hidden state of Equation (6). With the correct hidden state inferred, it is straightforward to find w and w_0 such that Equation (5) yields optimal (and correct) predictions.

As one might then expect, and as Figure 4 confirms, LSTMs tend to have rates that are close to the optimal (maximal) rate and predictive accuracies that are only slightly below the optimal predictive accuracy. RCs and GLMs tend to have higher rates and lower predictive accuracies, but they are still within ~13% of optimal. We can see this qualitatively just by examining the predictive rate-accuracy curve in Figure 4: the closer that a point is to the curve, the more efficiently that predictor predicts. Among the points on the curve, potentially the most desirable point is the one at the highest achievable accuracy, at the top right. The points from the LSTMs tend to be closer to the curve and closer to the point at the top right, followed by RCs, and followed by GLMs. Interestingly, the points from all processes lie on a one-dimensional curve, speaking to some hidden simplicity in the relationship between rate and accuracy that likely holds only for binary-valued processes.

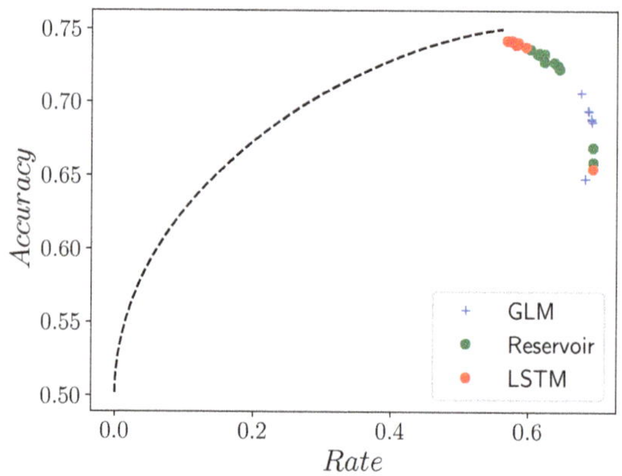

Figure 4. Predictive rate–accuracy curve for the Even Process in Figure 3, along with empirical predictive accuracies and rates of GLMs, RCs, and LSTMs of various sizes: orders range from 1–10 for GLMs, number of nodes range from 1–61 for RCs, and number of nodes range from 1–121 for LSTMs. Despite the Even Process' simplicity, there is a noticeable difference between the predictors' performances and between their performances and the optimal achievable performance.

As one might also expect, LSTMs and RCs with additional nodes and GLMs with higher orders (higher k) have higher predictive accuracies than LSTMs and RCs with fewer nodes and GLMs with lower orders. However, viewed another way, given the simplicity of the stimulus—indeed, given that a one-node LSTM can, in theory, learn the Even Process—the gap from the predictors' rates and accuracies to the optimal combinations of rate and accuracy is surprising. It is also surprising that none of the three predictors' rates fall below the maximal optimal rate.

Figure 5 introduces a similarly-simple three-state PDFA. If a 1 is observed after a 0, we are certain the machine is in state B; after state B, we know it will transition to state A; and then the parity of 0s following transition to state A tells us if it is in state A (even) or state B (odd). This PDFA is a combination of a Noisy Period-2 Process (between states A and B) and an Even Process (between states A and C).

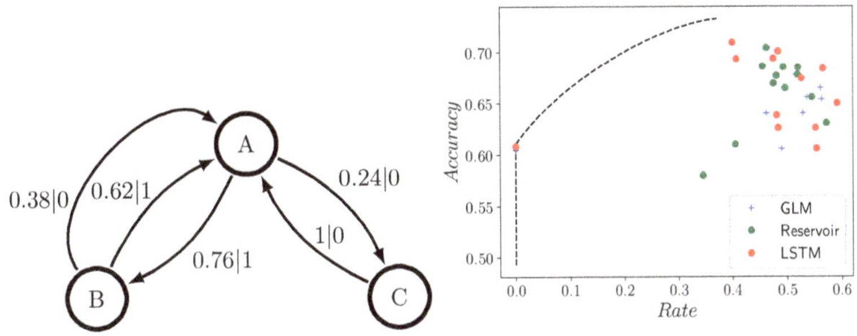

Figure 5. Predictive rate-accuracy curve for the Neven Process (PDFA shown at left), along with empirical predictive accuracies and rates of GLMs, RCs, and LSTMs of various sizes: orders range from 1–10 for GLMs, number of nodes range from 1–61 for RCs, and number of nodes range from 1–121 for LSTMs. Despite Neven Process' simplicity, there is a noticeable gap between the predictor's performance and the optimal performance achievable.

Given the Neven Process's simplicity, it is unsurprising that we can concoct an RNN that can infer the internal state. Let $h_t = (h_{t,A}, h_{t,B}, h_{t,C})$ be the hidden state that is $(1,0,0)$ if the internal state is A, $(0,1,0)$ if the internal state is B, and $(0,0,1)$ if the internal state is C. By inspection, we have:

$$\begin{aligned} h_{t+1,A} &= 1 - h_{t,A} \\ h_{t+1,B} &= x_t h_{t,A} \\ h_{t+1,C} &= (1 - x_t) h_{t,A} \,. \end{aligned}$$

One can straightforwardly find weights that lead to $p_{LSTM}(x_{t+1}|h_t)$ accurately reflecting the transmission (emission) probabilities. In other words, in theory a three-node RNN (and an equivalent three-node LSTM) can learn to predict the Neven process optimally.

However, the Neven Process' simplicity is belied by the gap between the predictors' accuracy and rate and the predictive rate-accuracy curve. In Figure 5, the point at zero rate implies that the predictor is spitting out the same symbol, regardless of input. The worst predictive accuracy falls short of the optimal by $\sim$15%, and none of the GLMs, RCs, or LSTMs get closer than $\sim$97% to optimal. Furthermore, almost all the rates surpass the maximal optimal predictor rate.

4.2. Comparing GLMs, RCs, and LSTMs

We now analyze the combined results obtained over all minimal PDFAs up to four states using two metrics. (Again, recall that they are 1338 unique machine topologies.) To compare across PDFAs, we first normalize the rate and accuracy by the rate and accuracy of the optimal predictor. Then, we find the distance from the predictor's rate and accuracy to the predictive rate-accuracy curve, which is similar in spirit to the metric of Ref. [44] and to the spirit of Ref. [8]. Note that this metric would have been markedly harder to estimate had we used nondeterministic probabilistic finite automata; that is, those without determinism (unifiliarity) in their transition structure [7].

Figure 6 showcases a histogram of the normalized distance to the predictive rate-accuracy curve, ignoring PDFAs for which the maximal optimal rate is 0 nats. The normalized distance for all three predictor types tends to be quite small, but even so, we can see differences in the three predictor types. LSTMs tend to have smaller normalized distances than RCs, and RCs tend to have smaller normalized distances to the predictive rate-accuracy curve than GLMs. In fact, LSTMs seem to be uniformly better lossy predictive feature extractors. Trained LSTMs on average have 0.8% normalized distance; RCs on average have 2.0% normalized distance; and GLMs on average have 4.5% normalized distance. When looking only at optimized LSTMs, RCs, and GLMs—meaning that the number of nodes or the order is chosen to minimize normalized predictive distortion—a few PDFAs still have high normalized predictive distortions of 4.6% for LSTMs, 9.7% for RCs, and 27.3% for GLMs.

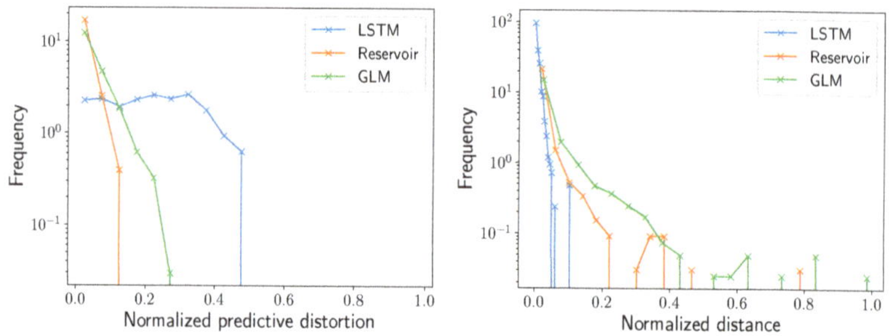

Figure 6. (**Left**) Histogram of normalized predictive distortions for LSTMs (blue), RCs (orange), and GLMs (green) using 798 distinct PDFAs. While LSTMs tend to have far higher predictive accuracies, they also have a much larger probability than reservoirs or GLMs do of having noticeable inaccuracies. Some recorded normalized predictive distortions were negative, indicating the effects of finite sample size. (**Right**) Histogram of normalized distances to the predictive rate-accuracy curve for LSTMs (blue), RCs (orange), and GLMs (green) using 798 distinct PDFAs. It is apparent that LSTMs are closer to the predictive rate-accuracy curves than reservoirs and GLMs.

The same trend holds for the percentage difference between the predictive accuracy and the maximal predictive accuracy, which we call the *normalized predictive distortion*, with a crucial modification. Trained LSTMs on average have 21.5% normalized predictive distortion; RCs on average have 1.8% normalized predictive distortion; and GLMs on average have 4.2% normalized predictive distortion. When looking only at optimized LSTMs, RCs, and GLMs—meaning that the number of nodes or the order is chosen to minimize normalized predictive distortion—a few PDFAs still have high normalized predictive distortions of 50% for LSTMs, 13.5% for RCs, and 25.5% for GLMs. However, perhaps the most interesting aspect of the Figure 6 is that LSTMs are far more likely than reservoirs or GLMs to have large normalized predictive distortions, surprisingly.

Unsurprisingly, increasing the GLM order and the number of nodes of the RCs and LSTMs tends to increase predictive accuracy and decrease the normalized distance.

Our final aim is to understand the PDFA characteristics that cause them to be harder to predict accurately and/or efficiently. We have two suspects, which are the most natural measures of process "complexity". This first is the generated process' entropy rate h_μ, the entropy of the next symbol conditioned on all previous symbols, which quantifies the intrinsic randomness of the stimulus. The second is the generated process' statistical complexity C_μ, the entropy of the causal states, which quantifies the intrinsic memory in the stimulus. The more random a stimulus, the harder it would be to predict; imagine having to find the optimal predictor for a biased coin whose bias is quite close to 1/2. The more memory in a stimulus, the more nodes in a network or the higher the order of the GLM required, it would seem. We performed a multivariate linear regression, trying to use h_μ and C_μ to predict the minimal normalized predictive distortion and minimal normalized distance. We find a small and positive correlation for LSTMs, reservoirs, and GLMs for predicting minimal deviations in accuracy from perfection, with an R^2 of 0.189, 0.134, and 0.132, respectively. For all three types of prediction algorithms, statistical complexity C_μ is positively correlated with deviations in accuracy. Entropy rate is positively correlated with deviations in accuracy for GLMs and reservoirs but, surprisingly, not LSTMs. Interestingly, the performance GLMs and RCs is impacted by increased randomness and increased memory in the stimulus, while the LSTMs' accuracy has little correlation with entropy rate and statistical complexity.

For the most part, we find that all three prediction methods–GLMs, RCs, and LSTMs—tend to learn to predict the PDFA outputs near-optimally, in that prediction accuracies differ from the optimal prediction accuracy by an average of roughly 5%. LSTMs outperform RCs, which outperform GLMs. However, we discovered simple PDFAs that cause the best LSTM to fail by as much as 5%, the best RC to fail by as much as 10%, and the best GLM to fail by as much as 27%.

Since none of the RNNs achieved perfect prediction accuracy, but the BSI method did [23], we conclude that existing methods for inferring causal states [6,23,25,33] are useful, despite the historically dominant reliance on RNNs. For example, as previously mentioned, Bayesian structural inference correctly infers the correct PDFAs almost 100% of the time, leading to essentially zero prediction error, on training sets that are orders of magnitude smaller than those used here [23].

5. Conclusions

We have known for a long time that reservoirs and RNNs can reproduce any dynamical system [15–17], and we have explicit examples of RNNs learning to infer the hidden states of a PDFA when shown the PDFA's output [18]. We revisited these examples to better understand if the finding of Ref. [18] is typical. How often do RNNs and RCs learn efficient and accurate predictors of PDFAs, especially given that BSI can yield an optimal predictor with orders-of-magnitude less training data?

We conducted a rather comprehensive search, analyzing 798 randomly-generated PDFAs with four states or less. For each PDFA, we trained GLMs, RCs, and RNNs of varying orders or varying numbers of nodes. Larger orders and larger numbers of nodes led to more accurate and more efficient predictors. On average, the various time series predictors have $\sim$5% predictive distortion. In other words, we are apparently better at classifying MNIST digits than sometimes predicting the output of a simple PDFA. Again, existing algorithms [23] can optimally predict the output of the PDFAs considered here with orders-of-magnitude less training data. (MNIST is a database of handwritten digits.) These findings lead us to conclude that algorithms that explicitly focus on inference of causal states [6,23–25] have a place in the currently RNN-dominated field of time series prediction.

More importantly, in this small data limit, overfitting is an issue for LSTMs but not RCs or GLMs. However, LSTMs are somehow excellent lossy predictive feature extractors nonetheless. The mechanism behind this is a subject for future research.

Perhaps most importantly, the predictive rate-accuracy framework that we introduce here or similar such frameworks could be useful for calibrating the performance of time series predictors. We have added a cost that comparatively little research has focused on: that of communicating the prediction. Implicitly, we are arguing that predictors which do not have maximal predictive accuracy but do have small communication costs might be useful nonetheless.

Author Contributions: S.E.M. and J.P.C. conceptualized the article and wrote the article, and S.E.M. performed the experiments. All authors have read and agreed to the published version of the manuscript.

Funding: This material is based upon work supported by, or in part by, the Air Force Office of Scientific Research under award number FA9550-19-1-0411 and the U. S. Army Research Laboratory and the U. S. Army Research Office under grants W911NF-18-1-0028 and W911NF-21-1-0048.

Institutional Review Board Statement: Not applicable.

Informed Consent Statement: Not applicable.

Data Availability Statement: Available upon reasonable request from the authors.

Conflicts of Interest: The authors declare no conflict of interest.

References

1. Schultz, W.; Dayan, P.; Montague, P.R. A neural substrate of prediction and reward. *Science* **1997**, *275*, 1593–1599. [CrossRef]
2. Montague, P.R.; Dayan, P.; Sejnowski, T.J. A framework for mesencephalic dopamine systems based on predictive Hebbian learning. *J. Neurosci.* **1996**, *16*, 1936–1947. [CrossRef]
3. Rao, R.P.; Ballard, D.H. Predictive coding in the visual cortex: A functional interpretation of some extra-classical receptive-field effects. *Nat. Neurosci.* **1999**, *2*, 79. [CrossRef]
4. Berger, T. *Rate Distortion Theory*; Prentice-Hall: New York, NY, USA, 1971.
5. Still, S.; Crutchfield, J.P.; Ellison, C.J. Optimal causal inference: Estimating stored information and approximating causal architecture. *Chaos Interdiscip. J. Nonlinear Sci.* **2010**, *20*, 037111. [CrossRef]
6. Still, S. Information bottleneck approach to predictive inference. *Entropy* **2014**, *16*, 968–989. [CrossRef]
7. Marzen, S.; Crutchfield, J.P. Predictive Rate-Distortion for Infinite-Order Markov Processes. *J. Stat. Phys.* **2014**, *163*, 1312–1338. [CrossRef]
8. Palmer, S.E.; Marre, O.; Berry, M.J.; Bialek, W. Predictive information in a sensory population. *Proc. Natl. Acad. Sci. USA* **2015**, *112*, 6908–6913. [CrossRef]
9. Tishby, N.; Zaslavsky, N. Deep Learning and the Information Bottleneck Principle. *arXiv* **2015**, arXiv:1503.02406.
10. Shwartz-Ziv, R.; Tishby, N. Opening the Black Box of Deep Neural Networks via Information. *arXiv* **2017**, arXiv:1703.00810.
11. Ash, R.B. *Information Theory*; John Wiley and Sons: New York, NY, USA, 1965.
12. Shalizi, C.R.; Crutchfield, J.P. Computational Mechanics: Pattern and Prediction, Structure and Simplicity. *J. Stat. Phys.* **2001**, *104*, 817–879. [CrossRef]
13. Bialek, W.; Nemenman, I.; Tishby, N. Predictability, complexity, and learning. *Neural Comput.* **2001**, *13*, 2409–2463. [CrossRef]
14. Crutchfield, J.P.; Feldman, D.P. Regularities Unseen, Randomness Observed: Levels of Entropy Convergence. *Chaos* **2003**, *13*, 25–54. [CrossRef] [PubMed]
15. Maass, W.; Natschläger, T.; Markram, H. Real-time computing without stable states: A new framework for neural computation based on perturbations. *Neural Comput.* **2002**, *14*, 2531–2560. [CrossRef]
16. Grigoryeva, L.; Ortega, J.P. Echo state networks are universal. *Neural Netw.* **2018**, *108*, 495–508. [CrossRef]
17. Doya, K. *Universality of Fully Connected Recurrent Neural Networks*; Technology Report; Deptartment of Biology, UCSD: La Jolla, CA, USA, 1993.
18. Cleeremans, A.; Servan-Schreiber, D.; McClelland, J.L. Finite state automata and simple recurrent networks. *Neural Comput.* **1989**, *1*, 372–381. [CrossRef]
19. Horne, B.G.; Hush, D.R. Bounds on the complexity of recurrent neural network implementations of finite state machines. In Proceedings of the 6th International Conference on Neural Information Processing Systems, Denver, CO, USA, 29 November–2 December 1993; pp. 359–366.
20. Schmidhuber, J.; Hochreiter, S. Long short-term memory. *Neural Comput.* **1997**, *9*, 1735–1780.
21. Collins, J.; Sohl-Dickstein, J.; Sussillo, D. Capacity and trainability in recurrent neural networks. *arXiv* **2016**, arXiv:1611.09913.
22. Nelder, J.A.; Wedderburn, R.W. Generalized linear models. *J. R. Stat. Stoc. A* **1972**, *135*, 370–384. [CrossRef]
23. Strelioff, C.C.; Crutchfield, J.P. Bayesian Structural Inference for Hidden Processes. *Phys. Rev. E* **2014**, *89*, 042119. [CrossRef] [PubMed]
24. Crutchfield, J.P.; Young, K. Inferring Statistical Complexity. *Phys. Rev. Let.* **1989**, *63*, 105–108. [CrossRef] [PubMed]
25. Pfau, D.; Bartlett, N.; Wood, F. Probabilistic deterministic infinite automata. *Adv. Neural Inf. Process. Syst.* **2010**, *23*, 1930–1938.
26. Littman, M.L.; Sutton, R.S. Predictive representations of state. *Adv. Neural Inf. Process. Syst.* **2002**, *14*, 1555–1561.
27. Creutzig, F.; Sprekeler, H. Predictive coding and the slowness principle: An information-theoretic approach. *Neural Comput.* **2008**, *20*, 1026–1041. [CrossRef] [PubMed]
28. Creutzig, F.; Globerson, A.; Tishby, N. Past-future information bottleneck in dynamical systems. *Phys. Rev. E* **2009**, *79*, 041925. [CrossRef] [PubMed]
29. Tishby, N.; Pereira, F.C.; Bialek, W. The information bottleneck method. *arXiv* **2000**, arXiv:physics/0004057.
30. Hopcroft, J.E.; Ullman, J.D. *Introduction to Automata Theory, Languages, and Computation*; Addison-Wesley: Reading, MA, USA, 1979.
31. James, R.G.; Mahoney, J.R.; Ellison, C.J.; Crutchfield, J.P. Many Roads to Synchrony: Natural Time Scales and Their Algorithms. *Phys. Rev. E* **2014**, *89*, 042135. [CrossRef] [PubMed]

32. Löhr, W. Models of Discrete-Time Stochastic Processes and Associated Complexity Measures. Ph.D. Thesis, University of Leipzig, Leipzig, Germany, 2009.
33. Shalizi, C.R.; Shalizi, K.L.; Crutchfield, J.P. Pattern discovery in time series, Part I: Theory, algorithm, analysis, and convergence. *J. Mach. Learn. Res.* **2002**, *10*, 60.
34. Csiszár, I. On the computation of rate-distortion functions (corresp.). *IEEE Trans. Inf. Theory* **1974**, *20*, 122–124. [CrossRef]
35. Johnson, B.D.; Crutchfield, J.P.; Ellison, C.J.; McTague, C.S. Enumerating Finitary Processes. *arXiv* **2010**, arXiv:1011.0036.
36. Crutchfield, J.P.; Young, K. Computation at the Onset of Chaos. In *Entropy, Complexity, and the Physics of Information*; Zurek, W., Ed.; SFI Studies in the Sciences of Complexity; Addison-Wesley: Reading, MA, USA, 1990; Volume VIII, pp. 223–269.
37. Packard, N.H. Adaptation toward the Edge of Chaos. In *Dynamic Patterns in Complex Systems*; Kelso, J.S., Mandell, A.J., Shlesinger, M.F., Eds.; World Scientific: Singapore, 1988; pp. 293–301.
38. Mitchell, M.; Crutchfield, J.P.; Hraber, P. Dynamics, Computation, and the "Edge of Chaos": A Re-Examination. In *Complexity: Metaphors, Models, and Reality*; Cowan, G., Pines, D., Melzner, D., Eds.; Santa Fe Institute Studies in the Sciences of Complexity; Addison-Wesley: Reading, MA, USA, 1994; Volume XIX, pp. 497–513.
39. Mitchell, M.; Hraber, P.; Crutchfield, J.P. Revisiting the Edge of Chaos: Evolving Cellular Automata to Perform Computations. *Complex Syst.* **1993**, *7*, 89–130.
40. Bertschinger, N.; Natschläger, T. Real-time computation at the edge of chaos in recurrent neural networks. *Neural Comput.* **2004**, *16*, 1413–1436. [CrossRef]
41. Boedecker, J.; Obst, O.; Lizier, J.T.; Mayer, N.M.; Asada, M. Information processing in echo state networks at the edge of chaos. *Theory Biosci.* **2012**, *131*, 205–213. [CrossRef] [PubMed]
42. Hochreiter, S. The vanishing gradient problem during learning recurrent neural nets and problem solutions. *Int. J. Uncertain. Fuzziness-Knowl.-Based Syst.* **1998**, *6*, 107–116. [CrossRef]
43. Kingma, D.P.; Ba, J. Adam: A method for stochastic optimization. *arXiv* **2014**, arXiv:1412.6980.
44. Zaslavsky, N.; Kemp, C.; Regier, T.; Tishby, N. Efficient compression in color naming and its evolution. *Proc. Natl. Acad. Sci. USA* **2018**, *115*, 7937–7942. [CrossRef] [PubMed]

entropy

MDPI

Article

Summarizing Finite Mixture Model with Overlapping Quantification

Shunki Kyoya * **and Kenji Yamanishi**

Graduate School of Information Science and Technology, The University of Tokyo, 7-3-1 Hongo, Bunkyo-ku, Tokyo 113-8656, Japan; yamanishi@mist.i.u-tokyo.ac.jp
* Correspondence: kyoya.shunki@plus-zero.co.jp

Abstract: Finite mixture models are widely used for modeling and clustering data. When they are used for clustering, they are often interpreted by regarding each component as one cluster. However, this assumption may be invalid when the components overlap. It leads to the issue of analyzing such overlaps to correctly understand the models. The primary purpose of this paper is to establish a theoretical framework for interpreting the overlapping mixture models by estimating how they overlap, using measures of information such as entropy and mutual information. This is achieved by merging components to regard multiple components as one cluster and summarizing the merging results. First, we propose three conditions that any merging criterion should satisfy. Then, we investigate whether several existing merging criteria satisfy the conditions and modify them to fulfill more conditions. Second, we propose a novel concept named clustering summarization to evaluate the merging results. In it, we can quantify how overlapped and biased the clusters are, using mutual information-based criteria. Using artificial and real datasets, we empirically demonstrate that our methods of modifying criteria and summarizing results are effective for understanding the cluster structures. We therefore give a new view of interpretability/explainability for model-based clustering.

Keywords: model-based clustering; merging mixture components; component overlap; interpretability

Citation: Kyoya, S.; Kenji, Y. Summarizing Finite Mixture Model with Overlapping Quantification. *Entropy* **2021**, *23*, 1503. https://doi.org/10.3390/e23111503

Academic Editor: Pasi Fränti

Received: 28 September 2021
Accepted: 8 November 2021
Published: 13 November 2021

1. Introduction

1.1. Motivation

Finite mixture models are widely used for modeling data and finding latent clusters (see McLachlan and Peel [1] and Fraley and Raftery [2] for overviews and references). When they are used for clustering, they are typically interpreted by regarding each component as a single cluster. However, the one-to-one correspondence between the clusters and mixture components does not hold when the components overlap. This is because the clustering structure then becomes more ambiguous and complex. Let us illustrate this using a Gaussian mixture model estimated for the Wisconsin breast cancer dataset in Figure 1 (details of the dataset and estimation are discussed in Section 8.2). A number of the components overlap with one another, which makes it difficult to estimate the shape of distribution or number of clusters. Therefore, we need an analysis of the overlaps to correctly interpret the models.

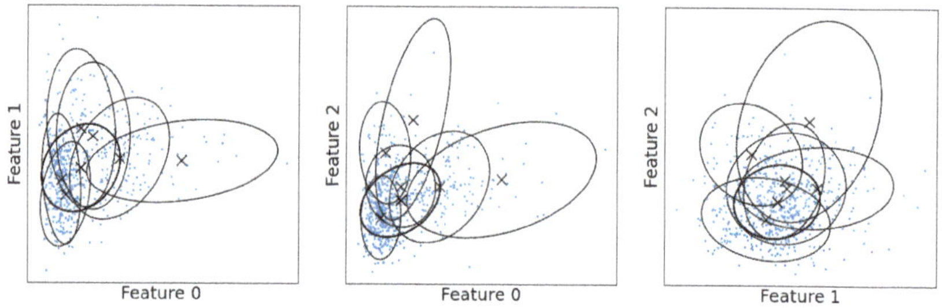

(a) feature 0, 1 (b) feature 0, 2 (c) feature 1, 2

Figure 1. Estimated Gaussian components for the Wisconsin breast cancer dataset [3].

We address this issue from two aspects. In the first aspect, we consider merging mixture components to regard several components as one cluster. We repeatedly select the most overlapping pairs of components to merge them. In this procedure, it is important how the degree of overlap is measured. A number of criteria for measuring cluster overlaps have been proposed [4–6], but they have not yet been compared theoretically. We give a theoretical framework for comparing merging criteria by defining three essential conditions that any method for merging clusters should satisfy. The more conditions any method satisfies, the better it is. From this viewpoint, we evaluate the existing criteria (entropy (Ent) [4], directly estimated misclassification (DEMP) [5] probability, mixture complexity (MC) [7]). We also modify these existing criteria so that they can satisfy more essential conditions.

In the second aspect, we consider how to summarize the merging results quantitatively. After merging mixture components, we obtain two types of clustering structures; those among the upper-components and those among sub-components within each upper-component, as illustrated in Figure 2. These structures might be still ambiguous because the upper-components are determined to be the different clusters, but they may overlap; the sub-components are determined to belong to the same cluster, but they may be scattered in the cluster. Therefore, we need to evaluate the degree to which the upper- and sub-components are discriminated as different clusters. We realize this using the notions of *mixture complexity* (MC) [7] and *normalized mixture complexity* (NMC). They give real-valued quantification of the number of effective clusters and the degree of their separation, respectively. We therefore develop a novel method for cluster summarization.

Our hypotheses in this paper are summarized as follows:

- Modifying merging criteria based on essential conditions can improve the ability to find cluster structures in the mixture model.
- Cluster summarization based on MC and NMC effectively describes the clustering structures.

We empirically verify them by experiments, using artificial and real datasets.

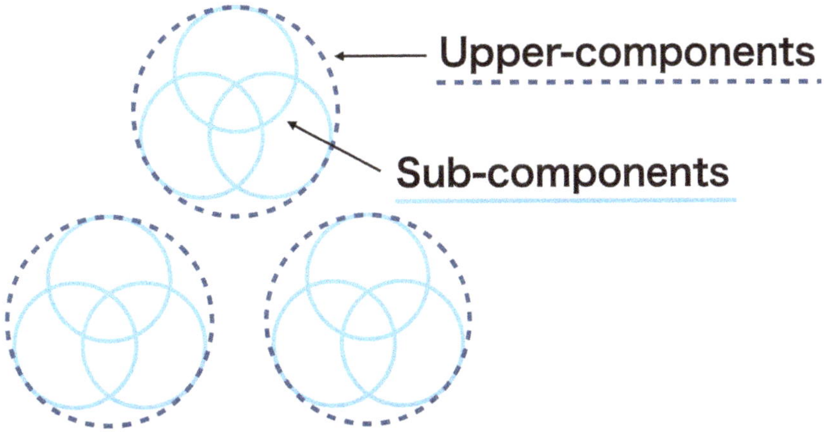

Figure 2. Upper-components and sub-components.

1.2. *Significance and Novelty of This Paper*

The significance and novelty of this paper is summarized below.

1.2.1. Proposal of Theoretical Framework for Evaluating Merging Criteria

We give a theoretical framework for evaluating merging methods by defining the *essential conditions*. They are necessary conditions that any merging criterion should satisfy: (1) the criterion should take the best value when the components are entirely overlapped, (2) it should take the worst value when the components are entirely separated, and (3) it should be invariant with respect to the scale of the weights. We empirically confirm that the more essential conditions any merging method satisfies, the better the clustering structure obtained in terms of larger interdistances and smaller intradistances.

1.2.2. Proposal of Quantitative Clustering Summarization

We propose a method for quantitatively summarizing clustering results based on MC and NMC. MC is an extended concept of the number of clusters into a real number from the viewpoint of information theory [7]. It quantifies the diversity among the components, considering their overlap and weight bias. NMC is defined by normalizing MC to remove the effects of weight bias. It quantifies the degree of the scatter of the components based only on their overlap. Furthermore, MC and NMC have desirable properties for clustering summarization: they are scale invariant and can quantify overlaps among more than two components. We empirically demonstrate that our MC-based method effectively summarizes the clustering structures. We therefore give a novel quantification of clustering structures.

2. Related Work on Finite Mixture Models and Model-Based Clustering

In this section, we present related work on finite mixture models and model-based clustering in four parts: roles of overlap, model, optimization, and visualization. The overlap has a particular impact on the construction of models.

2.1. *Roles of Overlap*

There has been widespread discussion about the roles of overlap in finite mixture models. One argues that the overlap is emerged to represent various distributions. While this flexibility is beneficial for modeling the data, various issues arise in applying them to clustering. For example, McLachlan and Peel [1] pointed out that some skew clusters required more than one Gaussian component to be represented. Moreover, Biernacki et al. [8] pointed out that the number of mixture components selected for estimating densities was typically more than that of clusters because of overlapping.

Model selection methods based on clustering (complete) likelihood, such as the integrated complete likelihood (ICL) [8], the normalized maximum likelihood (NML) [9,10], and the decomposed normalized maximum likelihood (DNML) [11,12], have been proposed to obtain less-overlapping mixtures so that one component corresponds to one cluster. However, they have problems in that they need to define the shape of the clusters in advance. This leads to a trade-off between shape flexibility and component overlap in model-based clustering.

Others argue that the overlap represents that the data belong to more than one cluster. For example, in clustering documents by their topics, the data may have several topics. Such issues have been widely discussed in the field of overlapping clustering. For example, Banerjee et al. [13] extended the mixture model to allow the data to belong to multiple clusters based on membership matrices. Fu and Banerjee [14] considered the product of cluster distributions to represent multiple memberships of the data. Xu et al. [15] proposed methods for describing more complex memberships by calculating correlation weights between the data and the cluster. While these methods allow complex relationships between the data and the clusters, cluster shapes become simple.

The overlap is also used for measuring the complexity of clustering structures in the concept of MC [7]. It is a non-integer valued quantity, which implies the uncertainty of determining the number of clusters. MC was introduced in the scenario of change detection in [7]. This paper gives a new application scenario of MC in the context of quantifying clustering structures. Moreover, this paper also newly introduces NMC as a variant of MC, which turns out to be most effective in this context.

2.2. Model

We discuss the issue of constructing models achieving both flexible cluster shapes and interpretability. Allowing each cluster to have complex shapes is a solution to tackle this. For example, mixtures of non-normal distributions have been proposed for this purpose, as reviewed by Lee and McLachlan [16]. Modeling each cluster as a finite mixture model, called the mixture of mixture model or multi-layer mixture model, has been considered in this regard. Various methods have been proposed to estimate such mixture models based on maximum likelihood estimation [17,18] and Bayesian estimation [19,20]. However, additional parameters are required for assigning sub-components to upper-clusters in many cases because changes of assignment do not change the overall distribution. Merging mixture components [4–6] is an alternative way of the composition of mixture models using single-layer estimations. In this approach, the criteria to measure the degree of component overlap have to be identified. Although various concepts have been developed to measure the degree of overlap, such as entropy [5], misclassification rate [4,6], and unimodality [4], they have not been satisfactorily compared yet.

2.3. Optimization

Merging components has also been discussed in the scenario of optimizing parameters in the mixture models. Ueda et al. [21] proposed splitting and merging mixture components to obtain better estimations, and Minagawa et al. [22] revised their methods to search the models with higher likelihoods. Zhao et al. [23] considered randomly swapping the mixture components during optimization, which allows a more flexible search than splitting and merging components. Because these methods aim only to optimize the models, there remains the problem of interpreting them.

We also refer to the agglomerative hierarchical clustering as a similar approach to merging components. Our methods are similar to the Bayesian hierarchical clustering methods [24,25] in that the number of merging is automatically decided. However, our approaches can not only create clusters, but also evaluate their shape and closeness under the assumption that the mixture models are given.

2.4. Visualization

Methods of interpreting clustering structures have been studied along with visualization methods. Visualizing the values of criteria with a dendrogram is useful for understanding cluster structures among sub-components [6]. Class-preserved projections [26] and parametric embedding [27] were proposed for visualizing structures among upper-clusters by reducing data dimension. We present a method to interpret both structures uniformly based on the MC and NMC.

3. Merging Mixture Components

We assume that data $x^N = x_1, \ldots, x_N$ and a finite mixture model are given. The probability distribution of the model f is written as follows:

$$f(x) = \sum_{k=1}^{K} \rho_k g_k(x),$$

where K denotes the number of components, $\rho_1, \ldots, \rho_K$ denote the mixture proportions of each component summing up to one, and $g(x|\theta_1), \ldots, g(x|\theta_K)$ denote the probability distributions. We assume that the data x^N are independently sampled from f. The random variable X following f is called an *observed variable*, because it can be observed as a data point. We also define the *latent variable* $Z \in \mathcal{Z} := \{1, \ldots, K\}$ as the index of the component from which X originated. The pair (X, Z) is called a *complete variable*. The distribution of the latent variable $P(Z)$ and the conditional distribution of the observed variable $P(X|Z)$ can be given by the following:

$$P(Z = k) = \rho_k, \quad P(X|Z = k) = g_k(X).$$

In the case that f is not known, we will replace f by its estimation $\hat{f}$ under the assumption that $\hat{f}$ is so close to f that x^N can be approximately regarded as samples from $\hat{f}$.

We discuss identifying cluster structures in x^N and f by merging mixture components as described below. First, we define a criterion function denoted as $\mathrm{Crit} : \mathcal{Z} \times \mathcal{Z} \to \mathbb{R}$, which measures the degree of overlap or closeness between two components. For simplicity, we change the sign of the original definitions as needed so that Crit takes smaller values as the components are closer. Then, we choose the closest two components that minimize the criterion and merge them. By repeating the merging process several times, we finally obtain clusters. We show the pseudo-code and computational complexity of this procedure in Appendix A.

4. Essential Conditions

In this section, we propose three *essential conditions* that the criteria should satisfy, so that the criteria can be compared in terms of the conditions. To establish the conditions, we restrict the criteria to those that can be calculated from the posterior probability of the latent variables $\{\gamma_k(x_n)\}_{k,n}$ defined as follows:

$$\gamma_k(x_n) := P(Z = k|X = x_n) = \frac{\rho_k g_k(x_n)}{f(x_n)},$$

where k is the index of the component. After merging the components i and j, the posterior probability can be easily updated as follows:

$$\gamma_{i \cup j}(x_n) := P(Z \in \{i, j\}|X = x_n) = \gamma_i(x_n) + \gamma_j(x_n).$$

Note that some other merging methods reestimate the distribution of the merged components as a single component [4]. We do not consider these in this study because they lack the benefit that the merged components can have complex shapes.

For later use, we define Best(Crit) and Worst(Crit) as the best and worst values that the criteria can take:

$$\mathrm{Best}(\mathrm{Crit}) := \min \mathrm{Crit}(i,j) \quad \text{w.r.t. } \{\gamma_{k,n}\}_{k,n},$$
$$\mathrm{Worst}(\mathrm{Crit}) := \max \mathrm{Crit}(i,j) \quad \text{w.r.t. } \{\gamma_{k,n}\}_{k,n},$$

where $\{\gamma_{k,n}\}_{k,n}$ is a set of $K \times N$ real values in $[0,1]$ that satisfies $\sum_k \gamma_{k,n} = 1$ for all n.

We formulate the three conditions. They provide natural and minimum conditions on the behaviors in the extreme cases that the components are entirely overlapped or separated and on the scale invariance of the criteria. The conditions for the moderate cases that the components partially overlap should be investigated in further studies.

First, we define the condition that a criterion should take the best value when the two components entirely overlap. It is formally defined as follows.

Definition 1. *If a criterion satisfies that*

$$(\forall n,\ g_i(x_n) = g_j(x_n)) \Rightarrow \mathrm{Crit}(i,j) = \mathrm{Best}(\mathrm{Crit}),$$

*then, we say that it satisfies the condition BO (*best in entirely overlap*).*

Next, we define the condition that the criterion should take the worst value when the two components are entirely separated.

Definition 2. *We consider that the sequence of the models $\{f_t = \sum_k \rho_{k,t} g_{k,t}\}_{t=1}^{\infty}$ satisfies the following:*

$$\forall n,\ g_{i,t}(x_n) g_{j,t}(x_n) \to 0 \tag{1}$$

as $t \to \infty$. We define $\mathrm{Crit}_t(i,j)$ as the criterion value based on f_t. Then, if (1) implies that

$$\lim_{t\to\infty} \mathrm{Crit}_t(i,j) \to \mathrm{Worst}(\mathrm{Crit}),$$

*we say that it satisfies the condition WS (*worst in entirely separate*).*

Note that this definition is written using limits in case that the distribution of the components has support in the entire space, such as the Gaussian distributions.

Finally, we define the condition that the value of the criterion should be invariant with the scale of mixture proportions.

Definition 3. *We consider that the components i and j are isolated from the other components, i.e., the sequence of the models $\{f_t = \sum_k \rho_{k,t} g_{k,t}\}_{t=1}^{\infty}$ satisfies the following:*

$$(g_{i,t}(x_n) + g_{j,t}(x_n)) g_{k,t}(x_n) \to 0$$

*for all $k \neq i,j$ and n as $t \to \infty$. In addition, we consider another sequence of the mixture model $\{\bar{f}_t = \sum_k \bar{\rho}_{k,t} g_{k,t}\}_{t=1}^{\infty}$ with different scales on the mixture proportions of the components i and j, i.e., $\bar{\rho}_{k,t} = a\rho_{k,t}$ $(k = i,j)$ holds for some $a > 0$. We define $\overline{\mathrm{Crit}}_t(i,j)$ as the criterion value based on $\bar{f}^{(t)}$. Then, we say that the criterion satisfies the condition SI (*Scale invariance*) if for any a, the following holds:*

$$\lim_{t\to\infty} \mathrm{Crit}_t(i,j) = \lim_{t\to\infty} \overline{\mathrm{Crit}}_t(i,j).$$

5. Modifying Merging Methods

In this section, we introduce the existing merging criteria and propose new criteria by modifying them so that they can satisfy more essential conditions.

5.1. Entropy-Based Criterion

First, we introduce the *entropy-based criterion* (Ent) proposed by Baudry et al. [5]. It selects the components that reduce the entropy of the latent variable the most. This criterion, denoted as $\mathrm{Crit}_{\mathrm{Ent}}$, is formulated as follows:

$$-\mathrm{Crit}_{\mathrm{Ent}}(i,j) := \sum_{n=1}^{N} \left(\Psi(\gamma_i(x_n)) + \Psi(\gamma_j(x_n)) - \Psi(\gamma_{i\cup j}(x_n))\right),$$

where $\Psi(x) := -x \log x$.

However, it violates the conditions BO and SI. Therefore, we propose to modify it in two regards. First, we correct the scale of the weights to make $\mathrm{Crit}_{\mathrm{Ent}}$ satisfy SI. We propose a new criterion $\mathrm{Crit}_{\mathrm{NEnt1}}$ defined as follows:

$$-\mathrm{Crit}_{\mathrm{NEnt1}}(i,j) := \frac{-\mathrm{Crit}_{\mathrm{Ent}}(i,j)}{N(\widetilde{\rho}_i + \widetilde{\rho}_j)},$$

where $\widetilde{\rho}_k := \sum_n \gamma_k(x_n)/N$. This satisfies the condition SI.

Next, we propose removing the effects of the weight biases to make $\mathrm{Crit}_{\mathrm{NEnt1}}$ satisfy BO. We further introduce a new criterion $\mathrm{Crit}_{\mathrm{NEnt2}}$ defined as follows:

$$\mathrm{Crit}_{\mathrm{NEnt2}}(i,j) := \frac{\mathrm{Crit}_{\mathrm{NEnt1}}(i,j)}{\widetilde{H}_{i,j}(Z)},$$

$$\widetilde{H}_{i,j}(Z) := \sum_{k\in\{i,j\}} \Psi\left(\frac{\widetilde{\rho}_k}{\widetilde{\rho}_i + \widetilde{\rho}_j}\right).$$

This satisfies all conditions: BO, WS, and SI.

5.2. Directly Estimated Misclassification Probabilities

Second, we introduce the criterion named directly estimated misclassification probabilities (DEMP) [4]. It selects the components with the highest misclassification probabilities. The criterion is formulated as follows:

$$-\mathrm{Crit}_{\mathrm{DEMP}}(i,j) := \max\left\{\widetilde{\mathcal{M}}_{j,i}, \widetilde{\mathcal{M}}_{i,j}\right\},$$

where

$$\widetilde{\mathcal{M}}_{j,i} := \widetilde{P}(\hat{z}(X) = j | Z = i) := \frac{\sum_n \gamma_i(x_n)\mathbf{1}(\hat{z}(x_n) = j)}{N\widetilde{\rho}_i},$$

$$\hat{z}(x) = \arg\max_{k=1,\ldots,K} \gamma_k(x).$$

However, this violates the condition BO when $\hat{z}(x_n)$ is not i or j for some n. Therefore, we modify it by restricting the choice of the latent variable to component i or j. We define $\hat{z}_{i,j}(x)$ as follows:

$$\hat{z}_{i,j}(x) := \arg\max_{k=i,j} \gamma_k(x_n)$$

and define $\mathrm{Crit}_{\mathrm{DEMP2}}$ by replacing $\hat{z}(x)$ with $\hat{z}_{i,j}(x)$ in the definition of $\mathrm{Crit}_{\mathrm{DEMP}}$. Then, this satisfies all essential conditions.

5.3. Mixture Complexity

Finally, we propose a new criterion based on mixture complexity (MC) [7]. MC is an extended concept of (the logarithm of) the number of clusters into a real value considering the overlap and bias among the components. It is defined based on information theory, and formulated as follows:

$$\mathrm{MC}\Big(\{\gamma_k(x_n)\}_{k,n}; \{w_n\}_n\Big) := \sum_{k=1}^{K} \Psi(\tilde{\rho}_k) - \sum_{n=1}^{N} \frac{w_n}{W} \sum_{k=1}^{K} \Psi(\gamma_k(x_n)),$$

where $\{w_n\}_n$ denotes the weights of the data x^N, $W := \sum_n w_n$ denotes their sum, and $\tilde{\rho}_k$ is redefined as $\tilde{\rho}_k := \sum_n w_n \gamma_k(x_n)/W$. Examples of MC for mixtures of two components are shown in Figure 3. In them, the exponential of the MCs take values between 1 and 2, according to the uncertainty in the number of clusters induced by the overlap or weight bias between the components.

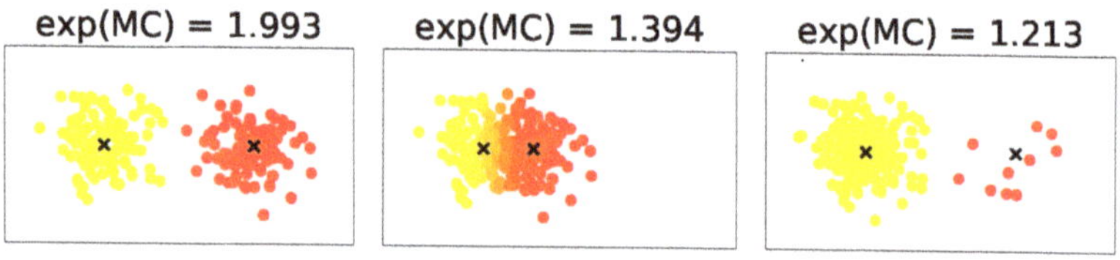

(a) (b) (c)

Figure 3. Examples of MC for mixtures of two components. Images are obtained from [7].

We first propose a new merging criterion $\mathrm{Crit}_{\mathrm{MC}}$ to select the components whose MCs are the smallest. It is defined as follows:

$$\mathrm{Crit}_{\mathrm{MC}}(i,j) := \mathrm{MC}\left(\left\{\frac{\gamma_k(x_n)}{\gamma_{i\cup j}(x_n)}\right\}_{k\in\{i,j\},n}; \{\gamma_{i\cup j}(x_n)\}_n\right).$$

However, this does not satisfy the condition WS because of the effects of the weight biases. Therefore, we modify it by removing the biases to propose a new criterion, which we call the *normalized mixture complexity* (NMC) $\mathrm{Crit}_{\mathrm{NMC}}$. The criterion is defined as follows:

$$\mathrm{Crit}_{\mathrm{NMC}}(i,j) := \frac{\mathrm{Crit}_{\mathrm{MC}}(i,j)}{\tilde{H}_{i,j}(Z)}.$$

It satisfies all conditions BO, WS, and SI. Note that it is equivalent to $\mathrm{Crit}_{\mathrm{NEnt2}}$ because $\mathrm{Crit}_{\mathrm{NMC}} = 1 + \mathrm{Crit}_{\mathrm{NEnt2}}$.

We summarize the relationships between the criteria and the essential conditions in Table 1. The modification led to the fulfillment of many conditions.

Table 1. Summary of the relationships between the criteria and the essential conditions. Check marks are attached to the conditions that are satisfied.

Before Modification				After Modification			
criterion	BO	WS	SI	criterion	BO	WS	SI
Ent		✓		NEnt1		✓	✓
DEMP	(✓)	✓	✓	DEMP2	✓	✓	✓
MC	✓		✓	NMC = NEnt2	✓	✓	✓

6. Stopping Condition

We also propose a new stopping condition based on NMC. First, we calculate the NMC for the (unmerged) mixture model f defined as follows:

$$\mathrm{NMC}_0 := \frac{\mathrm{MC}\left(\{\gamma_k(x_n)\}_{k,n}; \{1\}_n\right)}{\widetilde{H}(Z)}.$$

Since it represents the average degree of separation in the components of f, it can be used for the stopping condition for merging. Then, before merging components i and j, we compare $\mathrm{Crit}_{\mathrm{NMC}}(i,j)$ to NML_0. If $\mathrm{Crit}_{\mathrm{NMC}}(i,j) \geq \mathrm{NML}_0$, then the merging algorithm halts without merging components i and j. Otherwise, the algorithm merges components i and j and continues further.

Note that this stopping criterion can be applied when a criterion other than $\mathrm{Crit}_{\mathrm{NMC}}$ is used. In this case, we use the criterion to search the two closest components and use NMC to decide whether to merge them.

7. Clustering Summarization

In this section, we propose methods to quantitatively explain the merging results, using the MC and NMC.

We consider that a mixture model with K-component is merged into L upper-components. We define the sets $I_1, \dots, I_L$ that partition $\{1, \dots, K\}$ as the sets of the indices that are contained in each upper-component. Then, the MC and NMC among the upper-components, denoted as MC(up) and NMC(up), respectively, can be calculated as follows:

$$\mathrm{MC}(\mathrm{up}) := \mathrm{MC}\left(\left\{\sum_{k \in I_l} \gamma_k(x_n)\right\}_{l,n}, \{1\}_n\right),$$

$$\mathrm{NMC}(\mathrm{up}) := \frac{\mathrm{MC}(\mathrm{up})}{\sum_l \Psi(\widetilde{\tau}_l)},$$

where $\widetilde{\tau}_l$ denotes the weight of the upper-component l calculated as follows:

$$\widetilde{\tau}_l := \frac{1}{N} \sum_{k \in I_l} \gamma_k(x_n) = \sum_{k \in I_l} \widetilde{\rho}_k.$$

For each l, the MC and NMC in the sub-components within the upper-component l, written as $\mathrm{MC}(l)$ and $\mathrm{NMC}(l)$, respectively, can be calculated as follows:

$$\mathrm{MC}(l) := \mathrm{MC}\left(\left\{\frac{\gamma_k(x_n)}{\sum_{k' \in I_l} \gamma_{k'}(x_n)}\right\}_{k \in I_l, n}; \left\{\sum_{k' \in I_l} \gamma_k(x_n)\right\}_n\right),$$

$$\mathrm{NMC}(l) := \frac{\mathrm{MC}(l)}{\sum_{k \in I_l} \Psi\left(\widetilde{\rho}_l^{(k)}\right)},$$

where $\widetilde{\rho}_l^{(k)}$ denotes the relative weight of the sub-component $k \in I_l$ calculated as $\widetilde{\rho}_l^{(k)} := \widetilde{\rho}_k / \sum_{k' \in I_l} \widetilde{\rho}_{k'}$. NMC is undefined if the denominator is 0.

MC and NMC quantify the degree to which the components are regarded as clusters in different ways: larger values indicate that the components definitely look like different clusters. MC quantifies this by measuring (the logarithm of) the number of clusters continuously, considering the ambiguity induced by the overlap and weight bias among the components. It takes a value between 0 and the logarithm of the number of the components. In contrast, NMC measures the scattering of the components based only on their overlap. It takes a value between 0 and 1. They have also the desirable properties that they are scale invariant and can quantify overlaps among more than two components.

Therefore, we propose the summarization of clustering structures by listing MC(up), NMC(up), component weights, MC(l), and NMC(l) in a table, which we call the *clustering summarization*. The clustering summarization is useful for evaluating the confidence level of the clustering results.

We show an example of the clustering summarization using the mixture model illustrated in Figure 4. In this example, there are four Gaussian components as illustrated in Figure 4a, and two merged clusters on the left and right sides as illustrated in Figure 4b–d. The clustering summarization is presented in Table 2. For the upper-components, the exponential of MC is almost two, and the NMC is almost one. This indicates that two upper-components can be definitely regarded as different clusters. For both sub-components, the exponential of MC is larger than one. This indicates that they have more complex shapes than a single component. Moreover, the structures within Component 1 are more complex than those in 2, because the MC and NMC are larger.

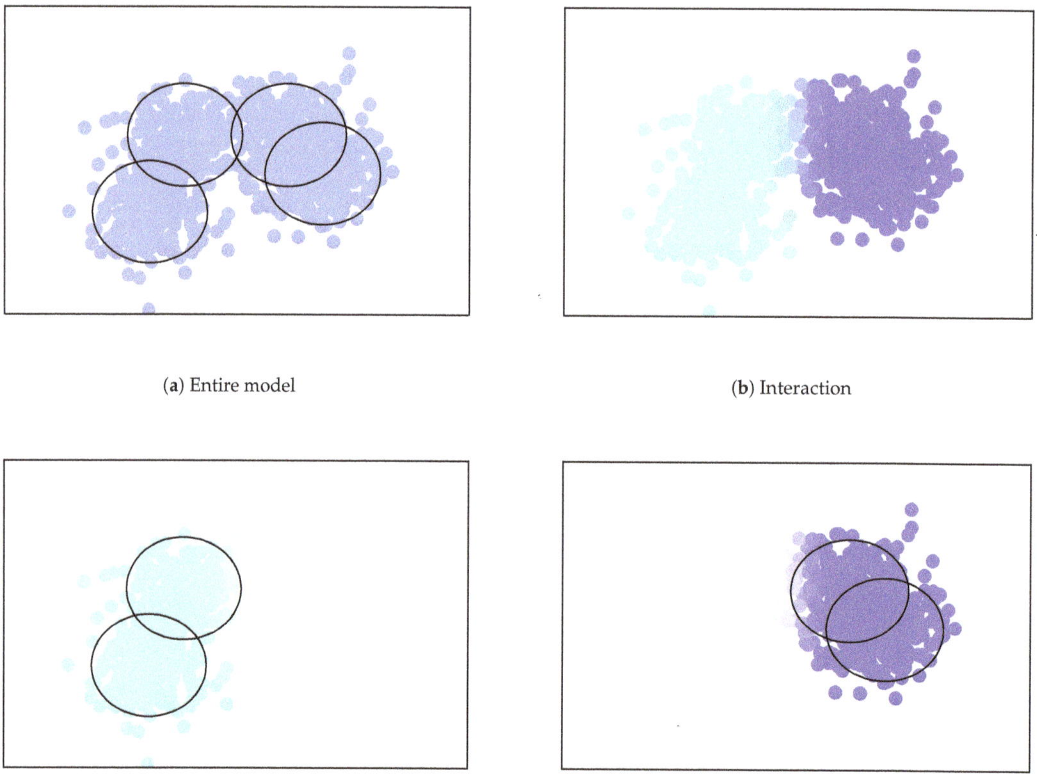

(**a**) Entire model (**b**) Interaction

(**c**) Cluster 1 (**d**) Cluster 2

Figure 4. Example of the merged mixture model. Images are obtained from [7].

Table 2. Example of a clustering summarization.

Upper-Components			
MC (exp):		0.647 (1.91)	
NMC:		0.933	
Component 1		**Component 2**	
Weight:	0.494	Weight:	0.506
MC (exp):	0.566 (1.76)	MC (exp):	0.324 (1.38)
NMC:	0.817	NMC:	0.467

8. Experiments

In this section, we present the experimental results to demonstrate the effectiveness of merging the mixture components and modifying the criteria.

8.1. Analysis of Artificial Dataset

To reveal the differences among the criteria, we conducted experiments with artificially generated Gaussian mixture models. First, we randomly created a two-dimensional Gaussian mixture model $f = \sum_{k=1}^{K} \rho_k \mathcal{N}(x; \mu_k, \Sigma_k)$ as follows:

$$\begin{aligned} K &:= 50, \\ (\rho_1, \ldots, \rho_K) &\sim \mathrm{Dir}(1, \ldots, 1), \\ \mu_1, \ldots, \mu_K &\overset{\text{i.i.d.}}{\sim} \mathcal{N}\left(\mu; [0,0], 3^2 \times I_2\right), \\ a_1, b_1, \ldots, a_K, b_K &\overset{\text{i.i.d.}}{\sim} U[0.5, 1.5], \\ \Sigma_k &:= [[a_k, 0], [0, b_k]] \quad (k = 1, \ldots, K), \end{aligned}$$

where $\mathrm{Dir}(\alpha, \ldots, \alpha)$ denotes the Dirichlet distribution, and $U[m, M]$ denotes the uniform distribution from m to M. Then, we sampled 5000 points x^{5000} from f, and ran the merging algorithms without stopping conditions. The algorithms were evaluated using the (maximum) intra-cluster distance $\mathcal{D}_{\text{intra}}$ and (minimum) inter-cluster distance $\mathcal{D}_{\text{inter}}$ defined as follows:

$$\begin{aligned} \mathcal{D}_{\text{intra}} &:= \max_{k=1,\ldots,K} \frac{\sum_n \gamma_k(x_n) \|x_n - \widetilde{\mu}_k\|^2}{\sum_{n'} \gamma_k(x_{n'})}, \\ \mathcal{D}_{\text{inter}} &:= \min_{1 \le i < j \le K} \|\widetilde{\mu}_i - \widetilde{\mu}_j\|^2, \end{aligned}$$

where $\widetilde{\mu}_1, \ldots, \widetilde{\mu}_K$ denote the centers of the components defined as

$$\widetilde{\mu}_k := \frac{\sum_n \gamma_k(x_n) x_n}{\sum_{n'} \gamma_k(x_{n'})}.$$

The clustering structure is said to be *better*, as $\mathcal{D}_{\text{intra}}$ is smaller and $\mathcal{D}_{\text{inter}}$ is larger. Both distances are measured with several K and compared among the algorithms with different criteria. Although we may obtain better results for these metrics by using them as merging criteria in a similar way as used in hierarchical clustering [28,29], we used them only for comparison rather than optimizing them.

The experiments were performed 100 times by randomly generating f and the data. Accordingly, the ranking of the criteria was calculated for each distance. Table 3 presents the average rank of each criterion. As seen from the table, the modifications of the criteria improved the rank. In addition, DEMP2 and NMC, satisfying all essential conditions, were always in the top three. These results indicate the effectiveness of the essential conditions.

Table 3. Average ranks of the criteria. For each K, the best rank is denoted in boldface.

			$\mathcal{D}_{\text{intra}}$		
K	40	30	20	10	5
Ent	6.00	6.00	6.00	5.94	5.58
NEnt1	3.67	4.15	3.93	4.15	4.18
DEMP	4.30	4.56	4.07	2.57	2.13
DEMP2	2.40	2.52	2.45	2.12	**1.91**
MC	**2.25**	2.04	3.21	4.53	4.94
NMC	2.37	**1.73**	**1.35**	**1.69**	2.27
			$\mathcal{D}_{\text{inter}}$		
K	40	30	20	10	5
Ent	5.02	5.14	5.38	5.50	5.19
NEnt1	4.29	4.23	4.26	4.56	3.99
DEMP	4.99	4.97	4.82	2.88	2.63
DEMP2	3.56	3.55	3.17	2.85	2.48
MC	2.09	2.06	2.35	3.90	4.86
NMC	**1.06**	**1.04**	**1.03**	**1.29**	**1.84**

To further investigate the relationships between the essential conditions and resulting cluster structures, we illustrated the cluster obtained in a trial where the intra-cluster distance was the largest in Figure 5. For the criterion Ent, one cluster continued to grow. This is because Ent lacks the condition SI, and is advantageous for larger clusters. For the criterion NEnt1, the growth of the larger clusters was mitigated by adding the condition SI to Ent. Nevertheless, the intra-cluster distances were still large because NEnt lacked the condition BO. It tended to create unnecessarily large clusters because it tended to merge larger and more distant components rather than smaller and closer components. The criterion NMC improved such a disadvantage by adding the condition BO to NEnt1. For the criterion MC, distant components were merged, as the condition WS was not satisfied. NMC overcame this by adding the condition WS to MC. The differences between DEMP and DEMP2 were unclear in Figure 5c,d, and both criteria elucidated the cluster structure well because they satisfied relatively many conditions. We conclude that the essential conditions are effective for obtaining better cluster structures.

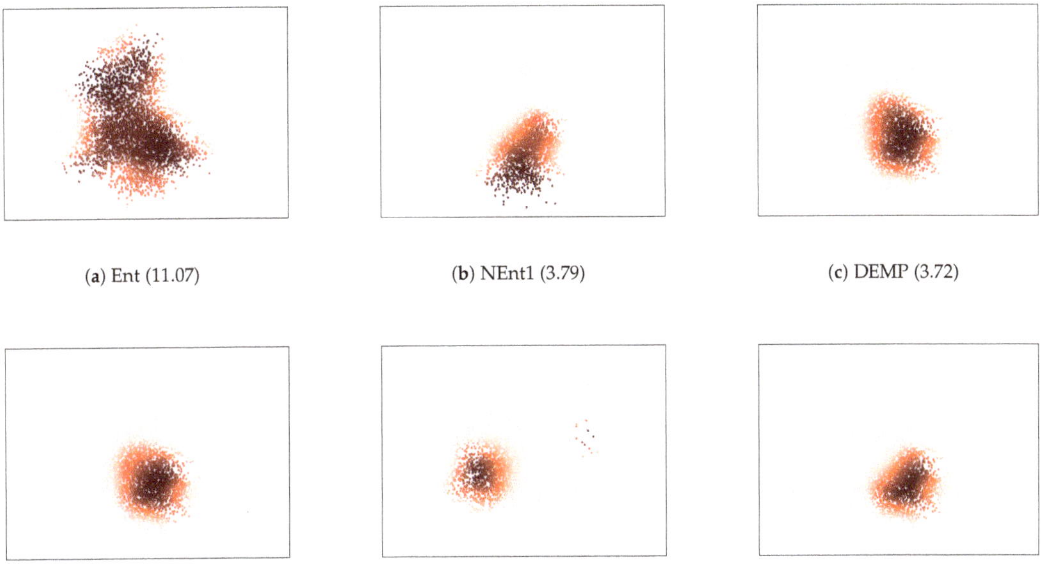

(**a**) Ent (11.07) (**b**) NEnt1 (3.79) (**c**) DEMP (3.72)

(**d**) DEMP2 (3.45) (**e**) MC (3.92) (**f**) NMC (2.79)

Figure 5. Scatter plots for the cluster with $K = 20$ whose intra-cluster distance is the largest. The thickness of the color corresponds to the posterior probabilities. The numbers in the parenthesis show $\mathcal{D}_{\text{intra}}$.

8.2. Analysis of Real Dataset

We discuss the results of applying the merging algorithms and clustering summarization to eight types of real datasets with true cluster labels. The details of the datasets and processing are described in Appendix B.

8.2.1. Evaluation of Clustering Using True Labels

First, we compared the clustering performance of the merging algorithms by measuring similarity between estimated and true cluster labels. Formally, given the dataset $\{x_n\}_n$ and the true labels $\{z_n^\star\}_n$, we first estimated the clustering structures using $\{x_n\}_n$ without seeing $\{z_n^\star\}_n$, and obtained the estimated labels $\{\hat{z}_n\}_n$. We define $K^\star$ and $\hat{K}$ as the number of the true and estimated clusters. Then, we evaluated the similarity between $\{z_n^\star\}_n$ and $\{\hat{z}_n\}_n$ using the adjusted Rand index (ARI) [30] and F-measure. ARI takes values between -1 and 1, and F-measure takes values between 0 and 1. Their larger value corresponds to better clustering. Both indices can be applied when the number of true and estimated clusters is different.

To run the merging algorithms, the mixture models should be estimated first. In our experiments, we estimated them by the variational Bayes Gaussian mixture model with $K = 20$ [31] implemented in the Scikit-learn package [32]; we adopted this, as it exhibited good performance in our experiments. We used the prior distributions of the mixture proportions as the Dirichlet distributions with $\alpha = 0.1$, and we set the other parameters for prior distributions as the default values in the package. For each dataset, we fitted the algorithm ten times with different initializations and used the best one.

We compared the merging algorithms with three types of model-based clustering algorithms based on the Gaussian mixture model, which are summarized in Table 4. First, we estimated the number of components, using BIC [33]. It selects a suitable model for describing the densities, and the mixture components tend to overlap. Nevertheless, it has been widely used for clustering by regarding each component as a cluster. Second, we estimated the number of clusters using DNML [11,12]. It selects a model whose components can be regarded as clusters by considering the description length of the latent and observed variables. Finally, we estimated the clusters as the mixture of Gaussian mixture models implemented by Malsiher-Walli et al, [20]. By fixing two integers K and L, K Gaussian mixture models were estimated with L components. The number of clusters was automatically adjusted by shrinking the redundant clusters. As in the original paper, we set $K = 30, L = 15$ (and some specific parameters in the paper) for the DLB dataset and $K = 10, L = 5$ for the other datasets.

Table 4. Overview of the comparison methods.

Abbreviation	Method	Reference
GMM + BIC	GMM and BIC criterion	[33]
GMM + DNML	GMM and DNML criterion	[11,12]
MixMix	Mixture of Gaussian mixture models	[20]

We estimated the models ten times and compared the average score among the methods. The average number of clusters are listed in Table 5, and F-measure and ARI are listed in Tables 6 and 7.

Two clusters that achieved the best score and that were obtained by the heuristics proposed in Section 6 are described. The best scores of the merging algorithms exceeded those of all other methods for six out of eight datasets. In particular, the merging methods satisfying many essential conditions, such as DEMP, DEMP2, and NMC, obtained high scores with a smaller number of clusters. Therefore, it can be said that the merging algorithms with more essential conditions are effective for elucidating the clustering structures. Moreover, the scores with NMC-based stopping conditions exceeded those of all other methods for four out of eight datasets.

Table 5. Estimated number of clusters. Merge (best F-measure) is the number of clusters when F-measure is highest. Merge (best ARI) is the number of clusters when ARI is highest. Merge (NMC) is the number of clusters obtained by the NMC-based stopping condition.

Dataset		**AIS**	**BTL**	**CRB**	**DLC**	**ECL**	**SDS**	**WSC**	**YST**
K^*		**2**	**3**	**4**	**4**	**5**	**3**	**2**	**2**
GMM + BIC		3.0	2.6	3.0	6.6	4.0	2.0	3.0	3.5
GMM + DNML		1.0	1.0	1.0	2.8	4.0	1.0	2.0	1.0
MixMix		2.7	1.2	1.0	7.4	4.5	3.2	2.1	2.8
Merge (Best F-measure)	Ent	18.0	20.0	19.1	17.6	19.0	19.4	19.2	18.6
	NEnt1	2.0	3.0	5.7	10.2	8.9	6.7	5.3	8.8
	DEMP	2.0	3.0	4.3	4.9	6.2	4.7	3.4	2.3
	DEMP2	2.0	3.0	4.3	4.9	5.7	4.1	2.7	2.0
	MC	2.0	3.0	5.6	3.6	4.9	7.1	2.7	2.0
	NMC	3.0	3.0	4.3	7.1	6.4	4.3	2.0	2.1
Merge (Best ARI)	Ent	19.0	20.0	19.4	17.6	19.0	19.4	19.2	18.6
	NEnt1	3.0	3.0	5.7	3.2	7.0	6.7	5.3	8.8
	DEMP	2.4	3.0	4.3	4.9	6.2	4.9	3.6	2.3
	DEMP2	2.0	3.0	4.3	5.0	6.1	4.5	2.7	2.0

Table 5. *Cont.*

Dataset K^*		AIS 2	BTL 3	CRB 4	DLC 4	ECL 5	SDS 3	WSC 2	YST 2
	MC	2.0	3.0	5.6	3.6	4.9	7.1	4.0	2.0
	NMC	4.0	3.0	4.3	7.0	6.4	4.4	2.0	2.1
Merge (NMC)	Ent	19.0	20.0	17.8	17.6	19.0	18.1	19.2	18.6
	NEnt1	4.0	3.0	3.6	6.9	4.7	4.9	5.6	8.9
	DEMP	4.8	3.0	3.3	9.1	5.7	5.3	4.4	8.0
	DEMP2	5.0	3.0	3.3	9.2	5.7	5.2	3.6	8.5
	MC	5.0	3.0	6.2	11.3	9.0	8.6	6.1	9.7
	NMC	4.0	3.0	3.1	6.4	4.3	3.2	2.9	6.9

Table 6. F-measure for the real datasets. For each merging algorithm, scores that exceed all comparison methods are denoted in boldface.

Dataset		AIS	BTL	CRB	DLC	ECL	SDS	WSC	YST
GMM + BIC		0.912	0.805	0.810	0.734	0.787	0.794	0.857	0.864
GMM + DNML		0.671	0.590	0.400	0.903	0.787	0.500	0.914	0.850
MixMix		0.925	0.578	0.400	0.761	0.829	0.849	0.947	0.826
Merge (Best)	Ent	0.916	**0.986**	**0.866**	**0.931**	**0.874**	**0.900**	0.897	**0.867**
	NEnt1	0.901	**0.986**	**0.889**	**0.922**	**0.860**	**0.928**	0.904	**0.868**
	DEMP	0.906	**0.986**	**0.877**	**0.952**	**0.874**	**0.908**	0.905	**0.942**
	DEMP2	0.906	**0.986**	**0.877**	**0.952**	**0.875**	**0.912**	0.905	**0.944**
	MC	**0.931**	**0.986**	**0.863**	**0.921**	**0.870**	**0.886**	0.886	**0.928**
	NMC	0.916	**0.986**	**0.893**	**0.949**	**0.875**	**0.938**	0.945	**0.942**
Merge (NMC)	Ent	0.892	**0.986**	**0.822**	**0.931**	**0.874**	**0.852**	0.897	**0.867**
	NEnt1	0.892	**0.986**	**0.828**	**0.905**	0.823	0.822	0.904	**0.868**
	DEMP	0.822	**0.986**	**0.823**	0.758	**0.867**	**0.886**	0.881	**0.870**
	DEMP2	0.805	**0.986**	**0.822**	0.754	**0.867**	**0.892**	0.880	0.820
	MC	0.803	**0.986**	**0.831**	0.706	**0.860**	**0.878**	0.858	0.771
	NMC	0.892	**0.986**	**0.828**	**0.916**	**0.848**	0.810	0.925	**0.878**

Table 7. ARI for the real datasets. For each merging algorithm, scores that exceed all comparison methods are denoted in boldface.

Dataset ($K^{\star}$)		AIS	BTL	CRB	DLC	ECL	SDS	WSC	YST
GMM + BIC		0.743	0.603	0.595	0.506	0.590	0.542	0.617	0.516
GMM + DNML		0.000	0.000	0.000	0.870	0.590	0.000	0.685	0.000
MixMix		0.751	0.110	0.000	0.501	0.673	0.623	0.799	0.589
Merge (Best)	Ent	0.700	**0.958**	**0.707**	**0.913**	**0.759**	**0.767**	0.688	0.508
	NEnt1	0.701	**0.958**	**0.739**	0.852	**0.719**	**0.810**	0.688	0.511
	DEMP	0.666	**0.958**	**0.732**	**0.934**	**0.763**	**0.782**	0.734	**0.769**
	DEMP2	0.657	**0.958**	**0.731**	**0.936**	**0.763**	**0.788**	0.732	**0.773**
	MC	0.741	**0.958**	**0.700**	0.849	**0.744**	**0.745**	0.664	**0.709**
	NMC	0.700	**0.958**	**0.748**	**0.928**	**0.769**	**0.832**	0.791	**0.760**
Merge (NMC)	Ent	0.700	**0.958**	**0.626**	**0.913**	**0.759**	**0.657**	0.688	0.508
	NEnt1	0.700	**0.958**	**0.642**	0.834	0.638	0.594	0.688	0.511
	DEMP	0.576	**0.958**	**0.640**	0.523	**0.754**	**0.728**	0.670	0.514
	DEMP2	0.545	**0.958**	**0.639**	0.521	**0.754**	**0.745**	0.670	0.402
	MC	0.534	**0.958**	**0.660**	0.452	**0.700**	**0.728**	0.633	0.313
	NMC	0.700	**0.958**	**0.643**	**0.878**	**0.725**	0.575	0.732	0.524

To further investigate the relationships between the performances of the algorithms and the shapes of the datasets, we estimated the proportion of outliers based on the k-nearest neighbor distances $\mathcal{D}_{\mathrm{nn}}^{(5)}$. We calculated the ratio of the 5-nearest neighbor distance $\mathcal{D}_{\mathrm{nn}}^{(5)}(x_n)$ and its average $(1/N)\sum_{n'}\mathcal{D}_{\mathrm{nn}}^{(k)}(x_{n'})$ for each data point, and we plotted the proportions for which the ratio exceeded 2.0, 3.0, 4.0, and 5.0 in Figure 6. As seen from the figure, the datasets where the merging methods did not work well, such as AIS, DLB, WSC, and YST, contained relatively many outliers. This is reasonable because the merging algorithms do not aim to merge distant clusters. We can conclude that the merging methods are particularly effective when the datasets have fewer outliers or when we want to find the aggregated clusters.

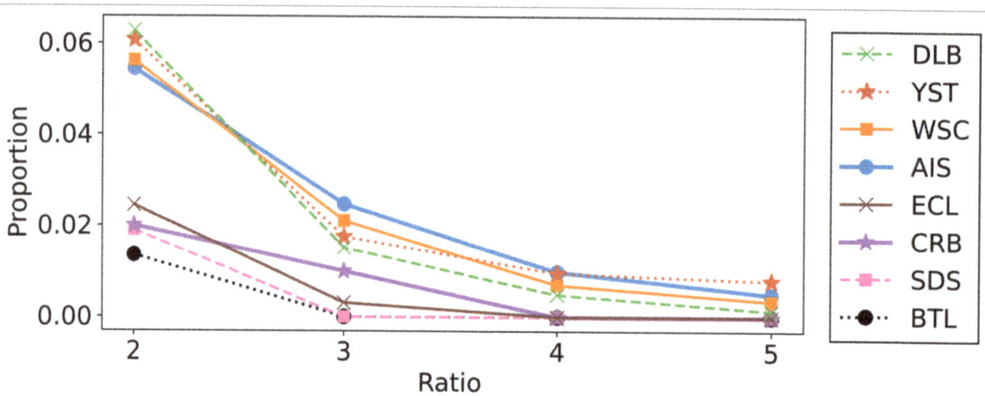

Figure 6. The proportions of the data x_n that satisfy $\mathcal{D}_{\mathrm{nn}}^{(5)}(x_n)/[(1/N)\sum_{n'}\mathcal{D}_{\mathrm{nn}}^{(5)}(x_{n'})] > 2.0, 3.0, 4.0, 5.0$.

8.2.2. Results of Clustering Summarization

Next, we analyzed the results of the merging methods using the clustering summarization proposed in Section 7. As examples, we show one result obtained using the NMC and NMC-based stopping conditions for the Flea beetles and Wisconsin breast cancer datasets. The clustering results are summarized in Tables 8 and 9, respectively. For the upper-components in Flea beetles dataset, the exponential of MC(up) was close to 3.0, and NMC(up) was close to 1.0; we see that the effective number of clusters was around three, and the clusters were well-separated. Components 2 and 3 were unmerged, and the exponentials of MC and NMC of Component 1 were close to 1.0 and 0.0, respectively. This indicates that each cluster can be represented by almost a single Gaussian distribution. Furthermore, the (exponentials of) MC and the NMC of the upper-components in the Wisconsin cancer dataset were 1.66 and 0.763, respectively. It can be expected that the situation was a partial overlap of the two clusters. For Components 1 and 2, NMCs were relatively large. This shows that partially separated components are needed to describe each component. MC of Component 2 was smaller than that of Component 1. Then, it is expected that Component 2 had simpler shapes than Component 1; however, the former seemed to have small components that might be outliers because NMC was larger. Plots of the predicted clusters are illustrated for the Flea beetles and Wisconsin breast cancer datasets in Figures 7 and 8, respectively. We observe that the predictions described previously match to the actual plots. Therefore, we can reveal significant information about the clustering structures by observing the clustering summarizations.

Table 8. Clustering summarization for the Flea beetles dataset.

Upper-Components					
MC (exp):			0.963 (2.62)		
NMC:			0.897		
Component 1		**Component 2**		**Component 3**	
Weight:	0.440	Weight:	0.268	Weight:	0.293
MC:	0.057	MC:	0.000	MC:	0.000
(exp)	(1.06)	(exp)	(1.00)	(exp)	(1.00)
NMC:	0.209	NMC:	-	NMC:	-

Table 9. Clustering summarization for the Wisconsin breast cancer dataset.

Upper-Components			
MC (exp):		0.509 (1.66)	
NMC:		0.763	
Component 1		**Component 2**	
Weight:	0.387	Weight:	0.613
MC (exp):	0.714 (2.04)	MC (exp):	0.270 (1.31)
NMC:	0.613	NMC:	0.676

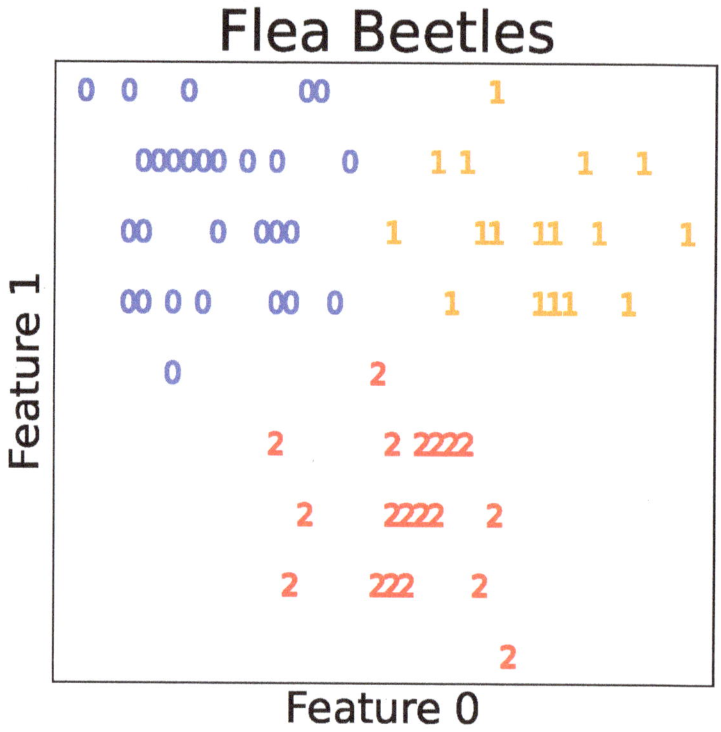

Figure 7. Predicted cluster labels for the Flea beetles dataset.

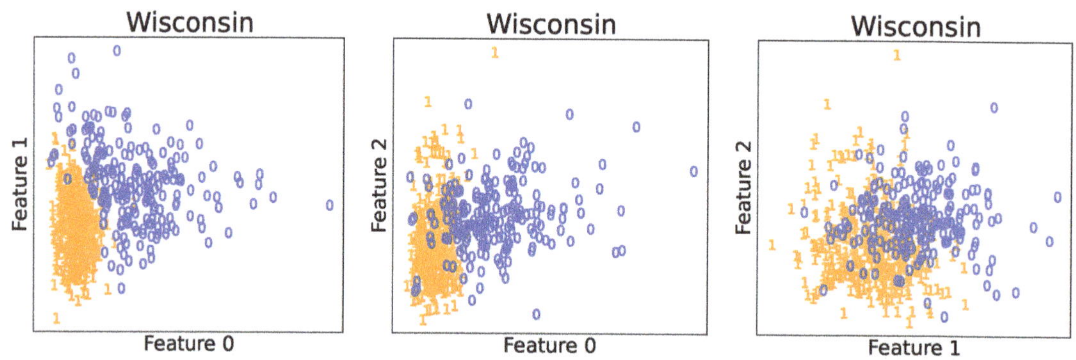

(**a**) feature 0, 1 (**b**) feature 0, 2 (**c**) feature 1, 2

Figure 8. Predicted cluster labels for the Wisconsin breast cancer dataset.

8.2.3. Relationships between Clustering Summarization and Clustering Quality

Finally, we confirmed that MC and NMC in the sub-components were also related to the quality of classification. To confirm this, we conducted additional experiments discussed below. First, we ran the merging algorithms until $K = 1$ without the stopping conditions. Then, for every merged clusters created at $K = K_{\text{start}}, \ldots, 1$, we counted the number of data points classified into them. We define $N_C^{(k)}$ as the number of points with true labels k classified into the merged cluster C. Then, we evaluated the quality of the cluster C using the entropy calculated as follows:

$$H_C = -\sum_{k=1}^{K^\star} \frac{N_C^{(k)}}{\sum_{k'} N_C^{(k)}} \log \frac{N_C^{(k)}}{\sum_{k'} N_C^{(k)}},$$

where the cluster C for $\sum_{k'} N_C^{(k)} = 0$ were omitted. This takes values between 0 and $\log K^\star$. Smaller values are preferred, because H_C becomes small when most of the points within the component share the same cluster label. We calculated the MC/NMC and H_C within the clusters for all datasets and merging algorithms, and we plotted the relationships between them in Figure 9. Note that the unmerged clusters were omitted because the NMC could not be defined. From the figure, it is evident that both MC and NMC had positive correlations with H_C. The correlation coefficients were 0.794 and 0.637 for MC and NMC, respectively. This observation is useful in applications. If the obtained cluster has smaller MC and NMC, then we can confirm that it contains only one group. Otherwise, we need to assume that it contains more than one group. Therefore, we conclude that MC and NMC indicate the confidence level of the cluster structures.

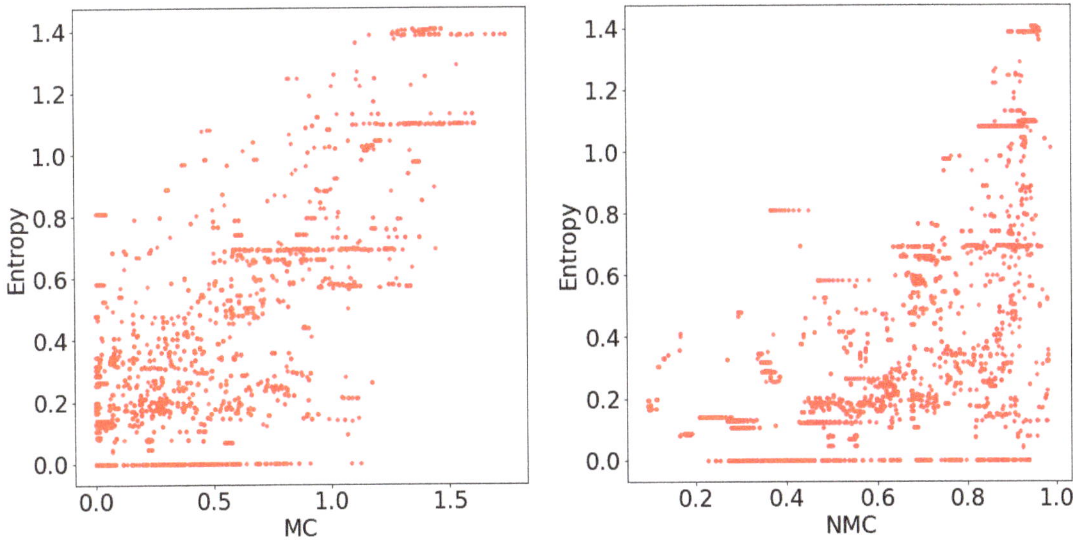

(a) (b)

Figure 9. Scatter plots of the MC/NMC and the entropy of the true cluster label.

9. Discussion

To improve the interpretability of the mixture models with overlap, we have established novel methodologies to merge the components and summarize the results.

For merging mixture components, we proposed essential conditions that the merging criteria should satisfy. Although there have been studies creating some rules in the clustering approach [34,35], they have not been applied to clustering by merging components. The proposed essential conditions for merging criteria contributed to comparing and modifying existing criteria. The limitation of our conditions is that they only provide the necessary conditions for extreme cases, where the components are entirely overlapped or separated. The conditions for the moderate cases that the components partially overlap should be investigated in further studies.

We also proposed a novel methodology to interpret the merging results based on clustering summarization. While previous studies [6,26,27] have focused on interpreting the structures among sub-components or upper-clusters only, our methods can quantify both structures uniformly based on the MC and NMC. They represented the overview of the structures in the mixture models by evaluating how much the components were distinguished based on the degree of overlap and weight bias.

We verified the effectiveness of our methods, using artificial and real datasets. In the artificial data experiments, we confirmed that the intra- and inter-cluster distances were improved corresponding to the modification of the criteria. Further, by observing the clusters with maximum intra-cluster distance, we found that the essential conditions were helpful to prevent the clusters from merging distant components or growing too much. In the real data experiments, we confirmed that the best scores of the proposed methods were better than the comparison methods for many datasets, and the scores obtained using the stopping condition were also better for the datasets containing relatively smaller outliers. In addition, we confirmed that the clustering summary was helpful to interpret the merging results. It was related to the shape of the clusters, weight biases, and the existence of the outliers. Further, we found that the MC and NMC within the components were also related to the quality of the classification. Therefore, the clustering summary also represented the confidence level of the cluster structures.

10. Conclusions

We have established the framework of theoretically interpreting overlapping mixture models by merging the components and summarizing merging results. First, we proposed three essential conditions for evaluating cluster-merging methods. They declared necessary properties that the merging criterion should satisfy. In this framework, we considered Ent, DEMP, and MC and their modifications to investigate whether they satisfied the essential conditions. The stopping condition based on NMC was also proposed.

Moreover, we proposed the clustering summarization based on MC and NMC. They quantify how overlapped the clusters are, how biased the clustering structure is, and how scattered the components are in a respective cluster. We can conduct this analysis from higher level clusters to lower level components to give a comprehensive survey of the global clustering structure. We then quantitatively explained the shape of the clusters, weight biases, and existence of the outliers.

In the experiments, we empirically demonstrated that the modification of the merging criteria improved the ability to find better clustering structures. We also investigated the merging order for each criterion and found that the essential conditions were helpful to prevent the clusters from merging distant components or growing too much. Further, we confirmed, using the real dataset, that the clustering summary revealed varied information in the clustering structure, such as the shape of the clusters, weight biases, the existence of the outliers, and even the confidence level of the cluster structures. We believe that this methodology gives a new view of the interpretability/explainability for model-based clustering.

We have studied how to interpret the overlapping mixture models after they were estimated. It remains for future study to apply merging criteria even in the phase of estimating mixture models.

Author Contributions: Conceptualization, S.K. and K.Y.; methodology, S.K. and K.Y.; software, S.K.; validation, S.K. and K.Y.; formal analysis, S.K.; investigation, S.K. and K.Y.; resources, S.K.; data curation, S.K.; writing—original draft preparation, S.K. and K.Y.; writing—review and editing, S.K. and K.Y.; visualization, S.K.; supervision, K.Y.; project administration, K.Y.; funding acquisition, K.Y. All authors have read and agreed to the published version of the manuscript.

Funding: This work was partially supported by JST KAKENHI JP19H01114 and JST-AIP JPMJCR19U4.

Institutional Review Board Statement: Not applicable.

Informed Consent Statement: Not applicable.

Data Availability Statement: All of the datasets used in this paper can be obtained in the manner described in the `README` file at https://github.com/ShunkiKyoya/summarize_cluster_overlap. Moreover, all of the experimental results can be reproduced by executing the .ipynb files.

Conflicts of Interest: The authors declare no conflict of interest.

Appendix A. Details of the Merging Algorithm

We show the pseudo-code and computational complexity of the merging algorithm. First, the pseudo-code of merging mixture components is shown in Algorithm A1.

Algorithm A1 Merging mixture components

Require: data x^N, finite mixture model f, criterion function Crit.
while (The number of components) > 1 **do**
  $i, j := \arg\min_{i<j} \mathrm{Crit}(i, j)$
  if a certain stopping condition is satisfied **then**
    return The current components.
  end if
  Merge components i and j.
end while
return The current components.

Next, we discuss the computational complexity in this algorithm given x^N and f below. First, the cost of calculating $\{\gamma_k(x_n)\}_{k,n}$ can be written as $\mathrm{O}(T_{\mathrm{dist}}NK)$, where T_{dist} is the cost to calculate $f(x)$ for a point. To merge components, it is needed to repeat updating $\{\mathrm{Crit}(i,j)\}_{i,j}$ and $\{\gamma_k(x_n)\}_{k,n}$ at most $(K-1)$ times. The cost for updating $\{\mathrm{Crit}(i,j)\}_{i,j}$ and $\{\gamma_k(x_n)\}_{k,n}$ are $\mathrm{O}(T_{\mathrm{crit}}K^2)$ and $\mathrm{O}(N)$, respectively, where T_{crit} is the cost to calculate $\mathrm{Crit}(i,j)$ for a pair of the components. Overall, we need $\mathrm{O}(K(T_{\mathrm{dist}} + T_{\mathrm{crit}}K^2 + N))$ to complete the algorithm.

For the criteria referred to in this section, their computational complexity T_{crit} are $\mathrm{O}(N)$ for Ent, NEnt1, DEMP2, MC, and NMC (NEnt2), and $\mathrm{O}(NK)$ for DEMP.

Appendix B. Details of the Datasets in the Real Data Experiment

The datasets used in the real data experiment are summarized in Table A1. We show the detail and preprocessing of them below. All variables in the datasets are normalized after they are selected.

Table A1. Summary of the real dataset, where N denotes the number of points, d denotes the number of features, and $K^{\star}$ denotes the number of true clusters.

Dataset	Abbreviation	(N, d)	$K^{\star}$
AIS	AIS	(202, 3)	2
Flea beatles	BTL	(74, 2)	3
Crabs	CRB	(200, 5)	4
DLBCL	DLB	(7932, 3)	4
Ecoli	ECL	(327, 6)	5
Seeds	SDS	(210, 7)	3
Wisconsin breast cancer	WSC	(569, 3)	2
Yeast	YST	(626, 3)	2

The AIS dataset [36] consists of the physical measurements of athletes who trained at the Australian Institute of Sport. Two cluster labels are male and female. As did Lee and McLachlan [16] and Malsiner-Walli et al. [20], we use three variables: BMI, LBM, and body fat percentage (BFat).

The Flea beetles dataset [37] consists of two physical measurements (width and angle) of flea beetles. Three cluster labels are the different species, named Concinna, Heikertingeri, and Heptapotamica.

The Crabs dataset [38] describes five morphological measurements (frontal lobe size, rear width, carapace length, carapace width, and body depth) of 200 crabs. Four cluster labels are formed by combining two color forms and two sexes (male and female).

The DLBCL dataset [39] contains fluorescent intensities of multiple conjugated antibodies (markers) on the cells derived from the lymph nodes of patients diagnosed with DLBCL (diffuse large B-cell lymphoma). As did Lee and McLachlan [40] and Malsiner-Walli et al. [20], we consider four labels corresponding to the cell populations.

The Ecoli dataset [41,42] contains cellular localization sites of proteins. We consider five variables named mcg, gvh, aac, alm1, and alm2. Binary attributes are omitted here. For the labels, we consider five localization sites named cp, im, imU, om, and pp. The other localization sites are omitted because there are little data assigned to them.

The Yeast dataset [41,42] also describes cellular localization sites of proteins. As did Franczac et al. [43] and Malsiner-Walli et al. [20], we select three variables and two cluster labels from the dataset. For the variables, we consider three attributes of proteins, named mcg, alm, and vac. For the labels, we consider two localization sites named CYT and ME3.

The Seeds dataset [44] consists of the seven geometric parameters of grains: area, perimeter, compactness, length of kernel, width of the kernel, asymmetry coefficient, and length of kernel groove. Three cluster labels are kernels belonging to different varieties of wheat: Kama, Rosa, and Canadian.

The Wisconsin breast cancer dataset [3] describes characteristics of the cell nuclei in the images of breast masses. Two cluster labels are benign and malignant. As did Fraley and Raftery [2] and Malsiner-Walli et al. [20], we select three variables: extreme area, extreme smoothness, and mean texture.

References

1. McLachlan, G.J.; Peel, D. *Finite Mixture Models*; Wiley Series in Probability and Statistics: New York, NY, USA, 2000.
2. Fraley, C.; Raftery, A.E. How Many Clusters? Which Clustering Method? Answers via Model-Based Cluster Analysis. *Comput. J.* **1998**, *41*, 578–588. [CrossRef]
3. Mangasarian, O.L.; Street, W.N.; Wolberg, W.H. Breast Cancer Diagnosis and Prognosis via Linear Programming. *Operat. Res.* **1995**, *43*, 570–577. [CrossRef]
4. Hennig, C. Methods for Merging Gaussian Mixture Components. *Adv. Data Anal. Class.* **2010**, *4*, 3–34. [CrossRef]

5. Baudry, J.P.; Raftery, A.E.; Celeux, G.; Lo, K.; Gottardo, R. Combining Mixture Components for Clustering. *J. Comput. Graph. Stat.* **2010**, *19*, 332–353. [CrossRef]
6. Melnykov, V. Merging Mixtute Components for Clustering Through Pairwise Overlap. *J. Comput. Graph. Stat.* **2016**, *25*, 66–90. [CrossRef]
7. Kyoya, S.; Yamanishi, K. Mixture Complexity and Its Application to Gradual Clustering Change Detection. *arXiv* **2020**, arXiv:2007.07467.
8. Biernacki, C.; Celeux, G.; Govaert, G. Assessing a Mixture Model for Clustering With the Integrated Completed Likelihood. *IEEE Trans. Patt. Anal. Mach. Intell.* **2000**, *22*, 719–725. [CrossRef]
9. Hirai, S.; Yamanishi, K. Efficient Computation of Normalized Maximum Likelihood Codes for Gaussian Mixture Models With Its Applications to Clustering. *IEEE Trans. Inform. Theory* **2013**, *59*, 7718–7727. [CrossRef]
10. Hirai, S.; Yamanishi, K. Correction to Efficient Cotampution of Normalized Maximum Likelihood Codes for Gaussian Mixture Models With Its Applications to Clustering. *IEEE Trans. Inform. Theory* **2019**, *65*, 6827–6828. [CrossRef]
11. Wu, T.; Sugawara, S.; Yamanishi, K. Decomposed Normalized Maximum Likelihood Codelength Criterion for Selecting Hierarchical Latent Variable Models. In Proceedings of the 23rd ACM SIGKDD International Conference on Knowledge Discovery and Data Mining, Halifax, NS, Canada, 13–17 August 2017; Association for Computing Machinery: New York, NY, USA, 2017.
12. Yamanishi, K.; Wu, T.; Sugawara, S.; Okada, M. The Decomposed Normalized Maximum Likelihood Code-Length Criterion for Selecting Hierarchical Latent Variable Models. *Data Mining Know. Discov.* **2019**, *33*, 1017–1058. [CrossRef]
13. Banerjee, A.; Krumpelman, C.; Basu, S.; Mooney, R.J.; Ghosh, J. Model-based Overlapping Clustering. In Proceedings of the 11th ACM SIGKDD International Conference on Knowledge Discovery and Data Mining, Chicago, IL, USA, 21–24 August 2005; Association for Computing Machinery: New York, NY, USA, 2005.
14. Fu, Q.; Banerjee, A. Multiplicative Mixture Models for Overlapping Clustering. In Proceedings of the 2008 Eighth IEEE International Conference on Data Mining, Pisa, Italy, 15–19 December 2008.
15. Xu, Y.; Yang, Y.; Wang, H.; Hu, J. An Overlapping Clustering Approach with Correlation Weight. In Proceedings of the International Joint Conference on Rough Sets, Olsztyn, Poland, 3–7 July 2017; Springer: Berlin/Heidelberg, Germany, 2017.
16. Lee, S.X.; McLachlan, G.J. Model-Based Clustering and Classification With Non-Normal Mixture Distributions. *Stat. Method Appl.* **2013**, *22*, 427–454. [CrossRef]
17. Li, J. Clustering Based on a Multi-layer Mixture Model. *J. Comput. Graph. Stat.* **2004**, *14*, 547–568. [CrossRef]
18. Di Zio, M.; Guarnera, U.; Rocci, R. A Mixture of Mixture Models For a Classifitation Problem. *Comput. Stat. Data Anal.* **2007**, *51*, 2573–2585. [CrossRef]
19. Yarebakan, H.Z.; Rajwa, B.; Dundar, M. The Infinite Mixture of Infinite Gaussian Mixtures. In *Advances in Neural Information Processing Systems*; Curran Associates Inc.: Montréal, QC, Canada, 2014; Volume 27.
20. Malsiner-Walli, G.; Frühwirth-Schnatter, S.; Grün, B. Identifying Mixtures of Mixtures Using Bayesian Estimation. *J. Comput. Graph. Stat.* **2017**, *26*, 285–295. [CrossRef]
21. Ueda, N.; Nakano, R.; Ghahramani, Z.; Hingon, G.E. SMEM Algorithm for Mixture Models. *Neur. Comput.* **2000**, *12*, 2109–2128. [CrossRef] [PubMed]
22. Minagawa, A.; Tagawa, N.; Tanaka, T. SMEM Algorithm Is Not Fully Compatible with Maximum-Likelihood Framework. *Neur. Comput.* **2002**, *14*, 1261–1266. [CrossRef]
23. Zhao, Q.; Hautamäki, V.; Kärkkäinen, I.; Fränti, P. Random Swap EM algorithm for Gaussian Mixture Models. *Pattern Recognit. Lett.* **2012**, *33*, 2120–2126. [CrossRef]
24. Heller, K.A.; Ghahramani, Z. Bayesian Hierarchical Clustering. In Proceedings of the 22nd International Conference on Machine Learning, Bonn, Germany, 7–11 August 2005; Association for Computing Machinery: New York, NY, USA, 2005.
25. Telgarsky, M.; Dasgupta, S. Agglomerative Bregman Clustering. In Proceedings of the 29th International Conference on Machine Learning, Edinburgh, UK, 26 June–1 July 2012; Association for Computing Machinery: New York, NY, USA, 2012.
26. Dhillon, I.S.; Modha, D.S.; Spangler, W.S. Class Visualization of High-Dimensional Data With Applications. *Comput. Stat. Data Anal.* **2002**, *41*, 59–90. [CrossRef]
27. Iwata, T.; Saito, K.; Ueda, N.; Stromsten, S.; Griffiths, T.L.; Tenenbaum, J.B. Parametric Embedding for Class Visualization. *Neural Comput.* **2007**, *19*, 2536–2556. [CrossRef]
28. Ward, J.H., Jr. Hierarchical Grouping to Optimize an Objective Function. *J. Am. Stat. Associat.* **1963**, *58*, 236–244. [CrossRef]
29. Sneath, P.H.A.; Sokal, R.R. *Numerical Taxonomy: The Principles and Practice of Numerical Classification*; W. H. Freeman: San Francisco, CA, USA, 1973.
30. Hubert, L.; Arabie, P. Comparing Partitions. *J. Class.* **1985**, *2*, 193–218. [CrossRef]
31. Bishop, C.M. *Pattern Recognition and Machine Learning*; Springer: New York, NY, YSA, 2006.
32. Pedregosa, F.; Varoquaux, G.; Gramfort, A.; Michel, V.; Thirion, B.; Grisel, O.; Blondel, M.; Prettenhofer, P.; Weiss, R.; Dubourg, V.; et al. Scikit-learn: Machine Learning in Python. *J. Mach. Learn. Res.* **2011**, *12*, 2825–2830.
33. Schwarz, G. Estimating the Dimension of a Model. *Annals Stat.* **1978**, *6*, 461–464. [CrossRef]
34. Puzicha, J.; Hofmann, T.; Buhman, J.M. A theory of proximity based clustering: structure detection by optimization. *Pattern Recognit.* **2000**, *33*, 617–634. [CrossRef]

35. Kleinberg, J. An Impossibility Theorem for Clustering. In *Advances in Neural Information Processing Systems 15*; Curran Associates Inc.: Vancouver, BC, Canada, 2002.
36. Cook, R.D.; Weisberg, S. *An Introduction to Regression Graphics*; Wiley-Interscience: New York, NY, USA, 1994.
37. Lubischew, A.A. On the Use of Discriminant Functions in Taxonomy. *Biometrics* **1962**, *18*, 455–477. [CrossRef]
38. Campbell, N.A.; Mahon, R.J. A Multivariate Study of Variation in Two Species of Rock Crab of The Genus Leptograpsus. *Austral. J. Zool.* **1974**, *22*, 417–425. [CrossRef]
39. Aghaeepour, N.; Finak, G.; Consortium, T.F.; Consortium, T.D.; Hoos, H.; Mosmann, T.R.; Brinkman, R.; Gottardo, R.; Scheuermann, R.H. Critical Assessment of Automated Flow Cytometry Data Analysis Techniques. *Nat. Meth.* **2013**, *10*, 228–243. [CrossRef] [PubMed]
40. Lee, S.X.; McLachlan, G.J. EMMIXuskew: An R Package for Fitting Mixtures of Multivariate Skew t Distributions via the EM Algorithm. *J. Stat. Softw.* **2013**, *55*, 1–22. [CrossRef]
41. Nakai, K.; Kanehisa, M. Expert System for Predicting Protein Localization Sites in Gram-Negative Bacteria. *Protein. Struct. Funct. Genet.* **1991**, *11*, 95–110. [CrossRef]
42. Nakai, K.; Kanehisa, M. A Knowledge Base for Predicting Protein Localization Sites in Eukaryotic Cells. *Genomics* **1992**, *14*, 897–911. [CrossRef]
43. Franczak, B.C.; Browne, R.P.; and, P.D.M. Mixtures of Shifted Asymmetric Laplace Distributions. *IEEE Trans. Patt. Anal. Mach. Intell.* **2014**, *36*, 1149–1157. [CrossRef] [PubMed]
44. Charytanowicz, M.; Niewczas, J.; Kulczycki, P.; Kowalski, P.A.; Łukasik, S.; Żak, S. Complete Gradient Clustering Algorithm for Features Analysis of X-Ray Images. *Informat. Technol. Biomed.* **2010**, *69*, 15–24.

entropy

MDPI

Article

Information-Corrected Estimation: A Generalization Error Reducing Parameter Estimation Method

Matthew Dixon [1,†] and Tyler Ward [2,*,†]

1 Department of Applied Mathematics, Illinois Institute of Technology, Chicago, IL 60616, USA; matthew.dixon@iit.edu
2 Department of Financial Engineering, NYU Tandon School of Engineering, New York, NY 11201, USA
* Correspondence: tw623@nyu.edu
† These authors contributed equally to this work.

Abstract: Modern computational models in supervised machine learning are often highly parameterized universal approximators. As such, the value of the parameters is unimportant, and only the out of sample performance is considered. On the other hand much of the literature on model estimation assumes that the parameters themselves have intrinsic value, and thus is concerned with bias and variance of parameter estimates, which may not have any simple relationship to out of sample model performance. Therefore, within supervised machine learning, heavy use is made of ridge regression (i.e., L2 regularization), which requires the the estimation of hyperparameters and can be rendered ineffective by certain model parameterizations. We introduce an objective function which we refer to as Information-Corrected Estimation (ICE) that reduces KL divergence based generalization error for supervised machine learning. ICE attempts to directly maximize a corrected likelihood function as an estimator of the KL divergence. Such an approach is proven, theoretically, to be effective for a wide class of models, with only mild regularity restrictions. Under finite sample sizes, this corrected estimation procedure is shown experimentally to lead to significant reduction in generalization error compared to maximum likelihood estimation and L2 regularization.

Keywords: generalization error; overfitting; information criteria; entropy

Citation: Dixon, M.; Ward, T. Information-Corrected Estimation: A Generalization Error Reducing Parameter Estimation Method. *Entropy* **2021**, *23*, 1419. https://doi.org/10.3390/e23111419

Academic Editors: Lizhong Zheng and Chao Tian

Received: 2 October 2021
Accepted: 25 October 2021
Published: 28 October 2021

1. Introduction

Kullback and Leibler [1] showed that minimizing a divergence $\rho_{KL}(f, g_\theta)$ between the truth, f, and a parametric model density, g_θ, is necessary and sufficient for making accurate predictions about data using the model defined by $\boldsymbol{\theta}$. Recent work [2] on Berk–Nash equilibria has shown the central role that KL divergence plays in game theoretic choice models such as multi-armed bandits and stochastic multi-party games. KL divergence thus plays a leading role in machine learning and neuroscience, with several inferential approaches developed in the information theory literature. Such approaches for minimizing KL divergence employ a range of methods, including data partitioning, Bayesian indirect inference and M-estimation [3–5]. These approaches are quite distinct from the standard penalized loss minimization framework and, as such, are non-trivial to combine with supervised learning methods such as neural networks.

It is well known that maximum likelihood estimation (MLE) introduces an asymptotic bias in the KL divergence minimizer which is problematic for both model estimation and model selection. For many models, where the parameters $\boldsymbol{\theta}$ are themselves important, this may be investigated as parameter bias and parameter variance. However, for models common in modern machine learning, the parameters themselves do not have any easily interpreted meaning. For these models, the parameters themselves are irrelevant and only the accuracy (in terms of KL divergence) of the model predictions matter. Within the information theory literature, this has often been referred to simply as bias (e.g., $b(G)$ from [6]). To distinguish it from parameter bias, one might refer to it as "prediction bias" or "generalization error". Generalization error is the more common terminology (see, for example, Equation 1.1.6 [7]) and will be used here.

Before the widespread use of machine learning, most models had interpretable parameters, and thus there is a large literature focused on reducing parameter bias. For instance, the jackknife [8] (leave-one-out cross-validation) estimator is an early example. More relevant to this paper is the approach of Firth [9] and later Kosmidis [10,11]. More recently, Pagui, Salvan, and Sartori [12] proposed a parameter bias reducing estimation methodology. An extensive review of the literature around this point can be found in [13]. Unfortunately, these approaches do not consider the impact on KL divergence-based generalization error and thus are not applicable to the field of machine learning where the parameters themselves are devoid of meaning. Heskes [14] shows that classifiers do have a notion of bias-variance decomposition for generalization error, but it is not computable from parameter bias and parameter variance. Therefore, parameter bias reducing formulations are not useful within machine learning unless it can be shown that they also reduce generalization error.

In fact, to seat the approach taken in this paper to generalization error, we recall much earlier and seminal work at the intersection of statistics and information theory. Akaike [15], and later Takeuchi [16], proposed information criteria (AIC and TIC, respectively) for model selection designed explicitly to reduce generalization error. Konishi and Kitagawa [6] extended the approach of Takeuchi to cases where MLE was not used to fit the underlying model, but still restricted themselves to the question of model selection. Stone [17] proved that Akaike's Information Criterion (AIC) is asymptotically equivalent to jackknifing when the estimator is finite. Takeuchi himself showed that TIC is an extension of AIC with fewer restrictions, and thus it too is equivalent to jackknifing whenever AIC would be valid.

For highly parameterized models, as are common in machine learning, model selection such as this is of limited utility. The parameter count may necessarily be very large, and thus none of the models fit using MLE may be acceptable. Then, merely choosing among them is unlikely to produce acceptable results. Within this field, typically L_2 or similar regularization is used to reduce generalization error. See Section 11.5.2 [18], for a typical example. For a more recent innovation, refer to [19]. Note that regularization schemes such as this often increase parameter bias while decreasing generalization error. Golub, Heath, and Wahba [20] showed that L_2 regularization is asymptotically equivalent to cross-validation for linear models, subject to certain assumptions. For nonlinear models, it has long been known that L_2 regularization is not always valid, and it is trivial to construct example models (See Section 4.1 for one such example) where this approach is always harmful in expectation.

Therefore, it is important to develop a method to reduce generalization error in model estimation analogous to the way that L_2 regularization would commonly be used for a highly parameterized model, but having applicability for a wider family of models, especially those for which L_2 regularization is not applicable. It is not the goal of this paper to perform a wide survey of generalization error reducing approaches, but we will rather propose an additional approach, investigate its properties, and show that it has superior performance when compared against L_2 regularization, which is currently the dominant generalization error reducing estimation procedure within the field of machine learning.

To this end, this paper introduces a generalization error reducing estimation approach referred to as Information Corrected Estimation (ICE). This estimator is proven to have a generalization error of only $O(n^{-\frac{3}{2}})$ instead of $O(n^{-1})$ as is the case for MLE, and is shown to be valid within a neighborhood around the MLE parameter estimate. Optimizing over this ICE objective function instead of the negative log likelihood thus produces parameters with superior out of sample performance.

Takeuchi's TIC and Firth's approach have never seen widespread use due to the computational and numerical issues that arise from the computation of this adjustment [21], and the ICE estimator in its raw form would have similar problems. Therefore, this paper also proposes an efficient approximation of this correction term, and shows through numerical experiments that the approximation is effective at improving model performance across a range of models.

2. Preliminaries

Let us assume that we have data $\boldsymbol{x}_n := \{x_1, \ldots, x_n\}$ generated from an unknown joint density function $f(\boldsymbol{x})$ of $X_n := \{X_1, \ldots, X_n\}$. Where necessary, we define Z_n to denote a second sample drawn from $f(\boldsymbol{x})$, independent of X_n, and $\boldsymbol{x}'_n$ is the observed realization of Z_n. We consider a model $\mathcal{M}_p$ given by a parametric family of densities $\mathcal{M}_p := \{g(\cdot|\boldsymbol{\theta}) \mid \boldsymbol{\theta} \in \boldsymbol{\Theta} \subseteq \mathbb{R}^p\}$, for some compact Euclidean parameter space $\boldsymbol{\Theta}$, which is misspecified and hence excludes the truth f. Henceforth, the distribution over x identified by $\boldsymbol{\theta}$ may be referred to as $g_{\boldsymbol{\theta}}(x) := g(x|\boldsymbol{\theta})$ where it is notationally convenient to do so.

Suppose that $\boldsymbol{\theta}_0$ is the quasi-true parameter of model $\mathcal{M}$, and $\hat{\boldsymbol{\theta}}(X_n)$ is the random variable representing the MLE of $\boldsymbol{\theta}_0$ fit on a dataset, $\boldsymbol{x}_n$. The negative log-likelihood of X_n under the distribution $g_{\boldsymbol{\theta}}$ is

$$-\ell(\boldsymbol{\theta}, X_n) := -\frac{1}{n}\sum_{i=1}^{n} \log g_{\boldsymbol{\theta}}(\boldsymbol{x}_i), \tag{1}$$

where $-\ell(\boldsymbol{\theta}, X_n)$ is written including a $\frac{1}{n}$ to make the expectation of this quantity $O(1)$ and asymptotically independent of n. Similarly, the minus sign is incorporated because $-\ell(\boldsymbol{\theta}, X_n)$ is a strictly non-negative quantity if $g_{\boldsymbol{\theta}}(\boldsymbol{x}_i)$ is a probability. The MLE, $\hat{\boldsymbol{\theta}}(X_n)$, minimizes the negative log likelihood of the data set with respect to the model:

$$\hat{\boldsymbol{\theta}}(\boldsymbol{x}_n) := \underset{\theta}{\operatorname{argmin}}[-\ell(\boldsymbol{\theta}, \boldsymbol{x}_n)]. \tag{2}$$

The expectation of $-\ell(\boldsymbol{\theta}, X_n)$ is the cross entropy between f and $g_{\boldsymbol{\theta}}$:

$$-\mathcal{L}(\boldsymbol{\theta}) := \mathbb{E}_{X_n}[-\ell(\boldsymbol{\theta}, X_n)]. \tag{3}$$

Here, the expectation is a function only of $\boldsymbol{\theta}$ and of the distribution f that generated the data X_n. As a function of the distribution f, this value is $O(1)$, but could be large for poorly conditioned f. The quasi-true parameter $\boldsymbol{\theta}_0$ is

$$\boldsymbol{\theta}_0 := \underset{\theta}{\operatorname{argmin}}[-\mathcal{L}(\boldsymbol{\theta})]. \tag{4}$$

Generalization Error in KL Divergence Based Loss Functions

Kullback and Leibler [1] viewed "information" as discriminating the sample data drawn from one distribution against another, and defined the KL-divergence ρ_{KL} between distributions in terms of the ability to make predictions about one by knowing the other. Here,

$$\rho_{KL}(f, g_{\boldsymbol{\theta}}) = \int \log[\frac{f(x)}{g_{\boldsymbol{\theta}}(x)}]f(x)dx. \tag{5}$$

This value is in general unknowable, but given a sample X_n from f, $-\ell(\boldsymbol{\theta}, X_n)$ will converge asymptotically to $\rho_{KL}(f, g_{\boldsymbol{\theta}})$ plus an additive constant that depends only on f. The convergence relies on White's regularity conditions [22].

A well known result by Stone [17] shows that the MLE is a biased estimator of the minimum KL-divergence:

$$\mathbb{E}_{X_n}[-\ell(\hat{\boldsymbol{\theta}}(X_n), X_n)] < \mathbb{E}_{X_n}[-\ell(\boldsymbol{\theta}_0, X_n)], \tag{6}$$

because it is evaluated on the data X_n which was used to fit $\hat{\boldsymbol{\theta}}$. Cross-validation was developed as a model selection technique to select a model from a group that actually minimizes $\mathbb{E}_{X_n}[\rho_{KL}(g_{\boldsymbol{\theta}_0}, g_{\hat{\boldsymbol{\theta}}(X_n)})]$ and not merely $\mathbb{E}_{X_n}[-\ell(\hat{\boldsymbol{\theta}}(X_n), X_n)]$ in the limit of large n. Takeuchi [16] and Akaike [15] explicitly modeled this bias (generalization error) of an estimation procedure $\boldsymbol{\theta}(X_n)$ as

$$b := \mathbb{E}_{X_n}\left[\ell(\boldsymbol{\theta}(X_n), X_n) - \mathbb{E}_{X_n'}[\ell(\boldsymbol{\theta}(X_n), X_n')]\right]. \tag{7}$$

Our goal is to obtain an estimate, b^*, of the generalization error b without using the MLE. We will then add this term to the objective function to develop the estimator $\boldsymbol{\theta}^*(X_n)$ so as to cancel the lower order terms of the generalization error. This estimator will then minimize $\mathbb{E}_{X_n}[\rho_{KL}(g_{\boldsymbol{\theta}_0}, g_{\boldsymbol{\theta}^*(X_n)})]$ more effectively than MLE, and potentially would in turn produce improved predictions from the model fitted over finite training sets.

Remark 1. *We note that under MLE, $b = O(\frac{1}{n})$ [16]. Equivalently, one could say that a particular realization of the generalization error $\ell(\boldsymbol{\theta}(X_n), X_n) - \mathbb{E}_{X_n'}[\ell(\boldsymbol{\theta}(X_n), X_n')]$ is itself $O_p(\frac{1}{n})$. Here, $O_p(\frac{1}{n})$ is used to indicate that the quantity is a random variable with finite variance, whose mean is $O(\frac{1}{n})$.*

3. Information Corrected Estimation (ICE)

We propose the following penalized likelihood function:

Definition 1 (ICE Objective).

$$-\ell^*(\boldsymbol{\theta}) = -\ell(\boldsymbol{\theta}) + \frac{1}{n}tr(I_{\boldsymbol{\theta}} J_{\boldsymbol{\theta}}^{-1}), \tag{8}$$

where $J_{\boldsymbol{\theta}}$ is the negative expected Hessian

$$J_{\boldsymbol{\theta}} := -\mathbb{E}_X[\partial^2_{\boldsymbol{\theta}} \log g(X|\boldsymbol{\theta})] = -\int f(x)\partial^2_{\boldsymbol{\theta}} \log g(x|\boldsymbol{\theta})dx, \tag{9}$$

and $I_{\boldsymbol{\theta}}$ is the Fisher Information matrix

$$I_{\boldsymbol{\theta}} := \mathbb{E}_X[\partial_{\boldsymbol{\theta}} \log g(X|\boldsymbol{\theta})\partial_{\boldsymbol{\theta}^T} \log g(X|\boldsymbol{\theta})]. \tag{10}$$

with $\hat{I}_{\boldsymbol{\theta}}$, $\hat{J}_{\boldsymbol{\theta}}$ being their estimates over the data.
Let $\boldsymbol{\theta}^*$ denote the minimizer of (8).

The trace term in Equation (8) will be familiar from Takeuchi [16]. However, Takeuchi showed only that this was the leading order of the bias for the MLE estimate $\hat{\boldsymbol{\theta}}$, and therefore the proof found there is not sufficient to justify a new estimator that will itself be the target of optimization, and is required to be valid away from $\hat{\boldsymbol{\theta}}$. As in Takeuchi, because I and J are unknowable, we will substitute their approximations computed from the training data, $\hat{I}_{\boldsymbol{\theta}}$ and $\hat{J}_{\boldsymbol{\theta}}$ during the actual computation of this objective. The numerical impact of this approximation will be examined in Section 4.2.1.

Remark 2. *Though AIC was developed before TIC, it is easily reproduced as a special case of TIC. Subject to certain conditions (guaranteed by the requirements of [15]), at least in expectation, $I_{\hat{\boldsymbol{\theta}}} = J_{\hat{\boldsymbol{\theta}}}$. Thus, the quantity within the TIC trace term, $I_{\hat{\boldsymbol{\theta}}} J_{\hat{\boldsymbol{\theta}}}^{-1}$, is the identity matrix. Therefore, its trace is equal to p, the parameter count of the model, recovering AIC. TIC itself can be derived using a proof that is similar to, though somewhat simpler than, the one we include in (A2), of which Takeuchi's proof is a special case that is valid only at the MLE estimate $\hat{\boldsymbol{\theta}}$.*

We also define $\hat{J}^*$ to be the negative hessian of $-\ell^*(\boldsymbol{\theta})$ rather than $-\ell(\boldsymbol{\theta})$, and similarly for $\hat{I}^*$, with expectations written as J^* and I^*. Analogously, $-\mathcal{L}^*(\boldsymbol{\theta})$ is the expectation of $-\ell^*(\boldsymbol{\theta})$ and $\boldsymbol{\theta}^*$ is the minimizer of $-\ell^*(\boldsymbol{\theta})$, while $\boldsymbol{\theta}_0^*$ is the minimizer of $-\mathcal{L}^*(\boldsymbol{\theta})$.

We refer to the estimation of $\boldsymbol{\theta}^*$, by minimization of this corrected likelihood function as Information-Corrected Estimation (ICE). As the terminology suggests, we depart from the corrective approach used in Information Criterion, by directly minimizing the bias corrected likelihood function. Note that unlike L_2 regularization, the correction term is parameter-free and thus would not require cross validation to estimate a hyperparameter such as the λ used by L_2.

General properties of this estimator are proved, and a set of regularity conditions are provided such that the estimator is asymptotically normal, and produces a bias that is $O_p(n^{-3/2})$ instead of the usual $O_p(n^{-1})$. Though this adds only a half-order to the bias correction, for most problems with reasonably large n, any increase in order is likely to greatly reduce bias. Experimental results demonstrate superior properties of ICE for linear models compared to MLE with and without L_2 regularization.

Remark 3. *For models satisfying White's regularity conditions (See [22]), it is known that J_{θ_0} is positive definite (thus non-singular) and continuous, and also that I_{θ_0} is continuous with respect to $\boldsymbol{\theta}$. Therefore, $\frac{1}{n}tr(I_{\theta_0}J_{\theta_0}^{-1})$ would always be well defined in an open region around $\boldsymbol{\theta}_0$. Similarly, the solution $\boldsymbol{\theta}^*$ would be expected to have the same properties, and hence (for large enough n) the estimate $\frac{1}{n}tr(\hat{I}_{\theta^*}\hat{J}_{\theta^*}^{-1})$ would be well defined when computed using the estimates $\hat{I}_{\theta^*}$ and $\hat{J}_{\theta^*}$.*

Remark 4. *N.B: Though $-\ell^*(\boldsymbol{\theta})$ is an estimator of $\mathcal{L}(\boldsymbol{\theta})$ accurate to within $O(n^{-\frac{3}{2}})$, that does not mean that $\mathcal{L}(\boldsymbol{\theta}^*)$ is reduced by any particular amount relative to $\mathcal{L}(\hat{\boldsymbol{\theta}})$. We expect that using this corrected objective will always (if it can be calculated accurately) generate some improvement by virtue of more accurately representing the true performance of the model out of sample, but there is no proof that this level of improvement has any particular form or asymptotic behavior.*

Our approach preserves the linear complexity of training with respect to n. However, the computation of $\hat{J}_{\theta^*}^{-1}$ at each iteration of the numerical solver requires the inversion of a symmetric positive definite matrix with a complexity of $O(p^3)$. Hence the approach is not suitable for high dimensional datasets without adjustment. See Section 5 for optimized approximations that are viable for larger parameter counts. Further exploration of large models based on this approach are beyond the scope of the present work.

Remark 5. *It is clear from inspection that if $-\ell(\boldsymbol{\theta})$ is strictly convex, then so too is $-\ell^*(\boldsymbol{\theta})$ for large enough n.*

We first provide a proof of asymptotic convergence of $\boldsymbol{\theta}^*$ under certain regularity conditions. With this convergence result in place, we then show that minimizing (8) leads to an $O(n^{-3/2})$ bias term, an improvement over the $O(\frac{1}{n})$ term produced by MLE.

Local Behavior of the ICE Objective

Suppose the following conditions hold:

1. $\mathcal{M}$ satisfies White's regularity conditions A1–A6 (see Appendix A.1 or [22]).
2. $\boldsymbol{\theta}_0$ is a global minimum of $-\mathcal{L}(\boldsymbol{\theta})$ in the compact space Θ defined in A2.
3. There exists a $\varepsilon > 0$ such that $-\mathcal{L}(\boldsymbol{\theta}_0) < -\mathcal{L}(\boldsymbol{\theta}_1) - \varepsilon$ for all other local minima $\boldsymbol{\theta}_1$.
4. For $k = 0,1,2,3,4,5$ the derivative $\partial_{\theta}^k\mathcal{L}(\boldsymbol{\theta})$ exists, is continuous, and bounded on an open set around $\boldsymbol{\theta}_0$.
5. For $k = 0,1,2,3,4,5$, the variance $\mathbb{V}[\partial_{\theta}^k\ell(\boldsymbol{\theta}, X_n)] \to 0$ as $n \to \infty$ on an open set around $\boldsymbol{\theta}_0$.

Then for sufficiently large n there exists a compact subset $U \subset \Theta$ containing $\boldsymbol{\theta}_0, \hat{\boldsymbol{\theta}}$, such that:

1. For $k = 0,1,2,3$ the derivative $\partial_{\theta}^k\ell^*(\boldsymbol{\theta}, x_n)$ exists, is continuous, and bounded on U, almost surely.
2. For $k = 0,1,2,3$, $\mathbb{V}[\partial_{\theta}^k\ell^*(\boldsymbol{\theta}, X_n)] \to 0$ as $n \to \infty$ on U, almost surely.
3. $\boldsymbol{\theta}^* \in U$ as $n \to \infty$ almost surely.
4. $\sqrt{n}(\boldsymbol{\theta}^* - \boldsymbol{\theta}_0^*) \to N(0, (\hat{J}_{\theta_0^*}^*)^{-1}\hat{I}_{\theta_0^*}^*(\hat{J}_{\theta_0^*}^*)^{-1})$ almost surely.
5. $-\mathcal{L}(\hat{\boldsymbol{\theta}}^*) = -\ell^*(\boldsymbol{\theta}^*(X_n), X_n) + O_p(n^{-3/2})$ almost surely.

Items (1–3) follow from Lemma A1 (see Appendix A.2). These are additional regularity conditions that are prerequisites for later theorems.

Item (4) follows from Theorem A1 in Appendix A.3. This states that the estimate $\boldsymbol{\theta}^*$ is asymtotically normal in a way that is analogous to classical asymptotic normality results for MLE. It is only true almost surely because results (1–3) upon which it relies are only true almost surely.

Item (5) follows from Theorem A2 in Appendix A.4. This item establishes the superior accuracy of the ICE objective compared to the MLE objective function in predicting out of sample errors. Like item (4) this is only true almost surely because intermediate results on which it relies are only true almost surely.

The reduction in generalization error seen arises from the optimization over the superior ICE objective function, analogous to the way that L_2 regularization is used for this purpose.

Remark 6. *The regularity conditions described here are only slightly more strict than the conditions described by White [22]. In particular, models having three continuous derivatives as required by White, but not 5 as needed here are thought to be very rare. Requirement (2) is just the definition of $\boldsymbol{\theta}_0$, which White labels differently, and requirement (3) excludes a pathological corner case, the further study of which is beyond the scope of this paper.*

Remark 7. *Note that as $-\ell(\boldsymbol{\theta}, x_n)$ is convex in the neighborhood of $\boldsymbol{\theta}_0$, so too is $-\ell^*(\boldsymbol{\theta})$ for large enough n because $-\ell^*(\boldsymbol{\theta}) \to -\ell(\boldsymbol{\theta})$. Thus it can be concluded that the local behavior of $-\ell^*$ in the neighborhood of $\boldsymbol{\theta}_0$ is not appreciably worse than the behavior of $-\ell$ if the problem is not too ill conditioned.*

4. Direct Computation Results

The following experiments have been designed to compare MLE, MLE with L_2 regularization, and ICE for regression. Each experiment involves simulation of training and test sets and is implemented in R. See the attached code to run each experiment.

Each of these experiments has been performed using the raw formula for ICE provided in Equation (8) with minimal adjustments. All gradients are computed using R's default finite difference approach. This means that for a model with p parameters, the objective function is dominated by the inversion of J, which costs $O(p^3)$ time and $O(p^2)$ space. The use of finite difference gradients further increases the time complexity to $O(p^4)$, compounding the problem. This approach is therefore viable for small models with few parameters, but not realistic for larger models. Optimizations to overcome this limitation will be considered in upcoming Section 5. The use of finite difference derivatives was not found to produce appreciable numerical differences in the final output, so analytic derivatives were not used for this analysis.

The code and results for this section is provided in [23]. Throughout this section, the following estimators will be compared.

MLE	$\hat{\boldsymbol{\theta}}(Z_n)$	$:= \operatorname{argmin}_{\boldsymbol{\theta}}[-\ell(\boldsymbol{\theta})]$
L_2 regularization	$\boldsymbol{\theta}^*_{L_2}(Z_n)$	$:= \operatorname{argmin}_{\boldsymbol{\theta},\lambda}[-\ell(\boldsymbol{\theta}) + \lambda\|\boldsymbol{\theta}\|_2^2]$
ICE	$\boldsymbol{\theta}^*_{ICE}(Z_n)$	$:= \operatorname{argmin}_{\boldsymbol{\theta}}[-\ell(\boldsymbol{\theta}) + \frac{1}{n}tr(\hat{I}_{\boldsymbol{\theta}}\hat{J}_{\boldsymbol{\theta}}^{-1})]$

4.1. Gaussian Error Model

We begin by considering the simplest case of univariate linear regression with Gaussian residuals. The advantage of this simple model is that the exact form of the correction term can be derived analytically and aids therefore in building intuition on its behavior. For such a toy model, $y \sim N(\mu, \sigma^2)$ and, for simplicity, the following example will consider μ to be a constant, but it is equally applicable if $\mu = \mu(x)$. Consider the parameters of the model to therefore be $\boldsymbol{\theta} := (\mu, \sigma)$ with their optimal values being $\boldsymbol{\theta}_0 := (\mu_0, \sigma_0)$. The the probability density function is

$$g(y, \boldsymbol{\theta}) = \frac{1}{\sqrt{2\pi}\sigma} e^{-\frac{(y-\mu)^2}{2\sigma^2}}. \quad (11)$$

It is known a priori that L_2 regularization cannot improve this model, as if $\mu_0 \neq 0$, any decrease in the magnitude of μ is likely to be systematically harmful. Similarly, a decrease in σ below $\hat{\sigma}$ results in a decrease in model distribution entropy, and hence would be generally making overfitting worse, and would generate a correspondingly higher KL-divergence than the MLE estimate. Consequently, we would expect any λ computed through cross-validation to be statistically indistinguishable from zero, and L_2 regularization to be generally harmful whenever $\lambda \neq 0$.

Generalization Error Analysis

The Gaussian model described was generated with $\mu_0 = 0.2$, $\sigma_0 = 0.2$, and $dy = 0.001$. For each of $n \in \{16, 32, 64, 128, 256, 512, 1024\}$, 500 independent simulations of the data $y_1, \ldots, y_n$ were performed, and then the parameters were fit from that data. In each simulation, $\boldsymbol{\theta}$ was computed using MLE, MLE with L_2 regularization, and ICE. The λ parameter for L_2 regularization was computed using 2-way cross-validation on the available data, and as expected, none of the computed values of λ were statistically different from zero.

For each estimate of $\boldsymbol{\theta}$, the KL-divergence $\rho_{KL}(f, g_{\boldsymbol{\theta}})$ was computed (using the known value of $\boldsymbol{\theta}_0$), and the results were compared. The ICE parameter estimation method showed statistically significant improvement over MLE at the 5-sigma level out to $n = 64$, and was improved by just under 1-sigma at $n = 1024$.

The KL-divergence results graphed against n on a log-log scale are shown in Figure 1. Every value of n is normalized by the average KL divergence of the MLE methodology to improve legibility. The L_2 series is statistically indifferent from the MLE series at 2 standard deviations beyond $n = 32$, and the two are not materially different for any n. The ICE series is at least 4.5 standard deviations below the MLE series until $n = 1024$.

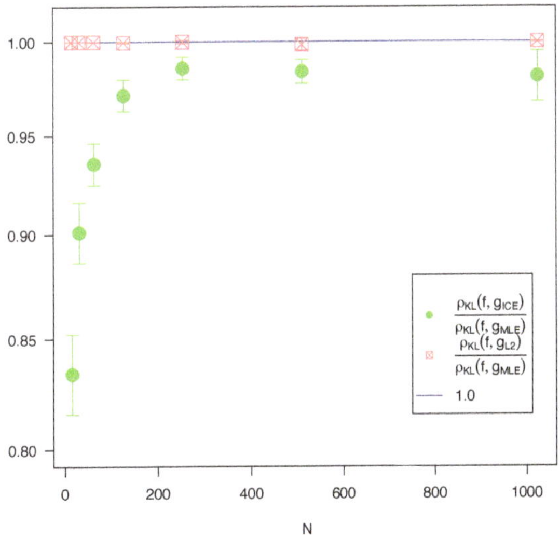

Figure 1. A comparison of the KL-divergence (y-axis) of various estimation methods against the number of training samples n. Each KL divergence value was divided by the average KL divergence of the MLE estimate for that value of n. The ICE and L_2 series are shown with 2 standard deviation error bars.

Remark 8. *In addition to the series shown in Figure 1, a series was computed using the true value of J, estimated from a much larger sample n = 1024 from the underlying distribution, and this series was indistinguishable from the series computed using $\hat{J}$ for every n, thus it was not graphed. This validates Takeuchi's approach of approximating J with $\hat{J}$ in this instance.*

As expected, the difference in μ between ICE and MLE is not statistically significant (at three standard deviations) for any n, but the ICE computed value of σ (shown in Figure 2) is considerably larger than the MLE estimate, especially for small values of n. This explains the greatly reduced KL-divergence noted in Figure 1.

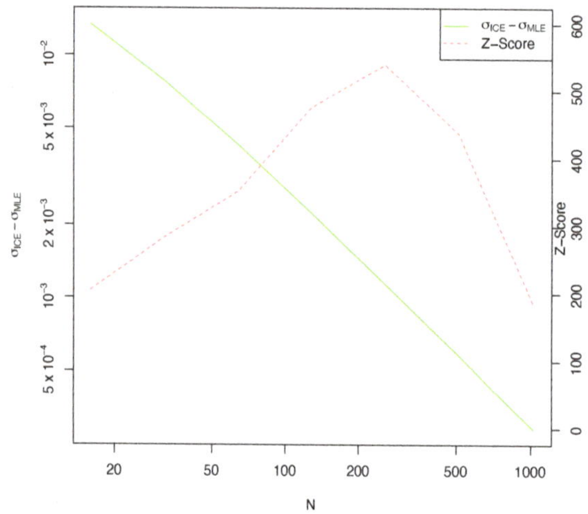

Figure 2. The error in the estimated $\hat{\sigma}_{ICE}$ and the Z-score of the estimate against the number of training samples n.

Note that the difference in estimated σ is always statistically significant when compared to the MLE value. This is because both MLE and ICE are fit on the same data, so ICE would always have a larger σ than MLE regardless of the actual data chosen from the distribution f. This is the cause of the large z-scores shown, always exceeding 200. We know from elementary statistics that correlation between the mean and std. deviation causes the MLE estimate of $\hat{\sigma}$ to be systematically low by a factor of $\frac{n-1}{n}$. Indeed, the ICE estimate of σ^* is closely tracking σ_0 whereas $\hat{\sigma}$ is closely tracking $\frac{\sigma_0(n-1)}{n}$ as expected. This is one example where reducing generalization error also reduces parameter bias as a side effect.

4.2. Friedman's Test Case

We now extend the example from Section 4.1 to the case where μ is no longer constants. For this example, we chose a standard regression test set, which is nonlinear in the features, based on Section 4.3 of [24]:

$$y_i = \mu_{\boldsymbol{\theta}}(x_i) + \varepsilon_i,\ \varepsilon \sim N(0, \sigma^2), \tag{12}$$

where the Friedman model is

$$\mu_{\boldsymbol{\theta}}(x_i) = \theta_0 sin(\pi x_{(i,0)} x_{(i,1)}) + \theta_1 (x_{(i,2)} - \theta_2)^2 + \theta_3 x_{(i,3)} + \theta_4 x_{(i,4)}. \tag{13}$$

The random features, X_j, are i.i.d. uniform random and the parameter values are fixed. The true parameter set, $\boldsymbol{\theta}_0 = (10.0, 20.0, 0.5, 10.0, 5.0, 1.0)$, reserves the last parameter (1.0) for the value of σ.

Note that here σ must be treated as an unknown parameter. To do otherwise implies that the modeler knows the amount of noise expected in the data. In the case of a known noise term, overfitting is impossible since overfitting arises when a model reduces the projected noise below its actual value, which can never arise when the noise level is known.

The model probability density $g(x, y|\boldsymbol{\theta})$ of y is given by

$$g(x_i, y_i|\boldsymbol{\theta}) = \frac{1}{\sqrt{2\pi\sigma^2}} e^{-\frac{(\mu_i - y_i)^2}{2\sigma^2}}. \tag{14}$$

Recall that in Section 4.1, the value of μ was considered to be a constant. This example is a natural extension of Section 4.1, and was chosen due to the well-explored difficulty of Friedman's problem.

We simulate 500 batches of equally sized training sets of length $n \in \{16, 32, 64, 128, 256, 512, 1024\}$. The test set is always of length 1024 to ensure accuracy for the smaller values of n. The starting point of the optimization is generated by adding a random perturbation, $\delta\boldsymbol{\theta} \sim N(0, 0.1)$, to each parameter. As before, the KL-divergence is computed between the distribution represented by the parameters and the true distribution, and these values are compared between estimation methods.

For each test sample, the KL divergence is computed using numerical integration with a dy increment of 0.01 over the interval containing $\mu \pm 10\sigma$ for both the true and model distributions. The computed probabilities are verified to numerically sum to unity within an error of $\pm 10^{-3}$.

In each simulation, $\boldsymbol{\theta}$ is computed using MLE, MLE with L_2 regularization, and ICE. The λ parameter for L_2 regularization is computed using 4-way cross validation on each batch of the training data.

As shown in Figure 3 and Table 1, L_2 is not effective for any value of n, and is is completely inactivated for $n > 32$. Where regularization is used (i.e., $\lambda \neq 0$), it generally underperforms MLE. ICE is effective across the entire data range, outperforming MLE for every n, and always by a statistically significant margin of at least 5 sigma.

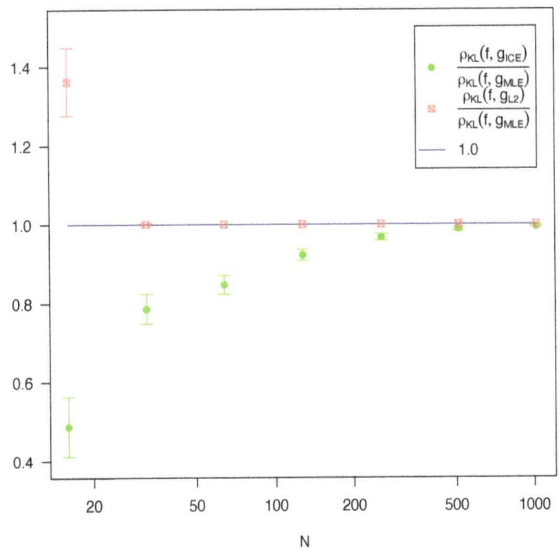

Figure 3. Comparison of the KL-divergence, averaged across 500 replications, of estimation methods against the number of training samples n. Each KL divergence value was divided by the average KL divergence of the MLE estimate for that value of n. The ICE and L_2 series are shown with 2 standard deviation error bars.

Table 1. Comparison of the average KL divergence across 500 replications for several model estimators given a fitting set size of n. For estimators other than $\hat{\theta}$, the values in parentheses denotes the t-statistic of the difference between this estimator and $\hat{\theta}$, with negative values indicating that the listed estimator has a lower KL divergence.

n	$\rho_{KL}(f, g_{\hat{\theta}})$	$\rho_{KL}(f, g_{\theta_{L2}})$	$\rho_{KL}(f, g_{\theta^*})$
16	6.19×10^{-1}	8.442×10^{-1} (8.40)	3.02×10^{-1} (−13.54)
32	1.74×10^{-1}	1.74×10^{-1} (0.16)	1.37×10^{-1} (−11.11)
64	6.85×10^{-2}	6.85×10^{-2} (0.0)	5.81×10^{-2} (−12.65)
128	3.82×10^{-2}	3.82×10^{-2} (0.0)	3.53×10^{-2} (−11.13)
256	2.19×10^{-2}	2.19×10^{-2} (0.0)	2.12×10^{-2} (−7.84)
512	1.53×10^{-2}	1.53×10^{-2} (0.0)	1.51×10^{-2} (−5.66)
1024	1.25×10^{-2}	1.25×10^{-2} (0.0)	1.24×10^{-2} (−5.03)

4.2.1. Impact of $\hat{J}$ Approximation

It was noted previously that Takeuchi used $\hat{J}$ (and likewise, $\hat{I}$) in place of the true value of J, and we do so here as well. Though there is no realistic way to avoid this approximation in the real world, and the optimized approach discussed in Section 5 has an entirely different set of approximations, the impact of this approximation will be briefly characterized here.

In Table 2, we revisit Table 1, but now drop the L_2 regualarization column, and add a new column where the ICE objective is allowed to use a much better approximated value of J, in this case approximated from 1024 independently drawn samples regardless of n.

Table 2. Comparison of the average KL divergence across 500 replications for ICE estimators with and without approximation of J given a fitting set size of n. For estimators other than $\hat{\theta}$, the values in parentheses denotes the t-statistic of the difference between this estimator and $\hat{\theta}$, with negative values indicating that the listed estimator has a lower KL divergence.

n	**MLE**	**ICE ($\hat{J}$)**	**ICE (J)**
16	6.19×10^{-1}	3.02×10^{-1} (−13.54)	6.21×10^{-1} (0.05)
32	1.74×10^{-1}	1.37×10^{-1} (−11.11)	1.51×10^{-1} (−3.74)
64	6.85×10^{-2}	5.81×10^{-2} (−12.65)	4.21×10^{-2} (−17.29)
128	3.82×10^{-2}	3.53×10^{-2} (−11.13)	2.99×10^{-2} (−11.82)
256	2.19×10^{-2}	2.12×10^{-2} (−7.84)	1.99×10^{-2} (−4.19)
512	1.53×10^{-2}	1.51×10^{-2} (−5.66)	1.48×10^{-2} (−1.62)
1024	1.25×10^{-2}	1.24×10^{-2} (−5.03)	1.24×10^{-2} (−0.72)

As can be seen from Table 2, using the true value of J is at most marginally helpful. In fact, for most values of n it displays slightly better average results, but slightly higher std. deviation of those results, and thus reduced T-statistics. Thus, we conclude that the Takeuchi's approximation, replacing J with $\hat{J}$ is reasonable. The same conclusion was reached in Section 4.1, see the remark there. We note also that the ICE estimator using $\hat{J}$ exhibits substantially better performance for very low sample sizes, but further investigation of this phenomenon is beyond the scope of the current paper.

In Table 3, we show the average matrix norms of J, $\hat{J}$, and also of the diagonal of $\hat{J}$, referred to as the matrix D. The matrix D will be examined further in Section 5, and is included here for completeness. We also show the norms of several matrix differences.

We note that the ICE objective values themselves exhibit much lower variation than the matrix norms show in Table 3. In particular though the matrix D is not actually converging to J as n increases, we see from the correction term it generates that this difference does not appear to have a material impact for larger n. We thus conclude that the major eigenvectors of $(\hat{J} - J)$ and $(D - J)$ are very nearly orthogonal to the gradient vectors used to construct $\hat{I}$ for large n.

Table 3. Mean matrix norms of J, its approximations, and differences from these approximations across 500 replications.

n	$\|\|J\|\|$	$\|\|\hat{J}\|\|$	$\|\|D\|\|$	$\|\|\hat{J}-J\|\|$	$\|\|D-J\|\|$	$\frac{1}{n}tr(IJ^{-1})$	$\frac{1}{n}tr(\hat{I}\hat{J}^{-1})$	$\frac{1}{n}tr(\hat{I}D^{-1})$
16	1423.29	925,489.26	1738.30	926,165.93	2905.64	0.1452	0.0027	6.9484
32	875.03	89,414.11	96.49	89,786.19	793.43	0.0521	0.0256	0.0291
64	214.05	4820.55	89.55	4646.51	125.02	0.0202	0.0160	0.0162
128	200.86	200.68	85.00	27.65	115.86	0.0097	0.0086	0.0086
256	194.36	191.81	82.70	19.12	111.67	0.0048	0.0045	0.0045
512	191.20	190.10	82.76	14.18	108.44	0.0023	0.0023	0.0023
1024	188.91	187.39	81.89	11.58	107.02	0.0012	0.0012	0.0012

It is not clear from examining the trace terms in Table 3 that D is a worse approximation of J than $\hat{J}$ is, even for small n where the impact of the ICE approach is most significant. A more complete investigation of the spectrum of these matrices is beyond the scope of the present work.

4.3. Multivariate Logistic Regression

The previous experiment is based on a well-known test case. In this second experiment, we assess the general performance of ICE under (i) varying dimensionality of the true data distribution, (ii) increasing misspecification, and (iii) increasing training set sizes. To achieve this goal, we generate a more exhaustive set of data from a more complex data generation process.

4.3.1. Data Generation Process

The synthetic data are designed to exhibit a number of characteristics needed to broadly evaluate the efficacy of ICE. First, the regressors should be sufficiently correlated so as to ensure that model selection is representative of typical datasets. However, we avoid multi-collinearity by ensuring the smallest eigenvalue is above a certain threshold. We additionally control the condition number of the covariance matrix Σ by randomly generating a symmetric positive definite covariance matrix $\Sigma \in \mathbb{R}^p$ using the eigen-decomposition

$$\Sigma = UDU^T, \tag{15}$$

where U is an orthogonal random matrix with elements $U_{ij} \sim N(0,1)$ and D is diagonal matrix of positive eigenvalues. The eigenvalues are uniformly distributed over the interval $[a,b]$ so that the condition number of Σ is b/a and the eigenvalues are kept distinct. Here, a is chosen to be 1×10^{-4} and b is chosen to be 0.1.

Using a Cholesky decomposition $\Sigma = \Gamma\Gamma^T$ and the random mean vector $\mu \sim N(0,1)$, we generate correlated gaussian vectors of dimension p with the properties

$$X_i = \mu + \Gamma_{ij} Z_j,\ Z_j \sim N(0,1), \forall j \in 1, \dots, p. \tag{16}$$

The data (x_n, y_n) are generated under a logistic regression

$$p(y=1|\boldsymbol{x}, \boldsymbol{\theta}_0) = f(\boldsymbol{x}|\boldsymbol{\theta}_0) = \frac{1}{1+e^{-\boldsymbol{x}\boldsymbol{\theta}_0}}. \tag{17}$$

A key challenge in assessing the efficacy of bias reduction is to avoid generating excessively low entropy distributions. In such cases, bias reduction will have marginal effect as the parameters are all nearly zero. To avoid such scenarios, the intercept parameter of the true model is adjusted a-posterior until the following conditions are met:

1. $c < \mathbb{E}_Z[p(Y=1|\mathbf{X}, \boldsymbol{\theta}_0)] < d$
2. $-\mathcal{L}(\boldsymbol{\theta}_0) > \epsilon$

where $c = 0.35$, $d = 0.65$, and $\epsilon = 0.2$. If these conditions can not be met, then the replication is discarded.

4.3.2. Model Performance Comparison

As in prior sections, KL divergence is computed between the estimated model and the true model for each of the estimation methods. The T-statistics of the difference with the corresponding MLE KL divergence are computed, with negative T-statistics showing that an approach is performing better than the MLE approach. For L_2 regularization in this section, the value of λ is computed via cross-validation, using two folds, on the provided fitting set.

Table 4 compares the KL divergences ρ_{KL} from the true distribution to the model distributions produced using various estimation approaches applied to misspecified data. Here m denotes the number of regressors that are not predictive, i.e., $\boldsymbol{\theta}_0$ contains m zeros. The experiment is replicated 300 times using the data generation process described above and the test set is fixed at 100,000 observations.

Table 4. Comparison of the KL divergence for the different estimation approaches applied to misspecified data. The values in parentheses denote the t-statistic relative to MLE. For $p = \{5, 10, 20\}$ there are $m = \{2, 4, 8\}$ non-explanatory variables added.

p	n	$\rho_{KL}(f, g_{\hat{\theta}})$	$\rho_{KL}(f, g_{\theta_{L_2}})$	$\rho_{KL}(f, g_{\theta^*})$
5	500	4.79×10^{-3}	3.01×10^{-3} (−13.43)	4.56×10^{-3} (−13.53)
5	1000	2.64×10^{-3}	1.76×10^{-3} (−10.94)	2.57×10^{-3} (−12.80)
5	2000	1.29×10^{-3}	1.09×10^{-3} (−6.15)	1.27×10^{-3} (−7.69)
5	5000	5.09×10^{-4}	4.60×10^{-4} (−4.69)	5.07×10^{-4} (−6.19)
10	500	9.79×10^{-3}	9.85×10^{-3} (0.16)	9.18×10^{-3} (−6.27)
10	1000	5.05×10^{-3}	5.13×10^{-3} (0.51)	4.83×10^{-3} (−4.90)
10	2000	2.50×10^{-3}	3.05×10^{-3} (5.99)	2.56×10^{-3} (1.70)
10	5000	1.06×10^{-3}	1.49×10^{-3} (7.72)	1.04×10^{-3} (−0.86)
20	500	2.18×10^{-2}	2.16×10^{-2}(−0.29)	1.95×10^{-2} (−8.71)
20	1000	1.13×10^{-2}	1.24×10^{-2} (3.79)	1.10×10^{-2} (−1.95)
20	2000	6.86×10^{-3}	7.52×10^{-3} (4.47)	6.72×10^{-3} (−1.67)
20	5000	3.57×10^{-3}	4.24×10^{-3} (6.56)	3.59×10^{-3} (0.45)

We observe that the t-statistic for $\boldsymbol{\theta}^*$ is most significant for relatively small sample sizes, particularly $n = 500$. For these small sizes, the improvement over MLE is greater, though noisier. There is uniform decay in improvement over $\hat{\boldsymbol{\theta}}$ as n grows, until for $p = 10$ and $p = 20$ the largest sizes are no longer statistically significant. This is expected, as both the MLE and ICE estimates are converging towards the true value of $\boldsymbol{\theta}_0$, and for large enough sample sizes the ICE correction would be dominated by numerical error, particularly the ill conditioning of J.

The L_2 estimate improves for small values of p, but then becomes progressively worse for large values of p. We observe that for dimensionality above $p = 5$, the L_2 regularization described here is no longer effective in reducing the KL-divergence. For low values of p the value of $\boldsymbol{\theta x}$ has comparatively low variance, and thus the logistic function is reasonably locally approximated as linear. For higher p this approximation is less realistic and the performance of L_2 regularization degrades.

For the ICE estimates, larger values of p show fluctuations that are often not statistically significant. It is apparent that larger p is increasing the variance of the ICE divergences, probably due to numerical errors and ill conditioning. Larger values of n reduce the absolute size of the divergence improvement whereas larger values of p seem to increase it.

Note that though the t-statistics are degrading for large n, the absolute magnitude of the differences is asymptotically small. For these sizes, the results are insignificant, but more importantly, immaterial.

4.3.3. Convergence Analysis for Large n

For 10 randomly chosen example problems, under which the model coefficients are now fixed, the convergence behavior for large n, the training set size, is explored. Note that the test set remains fixed at 100,000 observations for each problem. Table 5 compares the KL divergence (averaged over all 10 problems) under MLE ($\hat{\boldsymbol{\theta}}$), L_2 regularization, and ICE for progressively larger sample sizes. The divergences $\rho_{KL}(f, g_{\hat{\theta}})$ and $\rho_{KL}(f, g_{\theta^*_{ICE}})$ converge to zero as $n \to \infty$, as does $\rho_{KL}(f, g_{\theta^*_{L_2}})$.

Table 5. Comparison of the KL divergence under the MLE $\hat{\boldsymbol{\theta}}$, L_2 regularization and ICE regularization $\boldsymbol{\theta}^*_{ICE}$ against a large sample size for the case when $p = 10$ and $m = 4$.

n	$\mathcal{L}(\theta_0)$	$\rho_{KL}(f, g_{\hat{\theta}})$	$\rho_{KL}(f, g_{\theta_{L_2}})$	$\rho_{KL}(f, g_{\theta^*})$
500	0.5439	9.28×10^{-3}	7.92×10^{-3}	7.74×10^{-3}
1000	0.5439	5.50×10^{-3}	5.86×10^{-3}	4.81×10^{-3}
2000	0.5439	2.65×10^{-3}	3.67×10^{-3}	2.65×10^{-3}
5000	0.5439	1.85×10^{-3}	2.72×10^{-3}	1.35×10^{-3}
10,000	0.5439	5.75×10^{-4}	1.57×10^{-3}	9.32×10^{-4}
20,000	0.5439	5.84×10^{-4}	8.10×10^{-4}	6.11×10^{-4}
50,000	0.5439	3.83×10^{-4}	4.09×10^{-4}	3.64×10^{-4}
100,000	0.5439	1.67×10^{-4}	1.15×10^{-3}	1.86×10^{-4}

Generally the $\boldsymbol{\theta}^*_{ICE}$ estimates are seen to converge slightly faster than the $\hat{\boldsymbol{\theta}}$ estimates. The regularization in $\boldsymbol{\theta}^*_{L_2}$ is observed to be beneficial for very small sample sizes, but then becomes marginally detrimental for large n.

5. Optimized Computation Results

For any model satisfying White's Regularity Criteria, it is known that the matrix J is positive definite near the MLE optimum $\hat{\boldsymbol{\theta}}$. This implies that J is diagonally dominated, and indeed considering just its diagonal elements D, it is known that $tr(ID^{-1}) > 0$. Indeed $tr(ID^{-1})$ differs strongly from $tr(IJ^{-1})$ most strongly for models with strong regressor interactions. Therefore, using finite difference gradients, consider the following approximations for the ICE objective function:

1. $\boldsymbol{\theta}^*$: J is computed directly, ICE is implemented as written.
2. $\boldsymbol{\theta}^*_2$: J is taken to be constant w.r.t. $\boldsymbol{\theta}$: ($J_\theta = J_{\hat{\theta}}$).
3. $\boldsymbol{\theta}^*_3$: J is taken to be diagonal: ($J = D$).
4. $\boldsymbol{\theta}^*_4$: J is taken to be the identity: ($J = \mathcal{I}$).

Clearly, we expect that $\boldsymbol{\theta}^*_4$ above is the least accurate approximation, and items $\boldsymbol{\theta}^*_2$ and $\boldsymbol{\theta}^*_3$ have varying levels of accuracy depending on the problem at hand. The cost comparison of these approaches is shown in Table 6.

Table 6. The asymptotic computational cost (per iteration) of various proposed approximations as a function of parameter count p. Cost is amortized when $J_\theta = J_{\hat{\theta}}$ assuming that $n \approx p$. Note that a typical model will cost $O(p)$ in time and space for both the objective function and its gradients.

Approximation	Objective Cost (Space)	Objective Cost (Time)	Gradient Cost (Space)	Gradient Cost (Time)
Direct Computation	$O(p^2)$	$O(p^3)$	$O(p^2)$	$O(p^4)$
$J_\theta = J_{\hat{\theta}}$	$O(p^2)$	$O(p^2)$	$O(p^2)$	$O(p^3)$
$J = D$	$O(p)$	$O(p)$	$O(p)$	$O(p^2)$
$J = \mathcal{I}$	$O(p)$	$O(p)$	$O(p)$	$O(p^2)$

Remark 9. *When computing gradients for use in a solver, often approximation error will have only a marginal impact on the final result, though it may increase the number of iterations needed for convergence. Broyden's method [25] is a typical example of this approach in action. Efficient approximations of $[\partial_\theta \hat{f}]$ might similarly have only a minor effect on accuracy and iteration count. The construction of approximate analytical derivatives is beyond the scope of the present work.*

These approximations were computed and compared for the Friedman (see Section 4.2) problem, and the results are shown in Table 7 below.

Table 7. Comparison of the average KL divergence across 200 replications for MLE and several variants of ICE given a fitting set size of n. For estimators other than $\hat{\theta}$, the values in parentheses denotes the t-statistic of the difference between this estimator and $\hat{\theta}$, with negative values indicating that the listed estimator has a lower KL divergence.

n	$\rho_{KL}(f, g_{\hat{\theta}})$	$\rho_{KL}(f, g_{\theta^*})$	$\rho_{KL}(f, g_{\theta_2^*})$	$\rho_{KL}(f, g_{\theta_3^*})$	$\rho_{KL}(f, g_{\theta_4^*})$
8	$1.22 \times 10^{+1}$	$4.55 \times 10^{+0}$ (-4.89)	$5.70 \times 10^{+0}$ (-5.26)	$3.83 \times 10^{+0}$ (-5.22)	$1.28 \times 10^{+0}$ (-4.67)
16	6.68×10^{-1}	2.99×10^{-1} (-8.13)	3.53×10^{-1} (-10.56)	3.36×10^{-1} (-8.30)	5.47×10^{-1} (-2.11)
32	1.45×10^{-1}	1.14×10^{-1} (-6.90)	1.04×10^{-1} (-8.18)	1.08×10^{-1} (-10.16)	3.63×10^{-1} (19.60)
64	5.93×10^{-2}	4.80×10^{-2} (-10.42)	4.70×10^{-2} (-6.95)	4.81×10^{-2} (-9.81)	2.38×10^{-1} (39.08)
128	2.48×10^{-2}	2.24×10^{-2} (-6.38)	2.33×10^{-2} (-2.26)	2.26×10^{-2} (-6.00)	1.62×10^{-1} (61.85)
256	1.21×10^{-2}	1.16×10^{-2} (-4.28)	1.20×10^{-2} (-0.68)	1.16×10^{-2} (-4.11)	1.01×10^{-1} (68.82)
512	6.26×10^{-3}	6.10×10^{-3} (-2.41)	6.16×10^{-3} (-0.84)	6.10×10^{-3} (-2.39)	5.42×10^{-2} (63.84)
1024	3.05×10^{-3}	3.00×10^{-3} (-2.66)	3.04×10^{-3} (-0.37)	2.99×10^{-3} (-2.73)	2.61×10^{-2} (59.78)

From Table 7, it is apparent that approach θ_4^*, taking $J = \mathcal{I}$ is not effective. This is not surprising as the actual J matrix has dramatic differences in scale between regressors. Approximation θ_3^*, taking $J = D$ is accurate enough that it cannot be statistically distinguished from the direct computation of ICE by the test above. Approximation θ_2^* tends to underperform approximation (3).

Therefore, we propose taking $J = D$ as a more numerically stable approximation of the ICE objective.

6. Conclusions

Takeuchi [16] is believed to be the first to have proposed using an objective function similar to ICE in order to reduce generalization error, though it was applied via model selection. Firth [9] introduced a similar term to reduce parameter bias in model fitting, as opposed to model selection, though he derived it only for exponential model families and did not consider its effect on generalization error. It is not known why this approach did not find widespread use, but one may infer that the $O(p^4)$ computational cost and instability was enough to keep it from wider adoption.

In this paper, we reintroduce the objective function of [16] and provide a more general proof of its widespread applicability. We then show that efficient implementations costing only $O(p)$ are possible. Under finite sample sizes, this bias correction term is shown experimentally in several models to lead to significant reduction in bias compared to maximum likelihood estimation with and without L_2 regularization. ICE offers many advantages over L_2 penalized maximum likelihood estimation: (i) it's suitable for most nonlinear models, (ii) it's provably asymptotically convergent; and (iii) does not rely on any parameters which would need to be provided by the operator or deduced through cross-validation.

Author Contributions: Conceptualization, M.D. and T.W.; methodology, M.D. and T.W.; software, T.W.; validation, M.D. and T.W.; formal analysis, M.D. and T.W.; writing—original draft preparation, M.D. and T.W.; writing—review and editing, M.D. and T.W. All authors have read and agreed to the published version of the manuscript.

Funding: This research received no external funding.

Institutional Review Board Statement: Not applicable.

Informed Consent Statement: Not applicable.

Data Availability Statement: Data and code are available at https://doi.org/10.6084/m9.figshare.14312852.v1.

Conflicts of Interest: The authors declare no conflicts of interest.

Appendix A. Proofs

Appendix A.1. White's Regularity Conditions

Definition A1 (White's regularity conditions). *White [22] provides the following regularity conditions:*

A1: The independent random vectors, $X_i, i = 1, \dots, n$, have common joint distribution function F on Ω, a measurable Euclidean space, with measurable Radon–Nikodym density $f = dF/dx$.

A2: The family of distribution functions $G(x|\boldsymbol{\theta})$ has Radon–Nikodym densities $g(x|\boldsymbol{\theta}) = dG(x|\boldsymbol{\theta})/dx$ which are measurable in x for every $\boldsymbol{\theta} \in \Theta$, a compact subset of p-dimensional Euclidean space, and continuous in $\boldsymbol{\theta}$ for every $x \in \Omega$.

A3: (a) $\mathbb{E}[\log f(X)]$ exists and $|\log g(x|\boldsymbol{\theta})| \leq m(x)$, $\forall \boldsymbol{\theta} \in \Theta$, where m is integrable with respect to F; (b) $\rho_{KL}(f, g_{\boldsymbol{\theta}})$ has a unique minimum at $\boldsymbol{\theta}_0 \in \Theta$.

A4: $\partial_{\boldsymbol{\theta}}(\log g(x|\boldsymbol{\theta}))$ are measurable functions of x for each $\boldsymbol{\theta} \in \Theta$ and continuously differentiable functions of $\boldsymbol{\theta}$ for each $x \in \Omega$.

A5: $|\partial^2_{\boldsymbol{\theta}}(\log g(x|\boldsymbol{\theta}))|$ and $|\partial_{\boldsymbol{\theta}}(\log g(x|\boldsymbol{\theta})) \cdot \partial_{\boldsymbol{\theta}}(\log g(x|\boldsymbol{\theta})|)$ are dominated by functions integrable in x with respect to F for all $x \in \Omega$ and $\boldsymbol{\theta} \in \Theta$.

A6: (a) $\boldsymbol{\theta}_0$ is an interior point of the parameter space; (b) $\mathbb{E}[\partial_{\boldsymbol{\theta}}(\log g(x|\boldsymbol{\theta})) \cdot \partial_{\boldsymbol{\theta}}(\log g(x|\boldsymbol{\theta}))]$ is non-singular; (c) θ_0 is a regular point of $\mathbb{E}[\partial^2_{\boldsymbol{\theta}}(\log g(x|\boldsymbol{\theta}))]$.

Appendix A.2. Proof of Finite Variance

Lemma A1 (Finite variance). *Suppose the following conditions hold:*

1. *$\mathcal{M}$ satisfies White's regularity conditions A1–A6 (see Appendix A.1 or [22]).*
2. *$\boldsymbol{\theta}_0$ is a global minimum of $-\mathcal{L}(\boldsymbol{\theta})$ in the compact space Θ defined in A2.*
3. *There exists a $\varepsilon > 0$ such that $-\mathcal{L}(\boldsymbol{\theta}_0) < -\mathcal{L}(\boldsymbol{\theta}_1) - \varepsilon$ for all other local minima $\boldsymbol{\theta}_1$.*
4. *For $k = 0,1,2,3,4,5$ the derivative $\partial^k_{\boldsymbol{\theta}}\mathcal{L}(\boldsymbol{\theta})$ exists, is continuous, and bounded on an open set around $\boldsymbol{\theta}_0$.*
5. *For $k = 0,1,2,3,4,5$, the variance $\mathbb{V}[\partial^k_{\boldsymbol{\theta}}\ell(\boldsymbol{\theta}, X_n)] \to 0$ as $n \to \infty$ on an open set around $\boldsymbol{\theta}_0$.*

Then, for sufficiently large n there exists a compact subset $U \subset \Theta$ containing $\boldsymbol{\theta}_0, \hat{\boldsymbol{\theta}}$, such that

1. *For $k = 0,1,2,3$ the derivative $\partial^k_{\boldsymbol{\theta}}\ell^*(\boldsymbol{\theta}, x_n)$ exists, is continuous, and bounded on U, almost surely.*
2. *For $k = 0,1,2,3$, $\mathbb{V}[\partial^k_{\boldsymbol{\theta}}\ell^*(\boldsymbol{\theta}, X_n)] \to 0$ as $n \to \infty$ on U, almost surely.*
3. *$\boldsymbol{\theta}^* \in U$ as $n \to \infty$ almost surely.*

Proof. Assumptions (4) and (5) establish existence of some set, S, containing $\boldsymbol{\theta}_0$ such that $\mathcal{L}(\boldsymbol{\theta})$ is bounded on S, and its estimate, $\ell(\boldsymbol{\theta}, X_n)$ has finite variance. Therefore, $-\ell(\boldsymbol{\theta}, x_n)$ is also bounded on S almost surely. Similarly for the first 5 derivatives. White's criteria imply that $J_{\boldsymbol{\theta}_0}$ is positive definite on an open set around $\boldsymbol{\theta}_0$, and thus one can form a compact set $U \subset S$ containing an open set around $\boldsymbol{\theta}_0$ on which the minimum eigenvalue of $J_{\boldsymbol{\theta}}$ is bounded away from 0.

Note that $J_{\boldsymbol{\theta}}$ is three times differentiable on U by Assumption 4, as is $\hat{J}_{\boldsymbol{\theta}}$, as established above. Then $\hat{J}_{\boldsymbol{\theta}}^{-1}$ is also positive definite and bounded on U. It can be shown to also have three derivatives by using the well-known matrix relation

$$\partial_{\boldsymbol{\theta}} A^{-1} = -A^{-1}(\partial_{\boldsymbol{\theta}} A)A^{-1}. \tag{A1}$$

It follows that $\hat{J}_{\theta}^{-1}$ is also positive definite, nonsingular, and bounded on U. Similarly for $\hat{I}_{\theta}$, and thus $tr(\hat{I}_{\theta}\hat{J}_{\theta}^{-1})$ is bounded with finite variance on U. It also has three bounded derivatives with finite variance.

Therefore, $-\ell^*(\boldsymbol{\theta}, X_n) \to -\ell(\boldsymbol{\theta}, X_n)$, and $\boldsymbol{\theta}^* \to \boldsymbol{\theta}_0$ as $n \to \infty$, with the convergence being in probability. This means that U contains $\boldsymbol{\theta}_0$, $\hat{\boldsymbol{\theta}}$, and $\boldsymbol{\theta}^*$ almost surely for large enough n. Similarly, on U we have three continuous, bounded derivatives of $\partial_{\boldsymbol{\theta}}^k \ell^*(\boldsymbol{\theta}, \boldsymbol{x}_n)$ almost surely. □

Appendix A.3. Proof of Asymptotic Normality

Theorem A1 (Asymptotic Normality). *Provided the conditions hold in Lemma A1, namely, that*

1. *For $k = 0, 1, 2, 3$ the derivative $\partial_{\boldsymbol{\theta}}^k \ell^*(\boldsymbol{\theta}, \boldsymbol{x}_n)$ exists, is continuous, and bounded on U.*
2. *For $k = 0, 1, 2, 3$, $\mathbb{V}[\partial_{\boldsymbol{\theta}}^k \ell^*(\boldsymbol{\theta}, \boldsymbol{x}_n)] \to 0$ as $n \to \infty$ on U.*

Then

$$\sqrt{n}(\boldsymbol{\theta}^* - \boldsymbol{\theta}_0^*) \to N(0, (\hat{J}_{\boldsymbol{\theta}_0^*}^*)^{-1} \hat{I}_{\boldsymbol{\theta}_0^*}^* (\hat{J}_{\boldsymbol{\theta}_0^*}^*)^{-1}).$$

Proof. As the first derivatives of ℓ are continuous, the mean value theorem may be applied:

$$\partial_{\boldsymbol{\theta}} \ell^*(\boldsymbol{\theta}_0^*) = \partial_{\boldsymbol{\theta}} \ell^*(\boldsymbol{\theta}^*) + (\boldsymbol{\theta}^* - \boldsymbol{\theta}_0^*)\hat{J}_{\bar{\boldsymbol{\theta}}}^* = (\boldsymbol{\theta}^* - \boldsymbol{\theta}_0^*)\hat{J}_{\bar{\boldsymbol{\theta}}}^*. \tag{A2}$$

$\bar{\boldsymbol{\theta}}$ is between $\boldsymbol{\theta}^*$ and $\boldsymbol{\theta}_0^*$. Under the assumptions of Lemma A1, and given its finite variance, $\hat{J}_{\bar{\boldsymbol{\theta}}}$ is almost surely (in the large n limit) positive definite, and thus invertible as both $\boldsymbol{\theta}^*$ and $\boldsymbol{\theta}_0^*$ are in U, and $\bar{\boldsymbol{\theta}}$ is between them. Therefore,

$$(\boldsymbol{\theta}^* - \boldsymbol{\theta}_0^*) = (\hat{J}_{\bar{\boldsymbol{\theta}}}^*)^{-1} \partial_{\boldsymbol{\theta}} \ell^*(\boldsymbol{\theta}_0^*). \tag{A3}$$

Applying the mean value theorem a second time gives

$$\hat{J}_{\bar{\boldsymbol{\theta}}} = \hat{J}_{\boldsymbol{\theta}_0^*}^* + (\bar{\boldsymbol{\theta}} - \boldsymbol{\theta}_0^*)\hat{J}_{\boldsymbol{\theta}_1}^*, \tag{A4}$$

with $\boldsymbol{\theta}_1$ between $\bar{\boldsymbol{\theta}}$ and $\boldsymbol{\theta}_0^*$. If the order of $(\boldsymbol{\theta}^* - \boldsymbol{\theta}_0^*) = O_p(\delta)$, where $\delta := n^{-1/2}$, then

$$\hat{J}_{\bar{\boldsymbol{\theta}}}^* = \hat{J}_{\boldsymbol{\theta}_0^*}^* + O_p(\delta). \tag{A5}$$

As all of the $\hat{J}^*$ are bounded away from zero in probability, we have

$$(\hat{J}_{\bar{\boldsymbol{\theta}}}^*)^{-1} = (\hat{J}_{\boldsymbol{\theta}_0^*}^*)^{-1} + O_p(\delta), \tag{A6}$$

with the equality holding in probability. In the large n limit, $\delta \to 0$, and thus

$$(\boldsymbol{\theta}^* - \boldsymbol{\theta}_0^*) = (\hat{J}_{\boldsymbol{\theta}_0^*}^*)^{-1} \partial_{\boldsymbol{\theta}} \ell^*(\boldsymbol{\theta}_0^*). \tag{A7}$$

As $\partial_{\boldsymbol{\theta}} \ell^*(\boldsymbol{\theta}_0^*)$ is the sum of n independent vectors, it is asymptotically normally distributed by the central limit theorem, and its mean is 0 by the definition of $\boldsymbol{\theta}_0^*$. Its variance is therefore $\mathbb{V}[\partial_{\boldsymbol{\theta}^*} \ell^*(\boldsymbol{\theta}_0^*)] = \mathbb{E}[\partial_{\boldsymbol{\theta}} \ell^*(\boldsymbol{\theta}_0^*)(\partial_{\boldsymbol{\theta}} \ell^*(\boldsymbol{\theta}_0^*))^T] = \frac{1}{n}\hat{I}_{\boldsymbol{\theta}_0^*}$

Substituting this into Equation (A7) yields

$$\sqrt{n}(\boldsymbol{\theta}^* - \boldsymbol{\theta}_0^*) = N(0, (\hat{J}_{\boldsymbol{\theta}_0^*}^*)^{-1} \hat{I}_{\boldsymbol{\theta}_0^*}^* (\hat{J}_{\boldsymbol{\theta}_0^*}^*)^{-1}), \tag{A8}$$

establishing the result. □

Appendix A.4. Proof of Prediction Bias Order under ICE

Theorem A2 (Prediction Bias Estimation under ICE). *By minimizing $-\ell^*$ instead of $-\ell$, the first order terms of the prediction bias are cancelled leaving a $O_p(n^{-3/2})$ residual term*

$$-\mathcal{L}(\hat{\theta}^*) = -\ell(\theta^*(X_n), X_n) + O_p(n^{-3/2}). \tag{A9}$$

Proof. Note that in Takeuchi's proof [16], the use of $\hat{\theta}$ was prescribed. Therefore, this approach could be used only for model selection, and not for model fitting. Now consider the bias under the ICE estimator θ^*.

$$\begin{aligned} b(\theta^*(X_n), X_n) &= \mathbb{E}_{X_n}[\log g(X_n|\theta^*(X_n)) - \log g(X_n|\theta_0)] \\ &+ \mathbb{E}_{X_n}[\log g(X_n|\theta_0) - n\mathbb{E}_{Z_n}[\log g(Z_n|\theta_0)]] \\ &+ n\mathbb{E}_{X_n}[\mathbb{E}_{Z_n}[\log g(Z_n|\theta_0)] - \mathbb{E}_{Z_n}[\log g(Z_n|\theta^*(X_n))]]. \end{aligned}$$

As the second term is zero, this can be simplified to

$$\begin{aligned} b(\theta^*(X_n), X_n) &= -n\mathbb{E}_{X_n}[\ell(\theta^*(X_n), X_n) - \ell(\theta_0, X_n)] \\ &- n\mathbb{E}_{X_n}[\mathcal{L}(\theta_0) - \mathcal{L}(\theta^*(X_n))]. \end{aligned}$$

Define $\delta = \frac{1}{\sqrt{n}}$, and then recall from White [22] that $(\hat{\theta} - \theta_0)$ is $O_p(\delta)$. Similarly, recall from Theorem A1 that $(\theta^* - \theta_0^*)$ is also $O_p(\delta)$. As constructed, the error term $b(\theta^*(X_n), X_n)$ is $O_p(1)$ (actually $O(1)$ as it is an expectation), and terms of order $O_p(\delta)$ and higher will be dropped. Therefore, as in Takeuchi's derivation [16], the Taylor expansions below will be truncated at second order in δ, dropping terms of order $O_p(\delta^3)$ or higher. Additionally, we will occasionally drop indications of X_n where the meaning is clear and it greatly simplifies the notation.

With that truncation, recalling that θ_0 is a minimum of $\mathcal{L}(\theta)$ and thus has zero gradient:

$$n\mathbb{E}_{X_n}[\mathcal{L}(\theta_0) - \mathcal{L}(\theta^*(X_n))] = \frac{n}{2}\mathbb{E}_{X_n}[(\theta^* - \theta_0)^T J_{\theta_0}(\theta^* - \theta_0)] + O(\delta). \tag{A10}$$

Recall Theorem A1, and recall the that for the quadratic form:

$$\mathbb{E}[\varepsilon^T A\varepsilon] = tr(A\Sigma) + \mu^T A\mu, \qquad \mathbb{E}[\varepsilon] = \mu, \mathbb{V}[\varepsilon] = \Sigma. \tag{A11}$$

Therefore,

$$\begin{aligned} \frac{n}{2}\mathbb{E}_{X_n}[(\theta^* - \theta_0)^T J_{\theta_0}(\theta^* - \theta_0)] &= \frac{1}{2}tr(J_{\theta_0}[(J^*_{\theta_0^*})^{-1} I^*_{\theta_0^*}(J^*_{\theta_0^*})^{-1})]) \\ &+ \frac{n}{2}(\theta_0^* - \theta_0)^T J_{\theta_0}(\theta_0^* - \theta_0) \\ &+ O(\delta). \end{aligned}$$

Now note that the second term on the right is a constant, and therefore would take no part in any optimization. Therefore, it can be safely ignored and

$$n\mathbb{E}_{X_n}[\mathcal{L}(\theta_0) - \mathcal{L}(\theta^*(X_n))] = \frac{1}{2}tr(J_{\theta_0}[(J^*_{\theta_0^*})^{-1} I^*_{\theta_0^*}(J^*_{\theta_0^*})^{-1})]) + O(\delta). \tag{A12}$$

Addressing the first term, again taking a Taylor expansion, we find that

$$\ell(\theta_0) = \ell(\theta^*) + (\theta_0 - \theta^*)^T \partial_\theta \ell(\theta^*) + \frac{1}{2}(\theta_0 - \theta^*)^T \partial^2_\theta \ell(\theta^*)(\theta_0 - \theta^*) + O_p(\delta^3). \tag{A13}$$

Therefore, recalling that $nO_p(\delta^3) = O_p(\delta)$ gives

$$
\begin{aligned}
n\mathbb{E}_{\mathbf{X}_n}[\ell(\boldsymbol{\theta}^*(\mathbf{X}_n), X_n) - \ell(\boldsymbol{\theta}_0, X_n)] &= \frac{1}{2}tr(J_{\theta^*}[(J^*_{\theta^*_0})^{-1}I^*_{\theta^*_0}(J^*_{\theta^*_0})^{-1})]) \\
&+ \frac{n}{2}(\boldsymbol{\theta}^*_0 - \boldsymbol{\theta}_0)^T J_{\theta^*}(\boldsymbol{\theta}^*_0 - \boldsymbol{\theta}_0) \\
&+ n\mathbb{E}_{\mathbf{X}_n}[(\boldsymbol{\theta}^* - \boldsymbol{\theta}_0)^T \partial_{\boldsymbol{\theta}}\ell(\boldsymbol{\theta}^*)] \\
&+ O_p(\delta).
\end{aligned}
$$

Now examine the last term:

$$
\begin{aligned}
n\mathbb{E}_{\mathbf{X}_n}[(\boldsymbol{\theta}^* - \boldsymbol{\theta}_0)^T \partial_{\boldsymbol{\theta}}\ell(\boldsymbol{\theta}^*)] &= n\mathbb{E}_{\mathbf{X}_n}[(\boldsymbol{\theta}^* - \boldsymbol{\theta}^*_0)^T \partial_{\boldsymbol{\theta}}\ell(\boldsymbol{\theta}^*)] \\
&+ n\mathbb{E}_{\mathbf{X}_n}[(\boldsymbol{\theta}^*_0 - \boldsymbol{\theta}_0)^T \partial_{\boldsymbol{\theta}}\ell(\boldsymbol{\theta}^*)].
\end{aligned}
$$

However, $(\boldsymbol{\theta}^*_0 - \boldsymbol{\theta}_0)$ does not depend on $\mathbf{X}_n$, so it can be pulled out of the expectation, then substitute in a first order Taylor expansion

$$
\begin{aligned}
\mathbb{E}_{\mathbf{X}_n}[(\boldsymbol{\theta}^*_0 - \boldsymbol{\theta}_0)^T \partial_{\boldsymbol{\theta}}\ell(\boldsymbol{\theta}^*)] &= (\boldsymbol{\theta}^*_0 - \boldsymbol{\theta}_0)^T \mathbb{E}_{\mathbf{X}_n}[\partial_{\boldsymbol{\theta}}\ell(\boldsymbol{\theta}^*)] \\
&= (\boldsymbol{\theta}^*_0 - \boldsymbol{\theta}_0)^T \mathbb{E}_{\mathbf{X}_n}[\partial_{\boldsymbol{\theta}}\ell(\boldsymbol{\theta}^*_0) + (\boldsymbol{\theta}^* - \boldsymbol{\theta}^*_0)\partial^2_{\boldsymbol{\theta}}\ell(\boldsymbol{\theta}^*_0) + O_p(\delta^2)] \\
&= (\boldsymbol{\theta}^*_0 - \boldsymbol{\theta}_0)^T \partial_{\boldsymbol{\theta}}\mathcal{L}(\boldsymbol{\theta}^*_0) \\
&+ (\boldsymbol{\theta}^*_0 - \boldsymbol{\theta}_0)^T \mathbb{E}_{\mathbf{X}_n}[(\boldsymbol{\theta}^* - \boldsymbol{\theta}^*_0)]\partial^2_{\boldsymbol{\theta}}\mathcal{L}(\boldsymbol{\theta}^*_0) + O(\delta^3) \\
&= (\boldsymbol{\theta}^*_0 - \boldsymbol{\theta}_0)^T \partial_{\boldsymbol{\theta}}\mathcal{L}(\boldsymbol{\theta}^*_0) + O(\delta^3),
\end{aligned}
$$

with the last equality following from the fact that $\mathbb{E}_{\mathbf{X}_n}[(\boldsymbol{\theta}^* - \boldsymbol{\theta}^*_0)] = 0$ and the substitution of $(\boldsymbol{\theta}^*_0 - \boldsymbol{\theta}_0)^T O(\delta^2) = O(\delta^3)$. This term is therefore a constant, up to $O(\delta^3)$, and takes no part in optimization of $\boldsymbol{\theta}$, thus it can be dropped from further consideration. Therefore $n\mathbb{E}_{\mathbf{X}_n}[(\boldsymbol{\theta}^*_0 - \boldsymbol{\theta}_0)^T \partial_{\boldsymbol{\theta}}\ell(\boldsymbol{\theta}^*)] = O(\delta)$, and

$$
\begin{aligned}
n\mathbb{E}_{\mathbf{X}_n}[(\boldsymbol{\theta}^* - \boldsymbol{\theta}_0)^T \partial_{\boldsymbol{\theta}}\ell(\boldsymbol{\theta}^*)] &= n\mathbb{E}_{\mathbf{X}_n}[(\boldsymbol{\theta}^* - \boldsymbol{\theta}^*_0)^T \partial_{\boldsymbol{\theta}}\ell(\boldsymbol{\theta}^*)] \\
&+ O(\delta).
\end{aligned}
$$

Recombining these terms yields

$$
\begin{aligned}
n\mathbb{E}_{\mathbf{X}_n}[\ell(\boldsymbol{\theta}^*(\mathbf{X}_n), X_n) - \ell(\boldsymbol{\theta}_0, X_n)] &= \frac{1}{2}tr(J_{\theta^*}[(J^*_{\theta^*})^{-1}I^*_{\theta^*}(J^*_{\theta^*})^{-1})]) \\
&+ \frac{n}{2}(\boldsymbol{\theta}^*_0 - \boldsymbol{\theta}_0)^T J_{\theta^*}(\boldsymbol{\theta}^*_0 - \boldsymbol{\theta}_0) \\
&+ n\mathbb{E}_{\mathbf{X}_n}[(\boldsymbol{\theta}^* - \boldsymbol{\theta}^*_0)^T \partial_{\boldsymbol{\theta}}\ell(\boldsymbol{\theta}^*)] \\
&+ O(\delta).
\end{aligned}
$$

Thus the bias (neglecting the constant terms) is then

$$
\begin{aligned}
b(\boldsymbol{\theta}^*(\mathbf{X}_n), \mathbf{X}_n) &= -\frac{1}{2}tr(J_{\theta^*}[(J^*_{\theta^*})^{-1}I^*_{\theta^*}(J^*_{\theta^*})^{-1})]) \\
&- \frac{1}{2}tr(J_{\theta_0}[(J^*_{\theta^*_0})^{-1}I^*_{\theta^*_0}(J^*_{\theta^*_0})^{-1})]) \\
&- \frac{n}{2}(\boldsymbol{\theta}^*_0 - \boldsymbol{\theta}_0)^T J_{\theta^*}(\boldsymbol{\theta}^*_0 - \boldsymbol{\theta}_0) \\
&- n\mathbb{E}_{\mathbf{X}_n}[(\boldsymbol{\theta}^* - \boldsymbol{\theta}^*_0)^T \partial_{\boldsymbol{\theta}}\ell(\boldsymbol{\theta}^*)] \\
&+ O(\delta).
\end{aligned}
$$

Because $\ell^*(\boldsymbol{\theta}) = \ell(\boldsymbol{\theta}) + O_p(\delta^2)$, it follows that $J^*_{\theta} = J_{\theta} + O_p(\delta^2)$ and thus $J_{\theta^*}(J^*_{\theta^*})^{-1} = I + O(\delta^2)$. Similarly for J_{θ_0}. In addition $I_{\theta^*} = I_{\theta^*_0} + O(\delta)$, thus the two trace terms can be simplified and combined:

$$
\begin{aligned}
b(\boldsymbol{\theta}^*(\mathbf{X}_n), \mathbf{X}_n) &= -tr(I_{\boldsymbol{\theta}^*} J_{\boldsymbol{\theta}^*}^{-1}) \\
&- \frac{n}{2}(\boldsymbol{\theta}_0^* - \boldsymbol{\theta}_0)^T J_{\boldsymbol{\theta}^*}(\boldsymbol{\theta}_0^* - \boldsymbol{\theta}_0) \\
&- n\mathbb{E}_{\mathbf{X}_n}[(\boldsymbol{\theta}^* - \boldsymbol{\theta}_0^*)^T \partial_{\boldsymbol{\theta}} \ell(\boldsymbol{\theta}^*)] \\
&+ O(\delta).
\end{aligned}
$$

As $\boldsymbol{\theta}_0^*$ is a minimum of $\mathcal{L}^*$, begin by Taylor expanding the derivatives

$$\partial_{\boldsymbol{\theta}} \mathcal{L}(\boldsymbol{\theta}_0^*) = \partial_{\boldsymbol{\theta}} \mathcal{L}(\boldsymbol{\theta}_0) - (\boldsymbol{\theta}_0^* - \boldsymbol{\theta}_0) J_{\boldsymbol{\theta}_0} = -(\boldsymbol{\theta}_0^* - \boldsymbol{\theta}_0) J_{\boldsymbol{\theta}_0}. \tag{A14}$$

Moreover, show from the definition of $\mathcal{L}^*$ that

$$\partial_{\boldsymbol{\theta}} \mathcal{L}^*(\boldsymbol{\theta}_0^*) = 0 = \partial_{\boldsymbol{\theta}} \mathcal{L}(\boldsymbol{\theta}_0^*) + \frac{1}{n} \partial_{\boldsymbol{\theta}} tr(IJ^{-1}) \tag{A15}$$

Then, after Taylor expanding $\mathcal{L}(\boldsymbol{\theta}_0^*)$ around $\boldsymbol{\theta}_0$, it is seen that

$$(\boldsymbol{\theta}_0^* - \boldsymbol{\theta}_0) = \frac{1}{n} J_{\boldsymbol{\theta}_0}^{-1} \partial_{\boldsymbol{\theta}} tr(IJ^{-1}). \tag{A16}$$

Noting that $J_{\boldsymbol{\theta}}^{-1} = O(1)$ and $tr(IJ^{-1}) = O(1)$, it holds that $(\boldsymbol{\theta}_0^* - \boldsymbol{\theta}_0) = O(\delta^2)$. Therefore, $\frac{n}{2}(\boldsymbol{\theta}_0^* - \boldsymbol{\theta}_0)^T J_{\boldsymbol{\theta}^*}(\boldsymbol{\theta}_0^* - \boldsymbol{\theta}_0) = O(\delta^2)$, and can be neglected.
Then, the bias becomes

$$
\begin{aligned}
b(\boldsymbol{\theta}^*(\mathbf{X}_n), \mathbf{X}_n) &= -tr(I_{\boldsymbol{\theta}^*} J_{\boldsymbol{\theta}^*}^{-1}) \\
&- n\mathbb{E}_{\mathbf{X}_n}[(\boldsymbol{\theta}^* - \boldsymbol{\theta}_0^*)^T \partial_{\boldsymbol{\theta}} \ell(\boldsymbol{\theta}^*)] \\
&+ O(\delta).
\end{aligned}
$$

However, for the same reason, $\partial_{\boldsymbol{\theta}} \ell(\boldsymbol{\theta}^*) = O(\delta^2)$, so the last term $n\mathbb{E}_{\mathbf{X}_n}[(\boldsymbol{\theta}^* - \boldsymbol{\theta}_0^*)^T \partial_{\boldsymbol{\theta}} \ell(\boldsymbol{\theta}^*)] = O(\delta)$, and it too can be absorbed into the residual.

Therefore,

$$
\begin{aligned}
b(\boldsymbol{\theta}^*(\mathbf{X}_n), \mathbf{X}_n) &= -tr(I_{\boldsymbol{\theta}^*} J_{\boldsymbol{\theta}^*}^{-1}) \\
&+ O(\delta),
\end{aligned}
$$

and

$$-\mathcal{L}(\hat{\boldsymbol{\theta}}) = -\ell(\boldsymbol{\theta}^*(Z_n), Z_n) + \frac{1}{n} tr(I_{\hat{\boldsymbol{\theta}}} J_{\hat{\boldsymbol{\theta}}}^{-1}) + O_p(\delta^3) + C, \tag{A17}$$

where the constant C is composed of the neglected constant terms from earlier stages

$$
\begin{aligned}
C &= -\frac{1}{2}(\boldsymbol{\theta}_0^* - \boldsymbol{\theta}_0)^T J_{\boldsymbol{\theta}_0}(\boldsymbol{\theta}_0^* - \boldsymbol{\theta}_0) \\
&+ -(\boldsymbol{\theta}_0^* - \boldsymbol{\theta}_0)^T \partial_{\boldsymbol{\theta}} \mathcal{L}(\boldsymbol{\theta}_0^*).
\end{aligned}
$$

However, the last term is $O(\delta^3)$, and the first is $O(\delta^2)$, so this may be approximated as

$$C = -\frac{1}{2}(\boldsymbol{\theta}_0^* - \boldsymbol{\theta}_0)^T J_{\boldsymbol{\theta}_0}(\boldsymbol{\theta}_0^* - \boldsymbol{\theta}_0). \tag{A18}$$

Recalling again that $(\boldsymbol{\theta}_0^* - \boldsymbol{\theta}_0) = O(\delta^2)$, it is clear that $C = O(\delta^4)$, and may thus be absorbed into the $O(\delta^3)$ residual term. Therefore,

$$-\mathcal{L}(\hat{\boldsymbol{\theta}}) = -\ell(\boldsymbol{\theta}^*(Z_n), Z_n) + \frac{1}{n} tr(I_{\hat{\boldsymbol{\theta}}} J_{\hat{\boldsymbol{\theta}}}^{-1}) + O(\delta^3). \tag{A19}$$

Comparing this to the form of $\ell^*(\boldsymbol{\theta})$:

$$-\mathcal{L}(\hat{\boldsymbol{\theta}}) = -\ell^*(\boldsymbol{\theta}^*(Z_n), Z_n) + O_p(\delta^3). \tag{A20}$$

Thus, by minimizing $-\ell^*$ instead of $-\ell$, the first order terms of the prediction bias are canceled and, in expectation, a more accurate model is produced. □

References

1. Kullback, S.; Leibler, R.A. On Information and Sufficiency. *Ann. Math. Stat.* **1951**, *22*, 79–86. [CrossRef]
2. Esponda, I.; Pouzo, D. Berk–Nash Equilibrium: A Framework for Modeling Agents With Misspecified Models. *Econometrica* **2016**, *84*, 1093–1130. [CrossRef]
3. Jiang, B. Approximate Bayesian Computation with Kullback-Leibler Divergence as Data Discrepancy. In *Proceedings of the Twenty-First International Conference on Artificial Intelligence and Statistics, Playa Blanca, Spain, 9–11 April 2018*; Proceedings of Machine Learning Research; Storkey, A., Perez-Cruz, F., Eds.; PMLR: Playa Blanca, Lanzarote, 2018; Volume 84, pp. 1711–1721.
4. Nguyen, X.; Wainwright, M.J.; Jordan, M.I. Estimating Divergence Functionals and the Likelihood Ratio by Convex Risk Minimization. *IEEE Trans. Inf. Theory* **2010**, *56*, 5847–5861. [CrossRef]
5. Pérez-Cruz, F. Kullback-Leibler divergence estimation of continuous distributions. In Proceedings of the 2008 IEEE International Symposium on Information Theory, Toronto, ON, Canada, 6–11 July 2008; pp. 1666–1670.
6. Konishi, S.; Kitagawa, G. Generalised Information Criteria in Model Selection. *Biometrika* **1996**, *83*, 875–890. [CrossRef]
7. Roelofs, R. Measuring Generalization and Overfitting in Machine Learning. Ph.D. Thesis, University of California, Berkeley, CA, USA, 2019.
8. Miller, R.G. The Jackknife—A Review. *Biometrika* **1974**, *61*, 1–15.
9. Firth, D. Bias Reduction of Maximum Likelihood Estimates. *Biometrika* **1993**, *80*, 27–38. [CrossRef]
10. Kosmidis, I. Bias Reduction in Exponential Family Nonlinear Models. Ph.D. Thesis, University of Warwick, Coventry, UK, 2007.
11. Kosmidis, I.; Firth, D. A generic algorithm for reducing bias in parametric estimation. *Electron. J. Stat.* **2010**, *4*, 1097–1112. [CrossRef]
12. Kenne Pagui, E.C.; Salvan, A.; Sartori, N. Median bias reduction of maximum likelihood estimates. *Biometrika* **2017**, *104*, 923–938. [CrossRef]
13. Kosmidis, I. Bias in parametric estimation: Reduction and useful side-effects. *Wiley Interdiscip. Rev. Comput. Stat.* **2014**, *6*, 185–196. [CrossRef]
14. Heskes, T. Bias/Variance Decompositions for Likelihood-Based Estimators. *Neural Comput.* **1998**, *10*, 1425–1433. [CrossRef] [PubMed]
15. Akaike, H. Information Theory and an Extension of the Maximum Likelihood Principle. In Proceedings of the 2nd International Symposium on Information Theory, Tsahkadsor, Armenia, 2–8 September 1971; Petrov, B.N., Csaki, F., Eds.; Akademiai Kiado: Budapest, Hungary, 1973; pp. 267–281.
16. Takeuchi, K. Distribution of information statistics and validity criteria of models. *Math. Sci.* **1976**, *153*, 12–18.
17. Stone, M. An Asymptotic Equivalence of Choice of Model by Cross-Validation and Akaike's Criterion. *J. R. Stat. Soc. Ser. B Methodol.* **1977**, *39*, 44–47. [CrossRef]
18. Hastie, T.; Tibshirani R.; Friedman J. *The Elements of Statistical Learning*; Springer: New York, NY, USA, 2009.
19. Ross, A.; Doshi-Velez, F. Improving the Adversarial Robustness and Interpretability of Deep Neural Networks by Regularizing Their Input Gradients. In Proceedings of the AAAI Conference on Artificial Intelligence, New Orleans, LA, USA, 2–7 February 2018; Volume 32.
20. Golub, G.H.; Heath, M.; Wahba, G. Generalized Cross-Validation as a Method for Choosing a Good Ridge Parameter. *Technometrics* **1979**, *21*, 215–223. [CrossRef]
21. Yanagihara, H.; Tonda, T.; Matsumoto, C. Bias correction of cross-validation criterion based on Kullback–Leibler information under a general condition. *J. Multivar. Anal.* **2006**, *97*, 1965–1975. [CrossRef]
22. White, H. Maximum Likelihood Estimation of Misspecified Models. *Econometrica* **1982**, *50*, 1–25. [CrossRef]
23. Ward, T. Information Corrected Estimation: A Bias Correcting Parameter Estimation Method, Supplementary Materials, 2021. Available online: https://figshare.com/articles/software/Information_Corrected_Estimation_A_Bias_Correcting_Parameter_Estimation_Method_Supplementary_Materials/14312852/1 (accessed on 2 October 2021).
24. Friedman, J.H. Multivariate adaptive regression splines. *Ann. Stat.* **1991**, *19*, 1–67. [CrossRef]
25. Broyden, C.G. A Class of Methods for Solving Nonlinear Simultaneous Equations. *Math. Comput.* **1965**, *19*, 577–593. [CrossRef]

entropy

MDPI

Article

On Supervised Classification of Feature Vectors with Independent and Non-Identically Distributed Elements

Farzad Shahrivari and Nikola Zlatanov *

Electrical and Computer Systems Engineering, Monash University, Alliance Ln, Clayton, VIC 3168, Australia; Farzad.shahrivari@monash.edu
* Correspondence: Nikola.zlatanov@monash.edu

Abstract: In this paper, we investigate the problem of classifying feature vectors with mutually independent but non-identically distributed elements that take values from a finite alphabet set. First, we show the importance of this problem. Next, we propose a classifier and derive an analytical upper bound on its error probability. We show that the error probability moves to zero as the length of the feature vectors grows, even when there is only one training feature vector per label available. Thereby, we show that for this important problem at least one asymptotically optimal classifier exists. Finally, we provide numerical examples where we show that the performance of the proposed classifier outperforms conventional classification algorithms when the number of training data is small and the length of the feature vectors is sufficiently high.

Keywords: supervised classification; independent and non-identically distributed features; analytical error probability

Citation: Shahrivari, F.; Zlatanov, N. On Supervised Classification of Feature Vectors with Independent and Non-Identically Distributed Elements. *Entropy* **2021**, *23*, 1045. https://doi.org/10.3390/e23081045

Academic Editors: Lizhong Zheng and Chao Tian

Received: 26 July 2021
Accepted: 10 August 2021
Published: 13 August 2021

1. Introduction

1.1. Background

Supervised classification is a machine learning technique that maps an input feature vector to an output label based on a set of correctly labeled training data. There is no single learning algorithm that works best on all supervised learning problems, as shown by the no free lunch theorem in [1]. As a result, there are many algorithms proposed in the literature whose performance depends on the underlying problem and the amount of training data available. The most widely used algorithms in the literature are decision trees [2,3], Support Vector Machines (SVM) [4,5], Rule-Based Systems [6], naive Bayes classifiers [7], k-nearest neighbors (KNN) [8], logistic regressions, and neural networks [9,10].

1.2. Motivation

In the following, we discuss the motivation for this work.

1.2.1. Lack of Tight Upper Bounds on the Performance of Classifiers

In general, there are no tight upper bounds on the performance of the classifiers used in practice. Many of the previous works only provide experimental performance results. However, this approach has drawbacks. For example, one has to rely on the trial-and-error approach in order to develop a good classifier for a given problem, which impacts the reliability. Next, the algorithms whose performance has been verified only experimentally may work for a given problem, but may fail to work when applied to a similar problem. Finally, experimental results do not provide intuition into the underlying problem, whereas the analytical results provide the understanding of the underlying problem and the corresponding solutions.

Motivated by this, in the paper, we aim to investigate classifiers with analytical upper bounds on their performance.

1.2.2. Independent and Non-Identically Distributed Features

In general, we can categorize the statistical properties of the feature vectors, which are the input to the classifier, into three types. To this end, let $Y^n(X) = [Y_1(X), Y_2(X), \ldots, Y_n(X)]$ denote the input feature vector to the supervised classifier, where n is the length of the feature vector and X is the label to which the feature vector $Y^n(X)$ belongs. Then, we can distinguish the following three types of feature vectors depending on the statistics of the elements in the feature vector $Y^n(X)$.

The first type of feature vector is when the elements of $Y^n(X)$ are independent and identically distributed (i.i.d.). This is the simplest features model, but also the least applicable in practice. This model is identical to hypothesis testing, which has been well investigated in the literature [11–13]. As a result, tight upper bounds on the performance of supervised learning algorithms for this type of feature vector are available in the hypothesis testing literature. For instance, the authors in [11] showed that the posterior entropy and the maximum a posterior error probability decay to zero with the length of the feature vector at the identical exponential rate, where the maximum achievable exponent is the minimum Chernoff information. In [12], the authors determine the requirements for the length of the vector $Y^n(X)$ and the number of labels m in order to achieve vanishing exponential error probability in testing m hypothesis that minimizes the rejection zone. In [13], the authors provide an upper bound and a lower-bound on the error probability of Bayesian m-ary hypothesis testing in terms of conditional entropy.

The second type of feature vectors is when the elements of $Y^n(X)$ are mutually dependent and non-identically distributed (d.non-i.d.). This type of features model is the most general model and the most applicable in practice. However, it is also the most difficult to tackle analytically. As a result, supervised learning algorithms proposed for this features model lack analytical tight upper bounds on their performance [14–23]. This is because there are not any frameworks that produce closed-form results when deriving statistics of vectors with d.non-i.d. elements when the underlying distributions are unknown. Then how can we investigate analytically classifiers for practical scenarios when the feature vectors have d.non-i.d. elements? A possible approach leads us to the third type of feature vectors, explained in the following.

The third type of feature vectors is when the elements of $Y^n(X)$ are mutually independent but non-identically distributed (i.non-i.d.). This features model is much simpler than the d.non-i.d. features model and, more importantly, it is analytically tractable, as we show in this paper. Furthermore, this features model is applicable in practice. Specifically, there exists a class of algorithms, known as Independent Component Analysis (ICA), that transform vectors with d.non-i.d. elements into vectors with i.non-i.d. elements with a zero or a negligible loss of information [24–28]. The origins of ICA can be traced back to Barlow [29], who argued that a good representation of binary data can be achieved by an invertible transformation that transform vectors with d.non-i.d. elements into vectors with i.non-i.d. elements. Finding such a transformation with no prior information about the distribution of the data has been considered an open problem until recently [28]. Specifically, the authors in [28] show that this hard problem can be accurately solved with a branch and bound search tree algorithm, or tightly approximated with a series of linear problems. Thereby, the authors in [28] provide the first efficient set of solutions to Barlow's problem. So far, the complexity of the fastest such algorithm is $\mathcal{O}(n \times 2^n)$ [28]. Nevertheless, since there exist such invertible transformations (i.e., no loss of information) which can transform vectors with d.non-i.d. elements into vectors with i.non-i.d. elements, we can tackle the features model comprised of d.non-i.d. elements by first transforming it (without loss of information) into the features model comprised of i.non-i.d. elements and then tackling the i.non-i.d. features model.

Motivated by this, in this paper, we investigate supervised classification of feature vectors with i.non-i.d. elements.

1.2.3. Small Training Set

The main factor that impacts the accuracy of supervised classification is the amount of training data. In fact, most supervised algorithms are able to learn only if there is a very large set of training data available [30]. The main reason for this is the curse of dimensionality [31,32], which states that "the higher the dimensionality of the feature vectors, the more training data are needed for the supervised classifier" [33]. For example, supervised classification methods such as random forest [34,35] and KNN [36] suffer from the curse of dimensionality. However, having large training data sets is not always possible in practice. As a result, designing a supervised classification algorithm that exhibits good performance even when the training data set is extremely small is important.

Motivated by this, in this paper, we investigate supervised classifiers for the case when t training feature vectors per label are available, where $t = 1, 2, ...$

1.3. Contributions

In this paper, we propose an algorithm for supervised classification of feature vectors with i.non-i.d. elements when the number of training feature vectors per label is t, where $t = 1, 2, ...$ Next, we derive an upper bound on the error probability of the proposed classifier for uniformly distributed labels and prove that the error probability exponentially decays to zero when the length of the feature vector, n, grows, even when only one training vector per label is available, i.e., when $t = 1$. Hence, the proposed classification algorithm provides an asymptotically optimal performance even when the number of training vectors per label is extremely small. We compare the performance of the proposed classifier with the naive Bayes classifier and to the KNN algorithm. Our numerical results show that the proposed classifier significantly outperforms the naive Bayes classifier and the KNN algorithm when the number of training feature vectors per label is small and the length of the feature vectors n is sufficiently high.

The proposed algorithm is a form of the nearest neighbor classification algorithm, where the nearest neighbor is searched in the domain of empirical distributions. As a result, we refer to the algorithm as *the nearest empirical distribution*. The nearest empirical distribution algorithm is not new and, to the best of our knowledge, it was first proposed in [37] for the case when the elements of $Y^n(X)$ are i.i.d., i.e., for the equivalent problem of hypothesis testing. However, in this paper, we propose the nearest empirical distribution algorithm for the case when the elements of $Y^n(X)$ are i.non-i.d., which is much more complex than the problem of hypothesis testing where the elements of $Y^n(X)$ are i.i.d.

To the best of our knowledge, this is the first paper that investigates the important problem of classifying feature vectors with i.non-i.d. elements and provides an upper bound on its error probability. The novelty of this paper is not with the classifier itself, but rather in showing the importance of the problem of classifying feature vectors with i.non-i.d elements and in showing analytically that at least one classifier with an asymptotically optimal error probability exists when at least one training feature vectors per label is available.

The remainder of this paper is structured as follows. In Section 2, we formulate the considered classification problem. In Section 3, we provide our classifier and derive an upper bound on its error probability. In Section 4, we provide numerical examples of the performance on the proposed classifier. Finally, Section 5 concludes the paper.

2. Problem Formulation

The machine learning model is comprised of a label X, a feature vector $Y^n(X) = [Y_1(X), Y_2(X), \ldots, Y_n(X)]$ of length n mapped to the label X, and a learned label $\hat{X}$, as shown in Figure 1. In this paper, we adopt the information-theoretic style of notations and thereby random variables are denoted by capital letters and their realizations are denoted with small letters. The feature vector $Y^n(X)$ is the input to the machine learning algorithm whose aim is to detect the label X from the observed feature vector $Y^n(X)$.

The performance of the machine learning algorithm is measured by the error probability $\mathbb{P}_e = \Pr\{X \neq \hat{X}\}$.

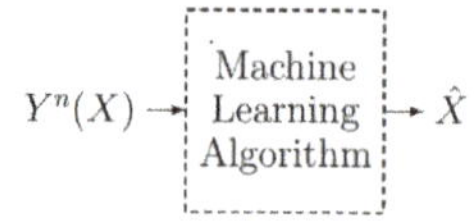

Figure 1. A typical structural modelling of the classification learning problem.

We adopt the modeling in [38–40] and represent the dependency between the label X and the feature vector $Y^n(X)$ via a joint probability distribution $p_{X,Y^n}(x,y^n)$. Now, in order to gain a better understanding of the problem, we include the joint probability distribution $p_{X,Y^n}(x,y^n)$ into the model in Figure 1. To this end, since $p_{X,Y^n}(x,y^n) = p_{Y^n|X}(y^n|x)p_X(x)$ holds, instead of $p_{X,Y^n}(x,y^n)$, we can include the conditional probability distribution $p_{Y^n|X}(y^n|x)$ and the probability distribution $p_X(x)$ into the model in Figure 1, and thereby obtain the model in Figure 2.

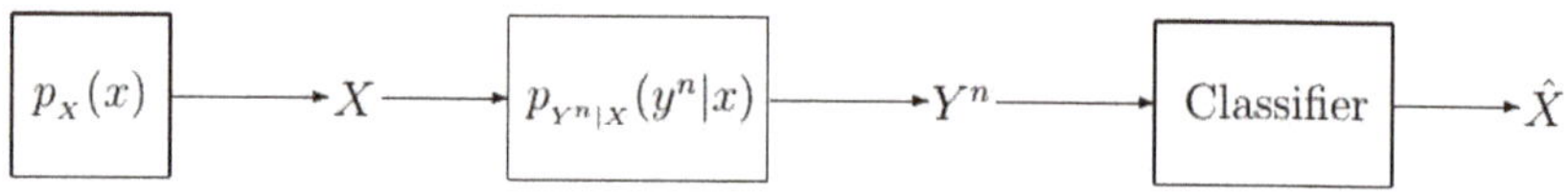

Figure 2. An alternative modeling of the classification learning problem.

Now, the classification learning model in Figure 2 is a system comprised of a label generating source X according to the distribution $p_X(x)$, a feature vector generator modelled by the conditional probability distribution $p_{Y^n|X}(y^n|x)$, a feature vector Y^n, a classifier that aims to detect X from the observed feature vector Y^n, and the detected label $\hat{X}$. Note that the system model in Figure 2 can be seen equivalently as a communication system comprised of a source X, a channel with input X and output Y^n, and a decoder (i.e., detector) that aims to detect X from Y^n. The notation used in this paper, letter X for labels and letter Y for features, is based on the notation used in information theory for modelling communication systems. In the classification model shown in Figure 2, we assume that the label X can take values from the set $\mathcal{X}$, according to $p_X(x) = 1/|\mathcal{X}|$, where $|\cdot|$ denotes the cardinality of a set. Next, we assume that the i-th element of the feature vector Y^n, Y_i, for $i = 1, 2, \ldots, n$, takes values from the set $\mathcal{Y} = \{y_1, y_2, \ldots, y_{|\mathcal{Y}|}\}$, according to the conditional probability distribution $p_{Y_i|X}(y_i|x)$.

Moreover, we assume that the elements of the feature vector Y^n are i.non-i.d. As a result, the feature vector Y^n takes values from the set $\mathcal{Y}^n$ according to the conditional probability distribution $p_{Y^n|X}(y^n|x)$ given by

$$p_{Y^n|X}(y^n|x) = p_{Y_1,Y_2,\ldots,Y_n|X}(y_1,y_2,\ldots,y_n|x) \stackrel{(a)}{=} \prod_{i=1}^{n} p_{Y_i|X}(y_i|x) \stackrel{(b)}{=} \prod_{i=1}^{n} p_i(y_i|x), \quad (1)$$

where (a) comes from the fact that elements in the feature vector Y^n are mutually independent and (b) is for the sake of notational simplicity, where p_i is used instead of $p_{Y_i|X}$. As a result of (1), the considered classification model in Figure 2 can be represented equivalently as in Figure 3.

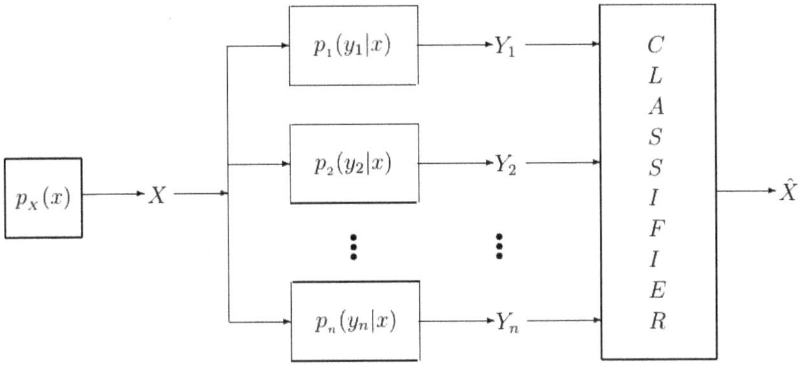

Figure 3. An alternative modelling of the classification learning problem when the elements of $Y^n(X)$ are mutually independent but non-identically distributed (i.non-i.d.).

Next, we assume that $p_i(y_i|x)$, $\forall i$, and thereby $p_{Y^n|X}(y^n|x)$ are unknown to the classifier. Instead, the classifier knows $\mathcal{X}$, $\mathcal{Y}$, and for each $x_i \in \mathcal{X}$, where $i = 1, 2, \ldots, |\mathcal{X}|$, it has access to a finite set of t correctly labelled input–output pairs $(x_i, \hat{y}^n_{i_1}), (x_i, \hat{y}^n_{i_2}), \ldots,$ $(x_i, \hat{y}^n_{i_t})$, denoted by $\mathcal{T}_i$, referred to as the training set for label x_i.

Finally, we assume that the following holds

$$\sum_{l=1}^{n} p_l(y|x_i) \neq \sum_{l=1}^{n} p_l(y|x_j), \text{ for } y \in \mathcal{Y} \text{ and } i \neq j. \tag{2}$$

The condition in (2) means that the distribution of the feature vectors $Y^n(X)$ for label $X = i$ is not a perturbation of distribution of the feature vectors $Y^n(X)$ for label $X = j$. As a result, the proposed classifier only applies to the subset of data vectors with i.non-i.d. elements that satisfy (2).

For the classification system model defined above and illustrated in Figure 3, we wish to propose a classifier that exhibits an asymptotically optimal error probability $\mathbb{P}_e = \Pr\{X \neq \hat{X}\}$ with respect to the length of Y^n, n, for any $t \geq 1$, i.e., for any $t \geq 1$, $\mathbb{P}_e \to 0$ as $n \to \infty$. Moreover, we wish to obtain an analytical upper bound on the error probability of the proposed classifier for a given t and n.

3. The Proposed Classifier and Its Performance

In this section, we propose our classifier, derive an analytical upper bound on its error probability, and prove that the classifier exhibits an asymptotically optimal performance when the length of the feature vector Y^n, n, satisfies $n \to \infty$. This is conducted in the following.

For a given vector $\mathrm{v}^n = (v_1, v_2, \ldots, v_n)$, let the Minkowski distance r be defined as

$$\|\mathrm{v}\|_r = \left(\sum_{i=1}^{n} v_i^r\right)^{(1/r)}. \tag{3}$$

Moreover, for a given feature vector $y^k = (y_1, y_2, , \ldots, y_k)$, let $\mathcal{I}[y^k = y]$ be a function defined as

$$\mathcal{I}[y^k = y] = \sum_{i=1}^{k} \mathcal{Z}[y_i = y], \tag{4}$$

where $\mathcal{Z}[y_i = y]$ is an indicator function assuming the value 1 if $y_i = y$ and 0 otherwise. Hence, $\mathcal{I}[y^k = y]$ counts the number of elements in Y^k that have the value y.

3.1. The Proposed Classifier

Let $\hat{y}_i^{nt}$ be a vector obtained by concatenating all training feature vectors for the input label x_i as

$$\hat{y}_i^{nt} = \left(\hat{y}_{i_1}^n, \hat{y}_{i_2}^n, \ldots, \hat{y}_{i_t}^n\right). \tag{5}$$

Let $P_{\hat{y}_i^{nt}}$ be the empirical probability distribution of the concatenated training feature vector for label x_i, $\hat{y}_i^{nt}$, given by

$$P_{\hat{y}_i^{nt}} = \left[\frac{\mathcal{I}[\hat{y}_i^{nt} = y_1]}{nt}, \frac{\mathcal{I}[\hat{y}_i^{nt} = y_2]}{nt}, \ldots, \frac{\mathcal{I}[\hat{y}_i^{nt} = y_{|\mathcal{Y}|}]}{nt}\right]. \tag{6}$$

Let y^n be the observed feature vector at the classifier whose label it wants to detect and let P_{y^n} denote the empirical probability distribution of y^n, given by

$$P_{y^n} = \left[\frac{\mathcal{I}[y^n = y_1]}{n}, \frac{\mathcal{I}[y^n = y_2]}{n}, \ldots, \frac{\mathcal{I}[y^n = y_{|\mathcal{Y}|}]}{n}\right]. \tag{7}$$

Using the above notations, we propose the following classifier.

Proposition 1. *For the considered system model, we propose a classifier with the following classification rule*

$$\hat{x} = x_i, \text{ where } i = \arg\min_i \left\|P_{y^n} - P_{\hat{y}_i^{nt}}\right\|_r, \tag{8}$$

where $r \geq 1$ and ties are resolved by assigning the label among the ties uniformly at random. (For example, if $\|P_{y^n} - P_{\hat{y}_i^{nt}}\|_r = \|P_{y^n} - P_{\hat{y}_j^{nt}}\|_r$ holds for, $i \neq j$, we set $\hat{x} = x_i$ or $\hat{x} = x_j$ uniformly at random).

As seen from (8), the proposed classifier assigns the label x_i if the empirical probability distribution of the concatenated training feature vector mapped to label x_i, $P_{\hat{y}_i^{nt}}$ is the closest, in terms of Minkowski distance r, to the empirical probability distribution of the observed feature vector P_{y^n}. In that sense, the proposed classifier can be considered as the nearest empirical distribution classifier.

3.2. Upper Bound on the Error Probability

The following theorem establishes an upper bound on the error probability of the proposed classifier.

Theorem 1. *Let $\bar{\mathrm{P}}_j$, for $j = 1, 2, \ldots, |\mathcal{X}|$, be a vector defined as*

$$\bar{\mathrm{P}}_j = \left[\bar{\mathrm{p}}(y_1|x_j), \bar{\mathrm{p}}(y_2|x_j), \ldots, \bar{\mathrm{p}}(y_{|\mathcal{Y}|}|x_j)\right], \tag{9}$$

where $\bar{\mathrm{p}}(y|x_j)$ is given by

$$\bar{\mathrm{p}}(y|x_j) = \frac{1}{n}\sum_{k=1}^{n} p_k(y|x_j). \tag{10}$$

Then, for a given $r \geq 1$, the error probability of the proposed classifier is upper bounded by

$$\mathbb{P}_e \leq 2|\mathcal{Y}|e^{-2n\epsilon^2} + 2|\mathcal{Y}|e^{-2nt^{1/3}\epsilon^2}, \tag{11}$$

where ϵ is given by

$$\epsilon = \min_{\substack{i,j \\ i\neq j}} \frac{||P_{\hat{y}_i^{nt}} - \bar{\mathrm{P}}_j||_r}{(2+t^{-1/3})|\mathcal{Y}|^{1/r}}. \tag{12}$$

Proof of Theorem 1. Without loss of generality we assume that x_1 is the input to $p_{Y^n|X}(y^n|x)$ and y^n is observed.

Let $\mathcal{A}_k^\epsilon$, for $1 \leq k \leq |\mathcal{Y}|$, be a set defined as

$$\mathcal{A}_k^\epsilon = \left\{ y^n : \left| \frac{\mathcal{I}[y^n = y_k]}{n} - \bar{\mathrm{p}}(y_k|x_1) \right| \leq \epsilon \right\}. \tag{13}$$

Furthermore, let $\mathcal{B}_k^\epsilon$, for $1 \leq k \leq |\mathcal{Y}|$, be a set defined as

$$\mathcal{B}_k^\epsilon = \left\{ \hat{y}^{nt} : \left| \frac{\mathcal{I}[\hat{y}^{nT} = y_k]}{nt} - \bar{\mathrm{p}}(y_k|x_1) \right| \leq \frac{\epsilon}{\sqrt[3]{t}} \right\}. \tag{14}$$

Let $\mathcal{A}^\epsilon = \bigcap_{k=1}^{|\mathcal{Y}|} \mathcal{A}_k^\epsilon$ and $\mathcal{B}^\epsilon = \bigcap_{k=1}^{|\mathcal{Y}|} \mathcal{B}_k^\epsilon$. Now, for any $y^n \in \mathcal{A}^\epsilon$, we have

$$\left(\sum_{k=1}^{|\mathcal{Y}|} \left| \frac{\mathcal{I}[y^n = y_k]}{n} - \bar{\mathrm{p}}(y_k|x_1) \right|^r \right)^{1/r} \overset{(a)}{\leq} \left(\sum_{k=1}^{|\mathcal{Y}|} \epsilon^r \right)^{1/r}, \tag{15}$$

where (a) follows from (13). Moreover, for $\hat{y}_1^{nt} \in \mathcal{B}^\epsilon$, we have

$$\left(\sum_{k=1}^{|\mathcal{Y}|} \left| \frac{\mathcal{I}[\hat{y}_1^{nt} = y_k]}{nt} - \bar{\mathrm{p}}(y_k|x_1) \right|^r \right)^{1/r} \overset{(a)}{\leq} \left(\sum_{k=1}^{|\mathcal{Y}|} \left(\frac{\epsilon}{\sqrt[3]{t}} \right)^r \right)^{1/r}, \tag{16}$$

where (a) follows from (14). Next, we have the following upper bound

$$\begin{aligned}
&\left(\sum_{k=1}^{|\mathcal{Y}|} \left| \frac{\mathcal{I}[y^n = y_k]}{n} - \frac{\mathcal{I}[\hat{y}_1^{nt} = y_k]}{nt} \right|^r \right)^{1/r} \\
&= \left(\sum_{k=1}^{|\mathcal{Y}|} \left| \frac{\mathcal{I}[y^n = y_k]}{n} - \bar{\mathrm{p}}(y_k|x_1) - \left(\frac{\mathcal{I}[\hat{y}_1^{nt} = y_k]}{nt} - \bar{\mathrm{p}}(y_k|x_1) \right) \right|^r \right)^{1/r} \\
&\overset{(a)}{\leq} \left(\sum_{k=1}^{|\mathcal{Y}|} \left| \frac{\mathcal{I}[y^n = y_k]}{n} - \bar{\mathrm{p}}(y_k|x_1) \right|^r \right)^{1/r} + \left(\sum_{k=1}^{|\mathcal{Y}|} \left| \frac{\mathcal{I}[\hat{y}_1^{nt} = y_k]}{nt} - \bar{\mathrm{p}}(y_k|x_1) \right|^r \right)^{1/r},
\end{aligned} \tag{17}$$

where (a) follows from the Minkowski inequality. Combining (15)–(17), we obtain

$$\left(\sum_{k=1}^{|\mathcal{Y}|} \left| \frac{\mathcal{I}[y^n = y_k]}{n} - \frac{\mathcal{I}[\hat{y}_1^{nt} = y_k]}{nt} \right|^r \right)^{1/r} \leq |\mathcal{Y}|^{1/r}\epsilon + |\mathcal{Y}|^{1/r}\frac{\epsilon}{\sqrt[3]{t}}. \tag{18}$$

Hence, the Minkowski distance between the empirical probability distribution of the observed vector y^n and the empirical probability distribution of the concatenated training vector for label x_1 is upper bounded by the right hand side of (18). We now derive a lower bound for $\hat{y}_i^{nt}$, where $i \neq 1$. For any x_i, such that $i \neq 1$, we have

$$\begin{aligned}
&\left(\sum_{k=1}^{|\mathcal{Y}|}\left|\frac{\mathcal{I}[y^n = y_k]}{n} - \frac{\mathcal{I}[\hat{y}_i^{nt} = y_k]}{nt}\right|^r\right)^{1/r} + \left(\sum_{k=1}^{|\mathcal{Y}|}\epsilon^r\right)^{1/r} \\
&\overset{(a)}{\geq} \left(\sum_{k=1}^{|\mathcal{Y}|}\left|\frac{\mathcal{I}[y^n = y_k]}{n} - \frac{\mathcal{I}[\hat{y}_i^{nt} = y_k]}{nt}\right|^r\right)^{1/r} + \left(\sum_{k=1}^{|\mathcal{Y}|}\left|\frac{\mathcal{I}[y^n = y_k]}{n} - \bar{\mathrm{p}}(y_k|x_1)\right|^r\right)^{1/r} \\
&\overset{(b)}{\geq} \left(\sum_{k=1}^{|\mathcal{Y}|}\left|\frac{\mathcal{I}[\hat{y}_i^{nt} = y_k]}{nt} - \bar{\mathrm{p}}(y_k|x_1)\right|^r\right)^{1/r}, \qquad (19)
\end{aligned}$$

where (a) follows from (15) and (b) is again due to the Minkowski inequality. The expression in (19), can be written equivalently as

$$\begin{aligned}
&\left(\sum_{k=1}^{|\mathcal{Y}|}\left|\frac{\mathcal{I}[y^n = y_k]}{n} - \frac{\mathcal{I}[\hat{y}_i^{nt} = y_k]}{nt}\right|^r\right)^{1/r} \\
&\geq \left(\sum_{k=1}^{|\mathcal{Y}|}\left|\frac{\mathcal{I}[\hat{y}_i^{nt} = y_k]}{nt} - \bar{\mathrm{p}}(y_k|x_1)\right|^r\right)^{1/r} - |\mathcal{Y}|^{1/r}\epsilon, \qquad (20)
\end{aligned}$$

where $i \neq 1$. Now, using the definitions of $P_{\hat{y}_i^{nt}}$ and $\bar{\mathrm{P}}_1$ given by (6) and (9), respectively, into (20) we can replace the expression in the right-hand side of (20) by $\|P_{\hat{y}_i^{nt}} - \bar{\mathrm{P}}_1\|_r$, and thereby for any $i \neq 1$ we have

$$\left(\sum_{k=1}^{|\mathcal{Y}|}\left|\frac{\mathcal{I}[y^n = y_k]}{n} - \frac{\mathcal{I}[\hat{y}_i^{nt} = y_k]}{nt}\right|^r\right)^{1/r} \geq \|P_{\hat{y}_i^{nt}} - \bar{\mathrm{P}}_1\|_r - |\mathcal{Y}|^{1/r}\epsilon. \qquad (21)$$

The expression in (21) represents a lower bound on the Minkowski r distance between the empirical probability distribution of the observed vector y^n and the empirical probability distribution of the concatenated training vector for any label x_i, where $i \neq 1$.

Using the bounds in (18) and (21), we now relate the left-hand sides of (18) and (21). As long as the following inequality holds for each $i \neq 1$,

$$|\mathcal{Y}|^{1/r}\epsilon\left(1 + \frac{1}{\sqrt[3]{t}}\right) < \|P_{\hat{y}_i^{nt}} - \bar{\mathrm{P}}_1\|_r - |\mathcal{Y}|^{1/r}\epsilon, \qquad (22)$$

which is equivalent to the following for $i \neq 1$

$$\epsilon < \frac{\|P_{\hat{y}_i^{nt}} - \bar{\mathrm{P}}_1\|_r}{(2 + t^{-1/3})|\mathcal{Y}|^{1/r}}, \qquad (23)$$

$$\begin{aligned}
\left(\sum_{k=1}^{|\mathcal{Y}|}\left|\frac{\mathcal{I}[y^n = y_k]}{n} - \frac{\mathcal{I}[\hat{y}_1^{nt} = y_k]}{nt}\right|^r\right)^{1/r} &\overset{(a)}{\leq} |\mathcal{Y}|^{1/r}\epsilon\left(1 + \frac{1}{\sqrt[3]{t}}\right) \\
&\overset{(b)}{<} \|P_{\hat{y}_i^{nt}} - \bar{\mathrm{P}}_1\|_r - |\mathcal{Y}|^{1/r}\epsilon \\
&\overset{(c)}{\leq} \left(\sum_{k=1}^{|\mathcal{Y}|}\left|\frac{\mathcal{I}[y^n = y_k]}{n} - \frac{\mathcal{I}[\hat{y}_i^{nt} = y_k]}{nt}\right|^r\right)^{1/r}, \qquad (24)
\end{aligned}$$

where (a), (b), and (c) follow from (18), (22), and (21), respectively. Thereby, from (24), we have the following for $i \neq 1$

$$\left(\sum_{k=1}^{|\mathcal{Y}|}\left|\frac{\mathcal{I}[y^n = y_k]}{n} - \frac{\mathcal{I}[\hat{y}_1^{nT} = y_k]}{nT}\right|^r\right)^{1/r} < \left(\sum_{k=1}^{|\mathcal{Y}|}\left|\frac{\mathcal{I}[y^n = y_k]}{n} - \frac{\mathcal{I}[\hat{y}_i^{nT} = y_k]}{nT}\right|^r\right)^{1/r}. \quad (25)$$

Note that the right- and left-hand sides of (25) can be replaced by the Minkowski distance of the vectors

$$\mathrm{v}_1 = \left[\frac{\mathcal{I}[y^n = y_1]}{n} - \frac{\mathcal{I}[\hat{y}_1^{nt} = y_1]}{nt}, \ldots, \frac{\mathcal{I}[y^n = y_{|\mathcal{Y}|}]}{n} - \frac{\mathcal{I}[\hat{y}_1^{nt} = y_{|\mathcal{Y}|}]}{nt}\right], \quad (26)$$

and

$$\mathrm{v}_2 = \left[\frac{\mathcal{I}[y^n = y_1]}{n} - \frac{\mathcal{I}[\hat{y}_i^{nt} = y_1]}{nt}, \ldots, \frac{\mathcal{I}[y^n = y_{|\mathcal{Y}|}]}{n} - \frac{\mathcal{I}[\hat{y}_i^{nt} = y_{|\mathcal{Y}|}]}{nt}\right], \quad (27)$$

respectively. Now, (26) and (27) can be replaced by $P_{y^n} - P_{\hat{y}_1^{nt}}$ and $P_{y^n} - P_{\hat{y}_i^{nt}}$, respectively, by the definitions of P_{y^n} and $P_{\hat{y}_i^{nt}}$ given by (7) and (6), respectively. Therefore, (25) can be written equivalently as

$$\|P_{y^n} - P_{\hat{y}_1^{nt}}\|_r < \|P_{y^n} - P_{\hat{y}_i^{nt}}\|_r. \quad (28)$$

Now, let us highlight what we have obtained. We obtained that there is an ϵ for which if (23) holds for $i \neq 1$, and for that ϵ there are sets $\mathcal{A}^\epsilon$ and $\mathcal{B}^\epsilon$ for which $y^n \in \mathcal{A}^\epsilon$ and $\hat{y}_1^{nt} \in \mathcal{B}^\epsilon$ then (28) holds for $i \neq 1$, and thereby our classifier will detect that x_1 is the correct label. Using this, we can upper bound the error probability as

$$\begin{aligned}\mathbb{P}_\mathrm{e} &= 1 - \Pr\{\hat{x}_1 = x_1\}\\ &\leq 1 - \Pr\Big\{(y^n \in \mathcal{A}^\epsilon) \cap (\hat{y}_1^{nt} \in \mathcal{B}^\epsilon) | \epsilon \in \mathcal{S}\Big\},\end{aligned} \quad (29)$$

where $\mathcal{S}$ is a set defined as

$$\mathcal{S} = \left\{\epsilon : \epsilon \leq \min_{\substack{i\\ i\neq 1}} \frac{\|P_{\hat{y}_i^{nt}} - \bar{\mathrm{P}}_1\|_r}{(2 + t^{-1/3})|\mathcal{Y}|^{1/r}}\right\}. \quad (30)$$

In the following, we derive the expression in (29). The right-hand side of (29) can be upper bounded as

$$\begin{aligned}1 - \Pr\Big\{(y^n \in \mathcal{A}^\epsilon) \cap (\hat{y}_1^{nt} \in \mathcal{B}^\epsilon) | \epsilon \in \mathcal{S}\Big\} &= \Pr\Big\{(y^n \notin \mathcal{A}^\epsilon) \cup (\hat{y}_1^{nt} \notin \mathcal{B}^\epsilon) | \epsilon \in \mathcal{S}\Big\}\\ &\overset{(a)}{\leq} \Pr\{y^n \notin \mathcal{A}^\epsilon | \epsilon \in \mathcal{S}\} + \Pr\Big\{\hat{y}_1^{nt} \notin \mathcal{B}^\epsilon | \epsilon \in \mathcal{S}\Big\},\end{aligned} \quad (31)$$

where (a) follows from Boole's inequality. Now, note that we have the following upper bound for the first expression in the right-hand side of (31)

$$\begin{aligned}\Pr\{y^n \notin \mathcal{A}^\epsilon | \epsilon \in \mathcal{S}\} &= \Pr\left\{ y^n \notin \bigcap_{k=1}^{|\mathcal{Y}|} \mathcal{A}_k^\epsilon \middle| \epsilon \in \mathcal{S} \right\} \\ &= \Pr\left\{ y^n \in \bigcup_{k=1}^{|\mathcal{Y}|} \overline{\mathcal{A}_k^\epsilon} \middle| \epsilon \in \mathcal{S} \right\} \\ &\overset{(a)}{\leq} \sum_{k=1}^{|\mathcal{Y}|} \Pr\{y^n \in \overline{\mathcal{A}_k^\epsilon} | \epsilon \in \mathcal{S}\} \\ &= \sum_{k=1}^{|\mathcal{Y}|} \Pr\left\{ \left| \frac{\mathcal{I}[y^n = y_k]}{n} - \bar{\mathrm{p}}(y_k|x_1) \right| > \epsilon \middle| \epsilon \in \mathcal{S} \right\} \\ &= \sum_{k=1}^{|\mathcal{Y}|} \Pr\left\{ \left| \sum_{j=1}^{n} \frac{\mathcal{Z}[y_j = y_k]}{n} - \bar{\mathrm{p}}(y_k|x_1) \right| > \epsilon \middle| \epsilon \in \mathcal{S} \right\}, \end{aligned} \tag{32}$$

where $\overline{\mathcal{A}_k^\epsilon}$ is the complement of $\mathcal{A}_k^\epsilon$ and (a) follows from Boole's inequality. Note that $\mathcal{Z}[y_1 = y_k], \mathcal{Z}[y_2 = y_k], \ldots, \mathcal{Z}[y_n = y_k]$ in (32) are n independent Bernoulli random variables with probabilities of success $p_1(y_k|x_1), p_2(y_k|x_1), \ldots, p_n(y_k|x_1)$, respectively. Let $\mathcal{W}[y_k]$ be a binomial random variable with parameters $(n, \bar{\mathrm{p}}(y_k|x_1))$. We proceed the proof by introducing the following well-known Hoefdding's Theorem from [41]. □

Theorem 2 (Hoeffding [41]). *Assume that $Z_1, Z_2, \ldots,$ and Z_n are n independent Bernoulli random variables with probabilities of success $p_1, p_2, \ldots,$ and p_n, respectively. Next, let Z be defined as $Z = Z_1 + Z_2 + \ldots + Z_n$ and, let $\bar{\mathrm{p}}$ be defined as $\bar{\mathrm{p}} = (p_1 + p_2 + \ldots + p_n)/n$. Let W be a binomial random variable with parameters $(n, \bar{\mathrm{p}})$. Then, for a given a and b, where $0 \leq a \leq n\bar{\mathrm{p}} \leq b \leq n$ holds, we have*

$$\Pr\{a \leq W \leq b\} \leq \Pr\{a \leq Z \leq b\}. \tag{33}$$

In other words, the probability distribution of W is more dispersed around its mean $n\bar{\mathrm{p}}$ than is the probability distribution of Z. Except in the trivial case when $a = b = 0$, the bound in (33) holds with equality if and only if $p_1 = \ldots = p_n = \bar{\mathrm{p}}$.

Proof of Theorem 2. Please refer to [41]. □

Setting $a = n(\bar{\mathrm{p}} - \delta)$ and $b = n(\bar{\mathrm{p}} + \delta)$ in (33), we obtain

$$\Pr\{n(\bar{\mathrm{p}} - \delta) \leq W \leq n(\bar{\mathrm{p}} + \delta)\} \leq \Pr\{n(\bar{\mathrm{p}} - \delta) \leq Z \leq n(\bar{\mathrm{p}} + \delta)\}. \tag{34}$$

Using (34), we have the following upper bound

$$\begin{aligned}\Pr\left\{ \left| \frac{Z}{n} - \bar{\mathrm{p}} \right| > \delta \right\} &= 1 - \Pr\{n(\bar{\mathrm{p}} - \delta) \leq Z \leq n(\bar{\mathrm{p}} + \delta)\} \\ &\overset{(a)}{\leq} 1 - \Pr\{n(\bar{\mathrm{p}} - \delta) \leq W \leq n(\bar{\mathrm{p}} + \delta)\} \\ &= \Pr\left\{ \left| \frac{W}{n} - \bar{\mathrm{p}} \right| > \delta \right\}, \end{aligned} \tag{35}$$

where (a) follows from (34).

We now turn to the proof of Theorem 1. According to Theorem 2, the probability distribution of $\mathcal{W}[y_k]$ is more dispersed around its mean $n\bar{\mathrm{p}}(y_k|x_1)$ than is the probability distribution of $\sum_{1\le j\le n}\mathcal{Z}[y_j=y_k]$. Therefore, we can upper bound the probability in the last line of (32) as

$$\Pr\left\{\left|\sum_{j=1}^{n}\frac{\mathcal{Z}[y_j=y_k]}{n}-\bar{\mathrm{p}}(y_k|x_1)\right|>\epsilon\middle|\epsilon\in\mathcal{S}\right\}\overset{(a)}{\le}\Pr\left\{\left|\frac{\mathcal{W}[y_k]}{n}-\bar{\mathrm{p}}(y_k|x_1)\right|>\epsilon\middle|\epsilon\in\mathcal{S}\right\},\tag{36}$$

where $\epsilon\in\mathcal{S}$ is defined in (30) and (a) follows from (35). Now, let us introduce another well-known Hoeffding's Theorem from [42].

Theorem 3 (Hoeffding's inequality [42]). *Let $W_1,W_2,\dots,W_n$ be n independent random variables such that for each $1\le i\le n$, we have $\Pr\{W_i\in[a_i,b_i]\}=1$. Then for S_n, defined as $S_n=\sum_{i=1}^{n}W_i$, we have*

$$\Pr\left\{S_n-\mathbb{E}\left[S_n\right]\ge\delta\right\}\le\exp\left(-\frac{2\delta^2}{\sum_{i=1}^{n}(b_i-a_i)^2}\right),\tag{37}$$

where $\mathbb{E}\left[S_n\right]$ is the expectation of S_n.

Proof of Theorem 3. Please refer to [42]. □

Back to (36), by using the result of (37) for $a_i=0$ and $b_i=1$ since the binomial random variable $\mathcal{W}[y_k]$ can take values 0 or 1, respectively, we have

$$\begin{aligned}\Pr\left\{\left|\sum_{j=1}^{n}\frac{\mathcal{Z}[y_j=y_k]}{n}-\bar{\mathrm{p}}(y_k|x_1)\right|>\epsilon\middle|\epsilon\in\mathcal{S}\right\}&\le 2\exp\left(-\frac{2n^2\epsilon^2}{\sum_{1\le i\le n}(1-0)^2}\right)\\&\le 2\mathrm{e}^{-2n\epsilon^2},\end{aligned}\tag{38}$$

where $\epsilon\in\mathcal{S}$ is defined in (30). Inserting (38) into (32), we obtain the following upper bound

$$\Pr\{y^n\notin\mathcal{A}^\epsilon|\epsilon\in\mathcal{S}\}\le 2|\mathcal{Y}|\mathrm{e}^{-2n\epsilon^2}.\tag{39}$$

Similarly, we have the following result for the second expression in the right-hand side of (31)

$$\begin{aligned}\Pr\left\{\hat{y}_1^{nt}\notin\mathcal{B}^\epsilon|\epsilon\in\mathcal{S}\right\}&=\Pr\left\{\hat{y}_1^{nt}\notin\bigcap_{k=1}^{|\mathcal{Y}|}\mathcal{B}_k^\epsilon\middle|\epsilon\in\mathcal{S}\right\}\\&=\Pr\left\{\hat{y}_1^{nt}\in\bigcup_{k=1}^{|\mathcal{Y}|}\overline{\mathcal{B}_k^\epsilon}\middle|\epsilon\in\mathcal{S}\right\}\\&\overset{(a)}{\le}\sum_{k=1}^{|\mathcal{Y}|}\Pr\{\hat{y}_1^{nt}\in\overline{\mathcal{B}_k^\epsilon}|\epsilon\in\mathcal{S}\}\\&=\sum_{k=1}^{|\mathcal{Y}|}\Pr\left\{\left|\frac{\mathcal{I}[\hat{y}_1^{nt}=y_k]}{nt}-\bar{\mathrm{p}}(y_k|x_1)\right|>\frac{\epsilon}{\sqrt[3]{t}}\middle|\epsilon\in\mathcal{S}\right\}\\&=\sum_{k=1}^{|\mathcal{Y}|}\Pr\left\{\left|\sum_{j=1}^{nt}\frac{\mathcal{Z}[y_j=y_k]}{nt}-\bar{\mathrm{p}}(y_k|x_1)\right|>\frac{\epsilon}{\sqrt[3]{t}}\middle|\epsilon\in\mathcal{S}\right\},\end{aligned}\tag{40}$$

where again (a) follows from Boole's inequality. Note that due to (5), for any integer number l such that $0\le l\le t-1$ the random variables $\mathcal{Z}[y_{nl+1}=y_k],\mathcal{Z}[y_{nl+2}=y_k],\dots,$ and $\mathcal{Z}[y_{nl+n}=y_k]$ in (40) are n independent Bernoulli random variables with the probabilities of success $p_1(y_k|x_1),p_2(y_k|x_1),\dots,$ and $p_n(y_k|x_1)$, respectively $(y_{nl+1},y_{nl+2},\dots,y_{nl+n}$ are elements of $\hat{y}_{1_{l+1}}^{n})$. In addition, note that

$$\bar{p}(y_k|x_1) = \frac{1}{n}\sum_{j=1}^{n} p_j(y_k|x_1)$$
$$= \frac{1}{nt}\left(\sum_{l=0}^{t-1}\sum_{j=1}^{n} p_j(y_k|x_1)\right). \quad (41)$$

Notice that for each $0 \leq l \leq t-1$, $p_1(y_k|x_1) + p_2(y_k|x_1) + \ldots + p_n(y_k|x_1)$ is the summation of the probabilities of success of the random variables $\mathcal{Z}[y_{nl+1} = y_k], \mathcal{Z}[y_{nl+2} = y_k], \ldots,$ and $\mathcal{Z}[y_{nl+n} = y_k]$. Thereby, the last expression on the right-hand side of (41) is the average probability of success of random variables $\mathcal{Z}[y_j = y_k]$ for $1 \leq j \leq nt$. Now, let $\mathcal{W}[y_k]$ be a binomial random variable with parameters $(nt, \bar{p}(y_k|x_1))$. Once again, according to Theorem 2, the probability distribution of $\mathcal{W}[y_k]$ is more dispersed around its mean $nt\bar{p}(y_k|x_1))$ than is the probability distribution of $\sum_{1\leq j\leq nt} \mathcal{Z}[y_j = y_k]$. Therefore, the probability in the last line of (40) can be upper bounded as

$$\Pr\left\{\left|\sum_{j=1}^{nt}\frac{\mathcal{Z}[y_j = y_k]}{nt} - \bar{p}(y_k|x_1)\right| > \frac{\epsilon}{\sqrt[3]{t}}\middle|\epsilon \in \mathcal{S}\right\} \overset{(a)}{\leq} \Pr\left\{\left|\frac{\mathcal{W}[y_k]}{nt} - \bar{p}(y_k|x_1)\right| > \frac{\epsilon}{\sqrt[3]{t}}\middle|\epsilon \in \mathcal{S}\right\}$$
$$\overset{(b)}{\leq} 2\exp\left(-\frac{2(nt)^2(t^{-1/3}\epsilon)^2}{\sum_{1\leq i\leq nt}(1-0)^2}\right)$$
$$\leq 2\mathrm{e}^{-2nt\left(t^{-2/3}\epsilon^2\right)}$$
$$= 2\mathrm{e}^{-2nt^{1/3}\epsilon^2}, \quad (42)$$

where $\epsilon \in \mathcal{S}$, defined in (30), (a) follows from (35) (in which n is replaced by nt), and (b) is the result of (37) for $a_i = 0$ and $b_i = 1$ since the binomial random variable $\mathcal{W}[y_k]$ can take values 0 or 1, respectively. Inserting (42) into (40), we have the following upper bound

$$\Pr\left\{\hat{y}_1^{nt} \notin \mathcal{B}^{\epsilon}\middle|\epsilon \in \mathcal{S}\right\} \leq 2|\mathcal{Y}|\mathrm{e}^{-2nt^{1/3}\epsilon^2}. \quad (43)$$

Inserting (39) and (43) into (31), and then inserting (31) into (29), we obtain the following upper bound for the error probability

$$\mathbb{P}_{\mathrm{e}} \leq 2|\mathcal{Y}|\mathrm{e}^{-2n\epsilon^2} + 2|\mathcal{Y}|\mathrm{e}^{-2nt^{1/3}\epsilon^2}, \quad (44)$$

where

$$\epsilon = \min_{\substack{i,j\\ i\neq j}} \frac{\|P_{\hat{y}_i^{nt}} - \bar{P}_j\|_r}{(2+t^{-1/3})|\mathcal{Y}|^{1/r}}, \quad (45)$$

which is the optimal value of ϵ that exhibits the tightest upper bound for the error probability $\mathbb{P}_{\mathrm{e}}$ given by (44). This completes the proof of Theorem 1.

The following corollary provides a simplified upper bound on the error probability when $t \to \infty$.

Corollary 1. *When the number of training vectors per label reaches infinity, i.e., when $t \to \infty$, which is equivalently to the case when the probability distribution $p(y^n|x)$ is known at the classifier, the error probability of the proposed classifier is upper bounded as*

$$\mathbb{P}_{\mathrm{e}} \leq 2|\mathcal{Y}|\mathrm{e}^{-2n\epsilon^2}, \quad (46)$$

where ϵ is given by

$$\epsilon = \min_{\substack{i,j \\ i \neq j}} \frac{\|\bar{\mathrm{P}}_i - \bar{\mathrm{P}}_j\|_r}{2|\mathcal{Y}|^{1/r}}. \tag{47}$$

Proof. The proof is straightforward. □

As can be seen from (8) and (11), the performance of the proposed classifier depends on r. We cannot derive the optimal value of r that minimizes the error probability since we do not have the exact expression of the error probability, we only have its upper bound. On the other hand, in practice, the optimal r with respect to the upper bound on the error probability also cannot be derived since the upper bound depends on $\bar{\mathrm{P}}_j$, which would be unknown in practice due to $p_{Y^n|X}(y^n|x)$ being unknown. As a result, for our numerical examples, we consider the Euclidean distance ($r = 2$), which is one of the most widely used distance metrics in practice.

The following corollary establishes the asymptotic optimality of the proposed classifier with respect to n.

Corollary 2. *The proposed classifier has an error probability that satisfies $\mathbb{P}_e \to 0$ as $n \to \infty$ if $|\mathcal{Y}| \leq \mathcal{O}(n^m)$, m is fixed, and $r > 2m$. Here, n^m indicates the dimension of our space, i.e., maximum number of alphabets each element in the feature vector y^n can take. Thereby, the proposed classifier is asymptotically optimal .*

Proof. For the proof, please see Appendix A. □

4. Simulation Results

In this section, we provide simulation results of the performance of the proposed classifier for $r = 2$ and compare it to benchmark schemes. The benchmark schemes that we adopt for comparison are the naive Bayes classifier and the KNN algorithm. We cannot adopt a classifier based on a neural network since neural networks require a very large training set, which we assume is not available. For the naive Bayes classifier, the probability distribution $p_{Y^n|X}(y^n|x)$ is estimated from the training vectors as follows. Let again $\hat{y}_i^{nt}$ be a vector obtained by concatenating all training feature vectors for the input label x_i as in (5). Then, the estimated probability distribution of $p(y_j = y|x_i)$, denoted by $\hat{p}(y_j = y|x_i)$, is found as

$$\hat{p}(y_j = y|x_i) = \frac{\mathcal{I}[\hat{y}_i^{nt} = y]}{nt}, \tag{48}$$

and the naive Bayes classifier decides according to

$$\hat{x} = \arg\max_{x_i} \prod_{k=1}^{n} \hat{p}(y_k|x_i). \tag{49}$$

The main problem of the naive Bayes classifier occurs when an alphabet $y_j \in \mathcal{Y}$ is not present in the training feature vectors. In that case, $\hat{p}(y_j|x_i)$ in (48) is $\hat{p}(y_j|x_i) = 0$, $\forall x_i \in \mathcal{X}$ and, as a result, the right hand side of (49) is zero since at least one of the elements in the product in (49) is zero. In this case, the naive Bayes classifier fails to provide an accurate classification of the labels. In what follows, we see that this issue of the naive Bayes classifier appears frequently when we have a small number of training feature vectors. On the other hand, the KNN classifier works as follows. For the observed feature vector y^n, the KNN classifier looks for the k nearest feature vectors to y^n, among all training feature vectors $\hat{y}_{r_s}^n$, for all $1 \leq r \leq |\mathcal{X}|$ and $1 \leq s \leq T$. Then by considering a set of K input–output pairs $(x_k, \hat{y}_{k_l}^n)$, for $k \in \{1, 2, \ldots, |\mathcal{X}|\}$ and $l \in \{1, 2, \ldots, |T|\}$, the KNN classifier decides a label which is the most frequent among x_k-s. The optimum value of k for $t = 1$ is $k = 1$.

In the following, we provide numerical examples where we illustrate the performance of the proposed classifier when $p_{Y^n|X}(y^n|x)$ is artificially generated.

4.1. The I.I.D. Case with One Training Sample per Label

In the following examples, we assume that the classifiers have access to only one training feature vector for each label, the elements of the feature vectors are generated i.i.d., and the alphabet size of the feature vector, $|\mathcal{Y}|$, is fixed.

In Figures 4 and 5, we compare the error probability of the proposed classifier with the naive Bayes classifier and the KNN algorithm for the case when $|\mathcal{Y}| = 6$ and $|\mathcal{Y}| = 20$, respectively. In both examples, we have two different labels, i.e., $|\mathcal{X}| = 2$. As a result, we have two different probability distributions $p_{Y^n|X_1}(y^n|x_1)$ and $p_{Y^n|X_2}(y^n|x_2)$. The probability distributions $p_{Y^n|X_1}(y^n|x_1)$ and $p_{Y^n|X_2}(y^n|x_2)$ are randomly generated as follows. We first generate two random vectors of length 6 and length 20 for Figures 4 and 5, respectively, where the elements of these vectors are drawn independently from a uniform probability distribution. Then we normalize these vectors such that the sum of their elements is equal to one. These two normalized randomly generated vectors then represent the two probability distributions $p_{Y_i|X_1}(y_i|x_1) = p_{Y|X_1}(y|x_1)$ and $p_{Y_i|X_2}(y_i|x_2) = p_{Y|X_2}(y|x_2)$, $\forall i$. Then, $p_{Y^n|X_k}(y^n|x_k)$ is obtained as $p_{Y^n|X_k}(y^n|x_k) = \prod_{i=1}^{n} p_{Y_i|X_k}(y_i|x_k)$, for $k = 1, 2$. The simulation is carried out as follows. For each n, we generate one training vector for each label, using the aforementioned probability distributions. Then, as test samples, we generate 1000 feature vectors for each label and pass these feature vectors through our proposed classifier, the naive Bayes classifier, and the KNN algorithm, and compute the errors. The length of the feature vector n is varied from $n = 1$ to $n = 100$. We repeat the simulation 5000 times and then plot the error probability. Figures 4 and 5 show that the proposed classifier outperforms both the naive Bayes classification and KNN. The main reason for this performance gain is because when only one training vector per label is available, the proposed classifier is more resilient to errors than the naive Bayes classifier, whereas the KNN algorithm has very poor performance because of the "curse of dimensionality". Specifically, the naive Bayes classifier cannot perform an accurate classification for small n compared to $|\mathcal{Y}|$ since the chance that an alphabet will not be present in one of the training feature vectors is close to 1. On the other hand, the KNN algorithm cannot perform an accurate classification for large n since the dimension of the input feature vector becomes much larger than the training data and the "curse of dimensionality" occurs.

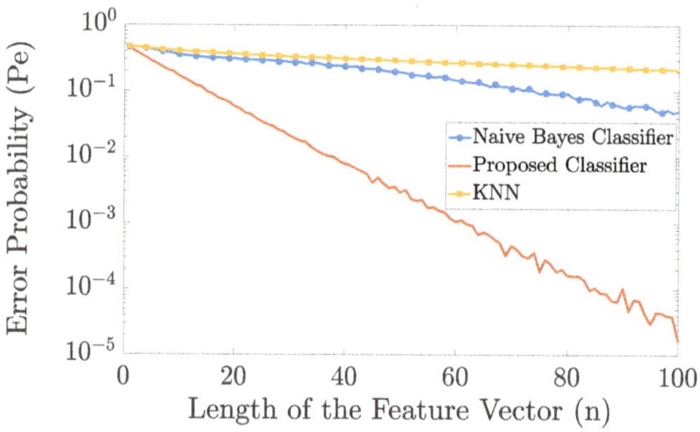

Figure 4. Comparison in error probability between the naive Bayes classifier, KNN, and the proposed classifier.

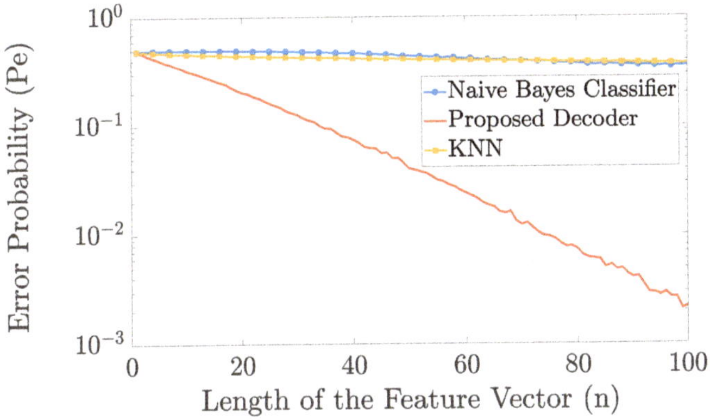

Figure 5. Comparison in error probability between the naive Bayes classifier, KNN, and the proposed classifier.

In Figure 6, we compare the performance of the proposed classifier for different values of r when $|\mathcal{Y}| = 6$ with the derived upper bounds. As can be seen, for this example, the derived theoretical upper bounds have similar slope as the exact error probabilities. Moreover, we can see that for this example, the optimal r is $r = 1$. However, this is not always the case and it depends on $p_{Y^n|X_k}(y^n|x_k)$, $|\mathcal{Y}|$, and $|\mathcal{X}|$.

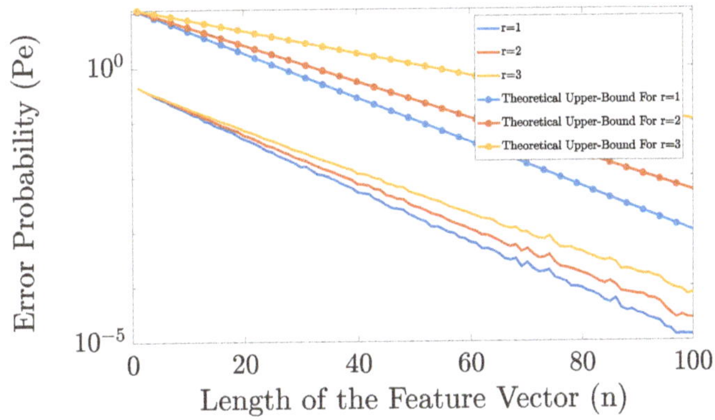

Figure 6. Comparison in error probability of the proposed classifier for different values of r when $|\mathcal{Y}| = 6$. The related theoretical upper bounds for each value of r are also given.

4.2. *The Overlapping I.Non-I.D. Case with One Training Sample per Label*

In this example, we consider the i.non-i.d. case where the probability distributions $p_i(y_i|x_k)$ are overlapping for all i, as shown in Figure 7. The small orthogonal lines on the x-axis in Figure 7 represent alphabets, i.e., the elements in $\mathcal{Y}$, and the probability of occurrence of an alphabet y_i is equal to the intersection between the corresponding orthogonal line to the represented probability distribution $p_i(y_i|x_k)$ for $k = 1, 2$. By "overlapping", we mean the following. Let $\mathcal{Y}_v$ and $\mathcal{Y}_u$ denote the set of outputs generated by $p_v(y_v|x_k)$ and $p_u(y_u|x_k)$, respectively. If for any v and u, $\mathcal{Y}_v \cap \mathcal{Y}_u \neq \emptyset$ holds, we say that the output alphabets are overlapping.

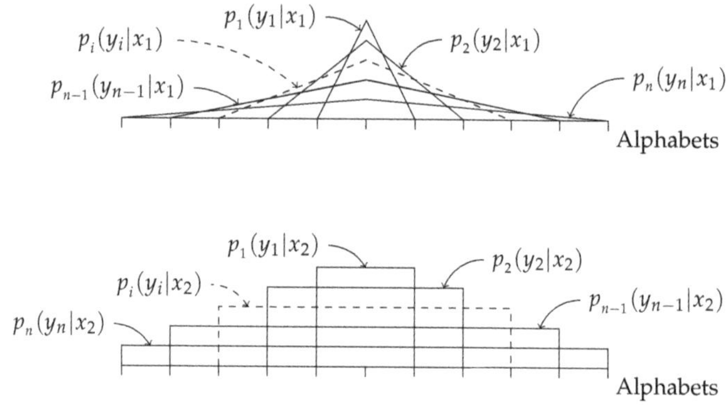

Figure 7. Illustration of the probability distributions $p_i(y_i|x_1)$ (upper figure) and $p_i(y_i|x_2)$ (lower figure), for $i = 1, 2, \ldots, n$.

To demonstrate the performance of our proposed classifier in the overlapping case, we assume that we have two different labels, $\mathcal{X} = \{x_1, x_2\}$, where the corresponding conditional probability distributions $p_i(y_i|x_1)$ and $p_i(y_i|x_2)$ are obtained as follows. For a given n, let $\mathcal{Y} = \{-n, -n+1, \ldots, 0, \ldots, n-1, n\}$ be the set of all alphabets. Note that the size of $\mathcal{Y}$ grows with n. Moreover, let $\mathbf{u}_i$ and $\mathbf{v}_i$ $(1 \le i \le n)$ be vectors of length $2n+1$, given by

$$\mathbf{u}_i = \left[0, \ldots, 0, \frac{1}{i(i+1)}, \frac{2}{i(i+1)}, \ldots, \frac{i}{i(i+1)}, \frac{i+1}{i(i+1)}, \frac{i}{i(i+1)}, \ldots, \frac{1}{i(i+1)}, 0, \ldots, 0\right], \tag{50}$$

$$\mathbf{v}_i = \left[0, \ldots, 0, \frac{1}{i(i+1)}, \frac{1}{i(i+1)}, \ldots, \frac{1}{i(i+1)}, \frac{1}{i(i+1)}, 0, \ldots, 0\right]. \tag{51}$$

The number of zeros in each side of the vectors $\mathbf{u}_i$ and $\mathbf{v}_i$ is $(n-i)$. To generate a feature vector from label $x_1(x_2)$, we generate the vector $y^n = (y_1, y_2, \ldots, y_n)$, where y_k takes values from the set $\mathcal{Y}$, with a probability distribution $p_i(y_i|x_1) = \mathbf{u}_i(1 + 2(n + y_i))$ $\left(p_i(y_i|x_2) = \mathbf{v}_i(1 + 2(n + y_i))\right)$.

The simulation is carried out as follows. For each n, we generate one training feature vector for each label. Then, we generate 1000 feature vectors for each label and pass them through our proposed classifier, the naive Bayes classifier, and the KNN algorithm and calculate the error probability. We change the length of the feature vector from $n = 1$ to $n = 100$ and repeat the simulation 1000 times and then plot the error probability.

As shown in Figure 8, there is a huge difference between the performance of the two benchmark classifiers and the proposed classifier. The error probability of the naive Bayes classifier is almost 0.5 for all shown values of n as it is susceptible to the problem of unseen alphabets in the training vectors. The error probability of the KNN classifier is also almost 0.5 for $n > 20$ as it is susceptible to the "curse of dimensionality". However, the error probability of our proposed classifier continuously decays as n increases.

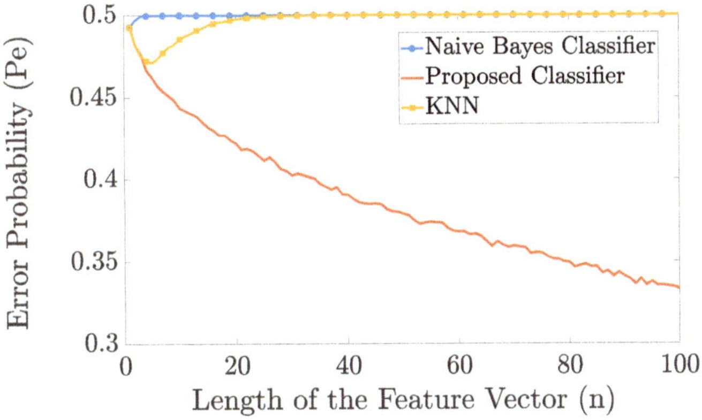

Figure 8. Comparison in error probability between the naive Bayes classifier, KNN, and the proposed classifier ($T = 1$).

In Figure 9, we run the same experiments as in Figure 8 but with $T = 100$, i.e., 100 training feature vectors per label. As can be seen from Figure 9, the performance of the proposed classifier is better than the naive Bayes classifier, for $n > 15$. Since $|\mathcal{Y}| = 2n + 1$, for small values of n, the naive Bayes classifier has access to many training samples and, thereby, its performance is very close to the case when the probability distribution $p_{Y^n|X}(y^n|x)$ is known, i.e., to the maximum-likelihood classifier, and hence it has the optimal performance. As n increases, the number of alphabets rises, i.e., $|\mathcal{Y}|$ rises, and due to the aforementioned issue of the naive Bayes classifier with unseen alphabets, our proposed classifier performs much better classification than the naive Bayes classifier. Furthermore, note that the error probability of our proposed classifier decays exponentially as n increases which is not the case with the naive Bayes classifier. Moreover, Figure 9 also shows the theoretical upper bound on the error probability we derived in (11).

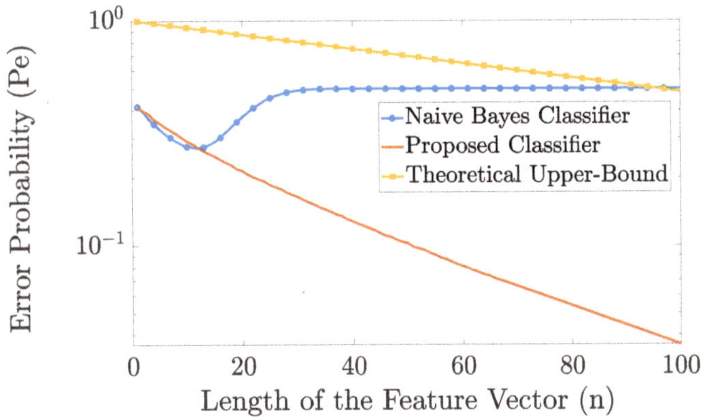

Figure 9. Comparison in error probability between the naive Bayes classifier and the proposed classifier ($T = 100$).

4.3. The Non-Overlapping I.Non-I.D. Case with One Training Sample for Each Label

In this example, we consider the i.non-i.d. case where the probability distributions $p_j(y_j|x_i)$ are non-overlapping for all j as shown in Figure 10, where we defined "overlapping" in Section 4.2. Hence, we test the other extreme in terms of possible distribution of the elements in the feature vectors Y^n.

To demonstrate the performance of our proposed classifier in the non-overlapping case, we assume that we have two different labels $\mathcal{X} = \{x_1, x_2\}$, the corresponding conditional probability distributions $p_i(y_i|x_1)$ and $p_i(y_i|x_2)$ are obtained as follows. For a given n, let $\mathcal{Y} = \{1, 2, 3, \dots, (n+1)^2 - 1\}$ be the set of all alphabets of the element in the feature vectors. Note again that the size of $\mathcal{Y}$ grows with n. in addition, let $\mathbf{u}_i$ and $\mathbf{v}_i$ for $(1 \leq i \leq n)$, be vectors of length $(n+1)^2 - 1$, given by

$$\mathbf{u}_i = \left[0, \dots, 0, \frac{1}{i(i+1)}, \frac{2}{i(i+1)}, \dots, \frac{i}{i(i+1)}, \frac{i+1}{i(i+1)}, \frac{i}{i(i+1)}, \dots, \frac{1}{i(i+1)}, 0, \dots, 0\right], \tag{52}$$

$$\mathbf{v}_i = \left[0, \dots, 0, \frac{1}{i(i+1)}, \frac{1}{i(i+1)}, \dots, \frac{1}{i(i+1)}, \frac{1}{i(i+1)}, 0, \dots, 0\right]. \tag{53}$$

The number of zeros in the left-hand sides of $\mathbf{u}_i$ and $\mathbf{v}_i$ is $i^2 - 1$. To generate a feature vector from the label $x_1(x_2)$, we generate the vector $y^n = (y_1, y_2, \dots, y_n)$, where y_k take values from the set $\mathcal{Y}$, with probability distribution $p_i(y_i|x_1) = \mathbf{u}_i(y_i)$ $(p_i(y_i|x_2) = \mathbf{v}_i(y_i))$.

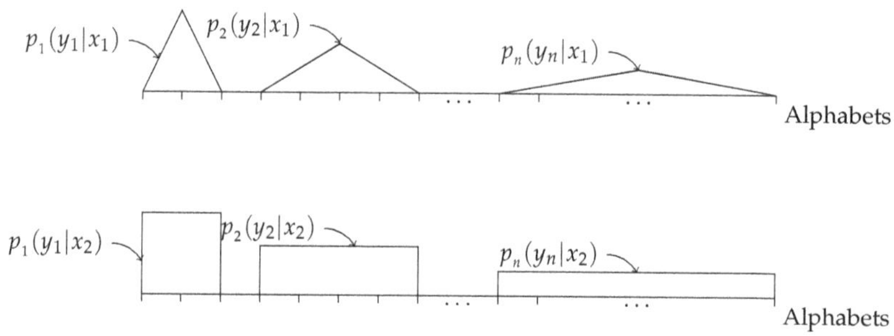

Figure 10. Illustration of the probability distributions $p_i(y_i|x_1)$ (upper figure) and $p_i(y_i|x_2)$ (lower figure), for $i = 1, 2, \dots, n$.

The simulation is carried out as follows. For each n, we generate one training feature vector for each label. Then we generate 250 feature vectors for each label and pass it through our proposed classifier, the naive Bayes classifier and KNN and calculate the error probabilities. We change the length of the vector from 1 to 80 and repeat the simulation 250 times and then plot the error probability. As shown in Figure 11, there is a huge difference between the performance of the proposed classifier and the two benchmark classifiers. The error probability of the naive Bayes classifier is almost 0.5 for all shown values of n as it is susceptible to the issue with unseen alphabets in the training feature vector. The error probability of the KNN classifier is almost 0.5 for all shown values of $n > 30$ as it becomes susceptible to the "curse of dimensionality". However, the error probability of our proposed classifier still decays continuously as n increases.

Note that, in our numerical examples, we compared our algorithm with the benchmark schemes on two extreme cases of i.non-i.d. vectors, referred to as "overlapping" and "non-overlapping". Any other i.non-i.d. vector can be represented as a combination of the "overlapping" and "non-overlapping" vectors. Since our algorithm works better than the benchmark schemes for small t on both these cases, it will work better than the benchmark schemes on any combination between "overlapping" and "non-overlapping" vectors, i.e., for any other i.non-i.d. vectors.

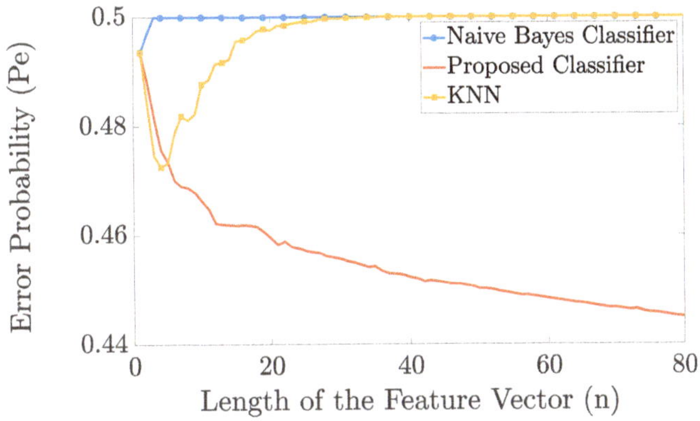

Figure 11. Comparison in error probability between the naive Bayes classifier and the proposed classifier ($T = 1$).

5. Conclusions

In this paper, we proposed a supervised classification algorithm that assigns labels to input feature vectors with independent but non-identically distributed elements, a statistical property found in practice. We proved that the proposed classifier is asymptotically optimal since the error probability moves to zero as the length of the input feature vectors grows. We showed that this asymptotic optimality is achievable even when one training feature vector per label is available. In the numerical examples, we compared the proposed classifier with the naive Bayes classifier and the KNN algorithm. Our numerical results show that the proposed classifier outperforms the benchmark classifiers when the number of training data is small and the length of the input feature vectors is sufficiency large.

Author Contributions: Methodology, F.S. and N.Z.; software, F.S.; formal analysis, F.S. and N.Z.; investigation, F.S.; supervision, N.Z.; writing—original draft preparation, F.S.; writing—review and editing, N.Z. All authors have read and agreed to the published version of the manuscript.

Funding: This research received no external funding.

Institutional Review Board Statement: Not applicable.

Informed Consent Statement: Not applicable.

Data Availability Statement: Data sharing is not applicable to this article as no new data were created or analyzed in this study.

Conflicts of Interest: The authors declare no conflict of interest.

Appendix A. Proof of Corollary 2

The proof is almost identical to the proof of Theorem 1; however, here we derive a looser upper-bound on the error-probability than that in (11), which is independent of $P_{\hat{y}_i^{nT}}$.

Without loss of generality we assume that x_1 is the input to $p_{Y^n|X}(y^n|x)$ and y^n is observed at the classifier.

Let $\mathcal{B}_{k,l}^{\epsilon}$, for $1 \leq k \leq |\mathcal{Y}|$ and $1 \leq l \leq |\mathcal{X}|$, be a set defined as

$$\mathcal{B}_{k,l}^{\epsilon} = \left\{ \hat{y}^{nt} : \left| \frac{\mathcal{I}[\hat{y}^{nt} = y_k]}{nt} - \bar{p}(y_k|x_l) \right| \leq \frac{\epsilon}{\sqrt[3]{t}} \right\}. \tag{A1}$$

Let $\mathcal{B}_l^\epsilon = \bigcap_{k=1}^{|\mathcal{Y}|} \mathcal{B}_{k,l}^\epsilon$. For $\hat{y}_1^{nt} \in \mathcal{B}_1^\epsilon$, we have

$$\left(\sum_{k=1}^{|\mathcal{Y}|} \left| \frac{\mathcal{I}[\hat{y}_1^{nt} = y_k]}{nt} - \bar{\mathrm{p}}(y_k|x_1) \right|^r \right)^{1/r} \overset{(a)}{\leq} \left(\sum_{k=1}^{|\mathcal{Y}|} \left(\frac{\epsilon}{\sqrt[3]{t}} \right)^r \right)^{1/r}, \tag{A2}$$

Using the same derivation as (18), for any $y^n \in \mathcal{A}^\epsilon$ and for $\hat{y}_1^{nt} \in \mathcal{B}_1^\epsilon$, we have:

$$\left(\sum_{k=1}^{|\mathcal{Y}|} \left| \frac{\mathcal{I}[y^n = y_k]}{n} - \frac{\mathcal{I}[\hat{y}_1^{nt} = y_k]}{nT} \right|^r \right)^{1/r} \leq |\mathcal{Y}|^{1/r}\epsilon + |\mathcal{Y}|^{1/r}\frac{\epsilon}{\sqrt[3]{t}}. \tag{A3}$$

On the other hand, the same as the derivation in (21), for each $i \neq 1$, we have:

$$\left(\sum_{k=1}^{|\mathcal{Y}|} \left| \frac{\mathcal{I}[y^n = y_k]}{n} - \frac{\mathcal{I}[\hat{y}_i^{nt} = y_k]}{nt} \right|^r \right)^{1/r} \geq \left\| P_{\hat{y}_i^{nt}} - \bar{\mathrm{P}}_1 \right\|_r - |\mathcal{Y}|^{1/r}\epsilon. \tag{A4}$$

Now, for any $\hat{y}_i^{nt} \in \mathcal{B}_i^\epsilon$, we have

$$\begin{aligned}
&\left\| P_{\hat{y}_i^{nt}} - \bar{\mathrm{P}}_1 \right\|_r + \left(\sum_{k=1}^{|\mathcal{Y}|} \left(\frac{\epsilon}{\sqrt[3]{t}} \right)^r \right)^{1/r} \\
&\overset{(a)}{\geq} \left(\sum_{k=1}^{|\mathcal{Y}|} \left| \frac{\mathcal{I}[\hat{y}_i^{nt} = y_k]}{nt} - \bar{\mathrm{p}}(y_k|x_1) \right|^r \right)^{1/r} + \left(\sum_{k=1}^{|\mathcal{Y}|} \left| \frac{\mathcal{I}[\hat{y}_i^{nt} = y_k]}{nt} - \bar{\mathrm{p}}(y_k|x_i) \right|^r \right)^{1/r} \\
&\overset{(b)}{\geq} \left(\sum_{k=1}^{|\mathcal{Y}|} \left| \bar{\mathrm{p}}(y_k|x_1) - \bar{\mathrm{p}}(y_k|x_i) \right|^r \right)^{1/r},
\end{aligned} \tag{A5}$$

where (a) follows from (A1) and (b) is again due to the Minkowski inequality. The expression in (A5), can be written equivalently as

$$\left\| P_{\hat{y}_i^{nt}} - \bar{\mathrm{P}}_1 \right\|_r \geq \left\| \bar{\mathrm{P}}_i - \bar{\mathrm{P}}_1 \right\|_r - |\mathcal{Y}|^{1/r}\frac{\epsilon}{\sqrt[3]{t}}. \tag{A6}$$

where $i \neq 1$. Using the bounds in (A6) and (A4), for any $i \neq 1$ we have

$$\left(\sum_{k=1}^{|\mathcal{Y}|} \left| \frac{\mathcal{I}[y^n = y_k]}{n} - \frac{\mathcal{I}[\hat{y}_i^{nt} = y_k]}{nt} \right|^r \right)^{1/r} \geq \left\| \bar{\mathrm{P}}_i - \bar{\mathrm{P}}_1 \right\|_r - |\mathcal{Y}|^{1/r}\epsilon\left(1 + \frac{1}{\sqrt[3]{t}}\right). \tag{A7}$$

Using the bounds in (A3) and (A7), we now relate the left-hand sides of (A3) and (A7) as follows. As long as the following inequality holds for each $i \neq 1$,

$$|\mathcal{Y}|^{1/r}\epsilon\left(1 + \frac{1}{\sqrt[3]{T}}\right) < \left\| \bar{\mathrm{P}}_i - \bar{\mathrm{P}}_1 \right\|_r - |\mathcal{Y}|^{1/r}\epsilon\left(1 + \frac{1}{\sqrt[3]{t}}\right), \tag{A8}$$

which is equivalent to the following for $i \neq 1$

$$\epsilon < \frac{\left\| \bar{\mathrm{P}}_i - \bar{\mathrm{P}}_1 \right\|_r}{2(1 + t^{-1/3})|\mathcal{Y}|^{1/r}}, \tag{A9}$$

we have the following for $i \neq 1$

$$\left(\sum_{k=1}^{|\mathcal{Y}|}\left|\frac{\mathcal{I}[y^n = y_k]}{n} - \frac{\mathcal{I}[\hat{y}_1^{nt} = y_k]}{nt}\right|^r\right)^{1/r} \overset{(a)}{\leq} |\mathcal{Y}|^{1/r}\epsilon\left(1+\frac{1}{\sqrt[3]{t}}\right)$$
$$\overset{(b)}{<} \|\bar{\mathrm{P}}_i - \bar{\mathrm{P}}_1\|_r - |\mathcal{Y}|^{1/r}\epsilon\left(1+\frac{1}{\sqrt[3]{t}}\right)$$
$$\overset{(c)}{\leq} \left(\sum_{k=1}^{|\mathcal{Y}|}\left|\frac{\mathcal{I}[y^n = y_k]}{n} - \frac{\mathcal{I}[\hat{y}_i^{nt} = y_k]}{nt}\right|^r\right)^{1/r}, \quad \text{(A10)}$$

where (a), (b), and (c) follow from (A3), (A8), and (A7), respectively. Thereby, from (A10), we have the following for $i \neq 1$

$$\left(\sum_{k=1}^{|\mathcal{Y}|}\left|\frac{\mathcal{I}[y^n = y_k]}{n} - \frac{\mathcal{I}[\hat{y}_1^{nt} = y_k]}{nt}\right|^r\right)^{1/r} \leq \left(\sum_{k=1}^{|\mathcal{Y}|}\left|\frac{\mathcal{I}[y^n = y_k]}{n} - \frac{\mathcal{I}[\hat{y}_i^{nt} = y_k]}{nt}\right|^r\right)^{1/r}, \quad \text{(A11)}$$

or equivalently as

$$\|P_{y^n} - P_{\hat{y}_1^{nt}}\|_r < \|P_{y^n} - P_{\hat{y}_i^{nt}}\|_r. \quad \text{(A12)}$$

Once again, we obtained that if there is an ϵ for which (A9) holds for $i \neq 1$ and for that ϵ there are sets $\mathcal{A}^\epsilon$ and $\mathcal{B}_i^\epsilon$ for which $y^n \in \mathcal{A}^\epsilon$ and $\hat{y}_j^{nt} \in \mathcal{B}_l^\epsilon$ for all $1 \leq l \leq |\mathcal{X}|$, then (A12) holds for $i \neq 1$, and thereby our classifier will detect that x_1 is the correct label. Using this, we can upper-bound the error probability as

$$\mathbb{P}_{\mathrm{e}} = 1 - \Pr\{\hat{x}_1 = x_1\}$$
$$\leq 1 - \Pr\left\{(y^n \in \mathcal{A}^\epsilon) \cap \left(\bigcap_{j=1}^{|\mathcal{X}|}\hat{y}_l^{nt} \in \mathcal{B}_l^\epsilon\right)\middle|\epsilon \in \mathcal{S}\right\}, \quad \text{(A13)}$$

where $\mathcal{S}$ is a set defined as

$$\mathcal{S} = \left\{\epsilon : \epsilon \leq \min_{\substack{i\\ i\neq 1}} \frac{\|\bar{\mathrm{P}}_i - \bar{\mathrm{P}}_1\|_r}{(2+t^{-1/3})|\mathcal{Y}|^{1/r}}\right\}. \quad \text{(A14)}$$

The right-hand side of (A13) can be upper-bounded as

$$1 - \Pr\left\{(y^n \in \mathcal{A}^\epsilon) \cap \left(\bigcap_{l=1}^{|\mathcal{X}|}\hat{y}_l^{nt} \in \mathcal{B}_j^\epsilon\right)\middle|\epsilon \in \mathcal{S}\right\} = \Pr\left\{(y^n \notin \mathcal{A}^\epsilon) \cup \left(\bigcup_{l=1}^{|\mathcal{X}|}\hat{y}_l^{nt} \notin \mathcal{B}_l^\epsilon\right)\middle|\epsilon \in \mathcal{S}\right\}$$
$$\overset{(a)}{\leq} \Pr\{y^n \notin \mathcal{A}^\epsilon | \epsilon \in \mathcal{S}\}$$
$$+ \sum_{l=1}^{|\mathcal{X}|}\Pr\left\{\hat{y}_l^{nt} \notin \mathcal{B}_l^\epsilon | \epsilon \in \mathcal{S}\right\}, \quad \text{(A15)}$$

Using the same derivation as (39), we have:

$$\Pr\{y^n \notin \mathcal{A}^\epsilon | \epsilon \in \mathcal{S}\} \leq 2|\mathcal{Y}|\mathrm{e}^{-2n\epsilon^2}. \quad \text{(A16)}$$

Similarly, we have the following result for the second expression in the right-hand side of (A15), which is the same as the derivation in (43)

$$\Pr\left\{\hat{y}_l^{nt} \notin \mathcal{B}_l^\epsilon | \epsilon \in \mathcal{S}\right\} \leq 2|\mathcal{Y}|\mathrm{e}^{-2nt^{1/3}\epsilon^2}. \quad \text{(A17)}$$

Inserting (A16) and (A17) into (A15), and then inserting (A15) into (A13), we obtain the following upper-bound for the error probability

$$\mathbb{P}_{\mathrm{e}} \leq 2|\mathcal{Y}|\mathrm{e}^{-2n\epsilon^2} + 2|\mathcal{X}||\mathcal{Y}|\mathrm{e}^{-2nt^{1/3}\epsilon^2}, \tag{A18}$$

where

$$\epsilon = \min_{\substack{i,j \\ i\neq j}} \frac{\|\bar{\mathrm{P}}_i - \bar{\mathrm{P}}_j\|_r}{2(1+t^{-1/3})|\mathcal{Y}|^{1/r}}, \tag{A19}$$

Now, if $|\mathcal{Y}| \leq n^m$, (A18) can be written as

$$\begin{aligned}
\mathbb{P}_{\mathrm{e}} &\leq 2|\mathcal{Y}|\mathrm{e}^{-2n\epsilon^2} + 2|\mathcal{X}||\mathcal{Y}|\mathrm{e}^{-2nt^{1/3}\epsilon^2} \\
&\leq 2n^m \exp\left(-2n \min_{\substack{i,j \\ i\neq j}} \frac{\|\bar{\mathrm{P}}_i - \bar{\mathrm{P}}_j\|_r^2}{2(1+t^{-1/3})^2 n^{2m/r}} \right) \\
&+ 2|\mathcal{X}|n^m \exp\left(-2nt^{1/3} \min_{\substack{i,j \\ i\neq j}} \frac{\|\bar{\mathrm{P}}_i - \bar{\mathrm{P}}_j\|_r^2}{2(1+t^{-1/3})^2 n^{2m/r}} \right) \\
&\leq \mathcal{O}\left(n^m \exp\left(-n^{1-\frac{2m}{r}} \right) \right).
\end{aligned} \tag{A20}$$

According to (A20), for a fixed $r > 2m$, the right-hand side of (A20) moves to zero as $n \to \infty$ and, thereby, the classifier is asymptotically optimal.

References

1. Shalev-Shwartz, S.; Ben-David, S. *Understanding Machine Learning: From Theory to Algorithms*; Cambridge University Press: Cambridge, MA, USA, 2014.
2. Quinlan, J.R. *Learning Efficient Classification Procedures and Their Application to Chess End Games*; Springer: Berlin/Heidelberg, Germany, 1983; pp. 453–482.
3. Breiman, L.; Friedman, J.H.; Olshen, R.A.; Stone, C.J. *Classification and Regression Trees*; Wadsworth and Brooks: Monterey, CA, USA, 1984.
4. Boser, B.E.; Guyon, I.M.; Vapnik, V.N. A training algorithm for optimal margin classifiers. In Proceedings of the Fifth Annual Workshop on Computational Learning Theory, San Jose, CA, USA, 27–29 July 1992; pp. 144–152.
5. Cortes, C.; Vapnik, V. Support-vector networks. *Mach. Learn.* **1995**, *20*, 273–297. [CrossRef]
6. Lallich, S.; Teytaud, O.; Prudhomme, E. *Association Rule Interestingness: Measure and Statistical Validation*; Studies in Computational Intelligence; Springer: Berlin, Germany, 2007; Volume 43, pp. 251–275.
7. Langley, P.; Iba, W.; Thompson, K. An analysis of bayesian classifiers. In Proceedings of the Tenth National Conference on Artificial Intelligence, San Jose, CA, USA, 14–16 July 1992; AAAI Press: San Jose, CA, USA, 1992; p. 223228.
8. Aha, D.W. Lazy learning. In *Artificial Intelligent Review*; Kluwer Academic Publishers: Cambridge, MA, USA, 1997; p. 710.
9. Bishop, C.M. *Neural Networks for Pattern Recognition*; Oxford University Press: Oxford, UK, 1995.
10. Hastie, T.; Tibshirani, R.; Friedman, J. *The Elements of Statistical Learning: Data Mining*; Springer Series in Statistics; Springer: New York, NY, USA, 2009.
11. Kanaya, F.; Han, T.S. The asymptotics of posterior entropy and error probability for bayesian estimation. *IEEE Trans. Inf. Theory* **1995**, *41*, 1988–1992. [CrossRef]
12. Gutman, M. Asymptotically optimal classification for multiple tests with empirically observed statistics. *IEEE Trans. Inf. Theory* **1989**, *35*, 401–408. [CrossRef]
13. Sason, I.; Verd, S. Arimotornyi conditional entropy and bayesian hypothesis testing. *IEEE Trans. Inf. Theory* **2018**, *64*, 4–25. [CrossRef]
14. Wang, C.; She, Z.; Cao, L. Coupled attribute analysis on numerical data. In Proceedings of the Twenty-Third International Joint Conference on Artificial Intelligence, Beijing, China, 2–9 August 2013; AAAI Press: Beijing, China, 2013; p. 17361742.
15. Wang, C.; Dong, X.; Zhou, F.; Cao, L.; Chi, C.H. Coupled attribute similarity learning on categorical data. *IEEE Trans. Neural Networks Learn. Syst.* **2015**, *26*, 781–797. [CrossRef] [PubMed]
16. Wang, C.; Cao, L.; Wang, M.; Li, J.; Wei, W.; Ou, Y. Coupled nominal similarity in unsupervised learning. In Proceedings of the 20th ACM International Conference on Information and Knowledge Management, Scotland, UK, 24–28 October 2011; Association for Computing Machinery: Glasgow, Scotland, UK, 2011; p. 973978.

17. Liu, C.; Cao, L. A coupled k-nearest neighbor algorithm for multi-label classification. In *PAKDD*; Springer International Publishing: Cham, Switzerland, 2015.
18. Liu, C.; Cao, L.; Yu, P. A hybrid coupled k-nearest neighbor algorithm on imbalance data. In Proceedings of the International Joint Conference on Neural Networks, Beijing, China, 6–11 July 2014; pp. 2011–2018.
19. Liu, C.; Cao, L.; Yu, P.S. Coupled fuzzy k-nearest neighbors classification of imbalanced non-iid categorical data. In Proceedings of the 2014 International Joint Conference on Neural Networks IJCNN, Beijing, China, 8–13 July 2014; pp. 1122–1129.
20. Cao, L. Non-IIDness Learning in Behavioral and Social Data. *Comput. J.* **2013**, *57*, 1358–1370. [CrossRef]
21. Cao, L. Coupling learning of complex interactions. *Inf. Process. Manag.* **2015**, *51*, 167–186. [CrossRef]
22. Cao, L.; Zhang, H.; Zhao, Y.; Luo, D.; Zhang, C. Combined mining discovering informative knowledge in complex data. *IEEE Trans. Syst. Man Cybern. Part B (Cybern.)* **2011**, *41*, 699–712.
23. Getoor, L.; Taskar, B. *Introduction to Statistical Relational Learning (Adaptive Computation and Machine Learning)*; The MIT Press: Cambridge, MA, USA, 2007.
24. Abed-Meraim, K.; Loubaton, P.; Moulines, E. A subspace algorithm for certain blind identification problems. *IEEE Trans. Inf. Theory* **2006**, *43*, 499–511. [CrossRef]
25. Almeida, L. Linear and nonlinear ica based on mutual information—The misep method. *IEEE Trans. Signal Process.* **2004**, *84*, 231–245. [CrossRef]
26. Hyvarinen, A.; Oja, E. Independent component analysis: Algorithms and applications. *Neural Netw.* **2000**, *13*, 411430. [CrossRef]
27. Yeredor, A. Independent component analysis over galois fields of prime order. *IEEE Trans. Inf. Theory* **2011**, *57*, 53425359. [CrossRef]
28. Nguyen, H.; Zheng, R. Binary independent component analysis with or mixtures. *IEEE Trans. Signal Processing* **2011**, *59*, 31683181. [CrossRef]
29. Barlow, H.B. Unsupervised learning. *Neural Comput.* **1989**, *1*, 295311. [CrossRef]
30. Nakhaeizadeh, G.; Taylor, C.C.; Kunisch, G. Dynamic Supervised Learning: Some Basic Issues and Application Aspects. In *Classification and Knowledge Organization*; Springer: Berlin/Heidelberg, Germany, 1997; pp. 123–135.
31. Koppen, M. *The Curse of Dimensionality*; SAGE: London, UK, 2000; pp. 4–8.
32. Duda, R.O.; Hart, P.E.; Stork, D.G. *Pattern Classification*, 2nd ed.; Wiley-Interscience: Hoboken, NJ, USA, 2000.
33. Jain, A.K.; Duin, R.P.W.; Mao, J. Statistical pattern recognition: A review. *IEEE Trans. Pattern Anal. Mach. Intell.* **2000**, *22*, 437. [CrossRef]
34. Xu, B.; Huang, J.; Williams, G.; Wang, Q.; Ye, Y. Classifying very high-dimensional data with random forests built from small subspaces. *Int. J. Data Warehous. Min.* **2012**, *8*, 44–63. [CrossRef]
35. Ye, Y.; Wu, Q.; Huang, J.Z.; Ng, M.K.; Li, X. Stratified sampling for feature subspace selection in random forests for high dimensional data. *Pattern Recognit.* **2013**, *46*, 769–787. [CrossRef]
36. Kouiroukidis, N.; Evangelidis, G. The effects of dimensionality curse in high dimensional knn search. In Proceedings of the 15th Panhellenic Conference on Informatics, Kastoria, Greece, 30 September–2 October 2011; pp. 41–45.
37. Matusita, K. Classification based on distance in multivariate gaussian cases. In *Proceedings of the Fifth Berkeley Symposium on Mathematical Statistics and Probability; Volume 1: Statistics*; University of California Press: Regents, CA, USA, 1967; pp. 299–304.
38. Collins, M.; Schapire, R.E.; Singer, Y. Logistic regression, adaboost and bregman distances. *Mach. Learn.* **2002**, *48*, 253285. [CrossRef]
39. Deisenroth, M.P.; Faisal, A.A.; Ong, C.S. *Mathematics for Machine Learning*; Cambridge University Press: Warsaw, Poland, 2020.
40. Lebanon, G.; Lafferty, J. Cranking: Combining rankings using conditional probability models on permutations. In Proceedings of the 19th International Conference on Machine Learning, San Francisco, CA, USA, 8–12 July 2002; pp. 363–370.
41. Hoeffding, W. On the distribution of the number of successes in independent trials. *Ann. Math. Statist.* **1956**, *27*, 713–721. [CrossRef]
42. Hoeffding, W. Probability inequalities for sums of bounded random variables. *J. Am. Stat. Assoc.* **1963**, *58*, 13–30. [CrossRef]

MDPI
St. Alban-Anlage 66
4052 Basel
Switzerland
Tel. +41 61 683 77 34
Fax +41 61 302 89 18
www.mdpi.com

Entropy Editorial Office
E-mail: entropy@mdpi.com
www.mdpi.com/journal/entropy

www.ingramcontent.com/pod-product-compliance
Lightning Source LLC
Chambersburg PA
CBHW040803250926
51850CB00007BA/1071
9783036553078